P9-DEU-883

God's Whisper, Creation's Thunder

*Echoes of Ultimate Reality
in the New Physics*

Brian Hines

My primeval god has been since the beginning of ages,

and no one knows His end;

And He is deep down within all, yea, He pervades everything:

of such Form has He been described.

His word rings through the universe:

the Lord's thunder is writ across all the skies.

—Namdev, a thirteenth century mystic

*Threshold Books is committed to publishing books
of spiritual significance and high literary quality.
All Threshold Books are printed on acid-free paper.
We will be happy to send you a catalog.*

Threshold Books, 139 Main Street, Brattleboro, Vermont 05301
802-254-8300
Cover by Threshold Design

copyright © 1996 by Brian Hines
ISBN 0-939660-48-2

1 2 3 4 5 6 7 8 9 10

Library of Congress Cataloging-in-Publication Data:

Hines, Brian, 1948-
God's whisper, creation's thunder:
echoes of ultimate reality in the new physics / Brian Hines
352. p.8.5"
Includes bibliographical references and indexes.
ISBN 0-939660-48-2 (pbk.:alk. paper)
1, Physics--Religious aspects. 2. Spiritual life. I. Title.
BL265,P4H56 1995
215'.3--dc20   95-30485 CIP

Contents

Acknowledgements

To *Maharaj Gurinder Singh*, for supporting me and this book from start to finish.

To *Faith Singh*, for serving as an invaluable advisor, editor, and shoulder-to-lean-on.

To *Laurel*, my wonderful wife, for staying awake (most of the time) when I read drafts to you in the evening, and for being so honest with your critiques and comments.

To *Tasha*, our dog, for faithfully listening to me when Laurel fell asleep.

To *Hans Lethe*, for sharpening my understanding of free will and karma through our spirited lunch-time discussions.

To *Kabir Helminski* and *Tom Goldberg* of Threshold Books, for making publishing such a positive experience. May all authors be so fortunate.

FOREWORD

John Davidson is the author of a series of books on science and mysticism. He is presently engaged in a series concerning the mysticism of Jesus, early Christianity and the ancient Middle East of which the first two books have so far been published. His most recent book is The Gospel of Jesus: In Search of His Original Teachings *(1995).*

The last century has seen a remarkable explosion in man's exploration and analysis of physical phenomena. The effects of scientific research have touched the lives of almost every human being—and most other creatures, too—upon the planet. It is apparent, however, that the technological development based upon this research has led to a massive destruction of the planetary environment and ecosystem. It is now arguably no longer a question of *whether*—but *when* and *how*—man will be forced by the extremity of the struggle for survival to curtail his activities. This external result of our scientific jamboree tells us something deeply significant about the nature of modern man and his science: man does not really know what he is doing nor does he genuinely understand the forces with which he is playing. He has been toying with surface phenomena with little or no comprehension of the far deeper underlying principles.

In brief, we have become extraordinarily selfish. Our goal has been entirely for material and personal gain. This is a measure of our superficiality and provides the underlying reason why our scientific adventure has resulted in such environmental and social problems. For if the motivation and basis of our understanding is awry, then that will surely be reflected in the outcome. As a consequence, while it is no doubt necessary for scientists and world leaders to gather together and determine practical means by which they can help to extricate us from our self-induced predicament, the most effective approach to solving our problems is to change the understanding that we have of science and of life. In a word, we have to change our attitude. For as long as our attitude remains self-oriented and material, our problems will continue.

It seems clear that it is the seeming success of our scientific achievements which has led us to adopt an increasingly materialistic philosophy regarding the whole of life. In many respects, this is as damaging to our hopes of reconstitution as our selfishness. For we have become so convinced of the rightness of our material "progress" that materiality has become our religion— while spirituality, ethics, morality and self-discipline have become almost outmoded concepts in many quarters. Generally, people have come to believe, without really considering the matter, that scientific understanding of the physical universe belies not only the existence of a God, but also of a spiritual side to life. Our children are even taught this in our schools. Education is being gradually restructured so that only those subjects which will help to make a person an economically productive citizen are given prominence. Basic human qualities are largely ignored. All this undermines a person's inherent sense of morality, unselfishness, and decency, leading inevitably to a decline in the social order and to a misuse of scientific understanding.

It is for this kind of reason that the last decade in particular has seen the publication of a number of books which have sought to rethink the philosophy upon which our science is based. In essence, they have attempted to relocate our thinking processes on a higher level than that of the purely material. *God's Whisper, Creation's Thunder* is a further advance along this road and is hence to be greatly welcomed. Like others before him, Brian Hines sets out to demonstrate that even apparently material science holds keys and pointers to a more spiritual reality. The reasoning is simple. If we could only realize that scientific understanding actually demonstrates the existence of a higher and more spiritual dimension, rather than negating it, then not only would our science make great strides, but the fruits of that research in daily life would be reflected in a more wholesome manner. Moreover, the individual himself might be stimulated into seeking for that spiritual estate within himself and so find a higher purpose to his life. But how can a deeper perception of the nature of the physical universe lead to a more spiritual outlook? To understand this, it is necessary to consider the fundamental nature of scientific research.

Observing and analyzing the phenomena of the physical creation, scientists have determined certain basic empirical laws by which the universe may be described and—to some extent—comprehended. The actual origin of these natural principles, however, still remains a mystery. Why there are three dimensions to space and not two or four or some other number; why light travels at the speed it does; why electricity behaves in the way it does; what electromagnetism and gravity actually are; how the bodies of living creatures are so intricately and complexly organized—these and many other conundrums of physical life remain a mystery.

Science can measure, describe, and relate one thing to another—but it cannot provide fundamental answers as to the intrinsic nature of life and existence. If we think it does, then we have mistaken a shadow for reality. In fact, the more science unravels the details of nature's order, organization, and integration—while lacking any understanding of its primal origins—the more it seems to demonstrate that there must be a supreme designer who has somehow brought everything into being and who holds all the ultimate secrets. This is not a new perspective, of course. It has been maintained by Socrates, Plato, Aristotle, Cicero, Marcus Aurelius, St Paul, and many others throughout history; and therein lies the first seeds of a shift in perspective. Scientists have not disproved the existence of a divine intelligence; they have helped to demonstrate it.

As the author of the present book points out, there is another group of scientists who also seek understanding of the universe and of life. And it appears that they have existed for as long as man can trace his history and that they have a methodology in common which seems to transcend all cultural barriers—for they have been present in every era, culture, religion, and race. These scientists are the mystics who have unanimously agreed (without conferring with each other) that there lies a world of experience beyond that of the five physical senses and the faculties of the human mind to which we are normally bound.

Experience of this super-sensory world, they all say, brings with it an automatic understanding, not only of this world, but also of higher principles in creation. And this understanding *is* undoubtedly linked to a deep and primordial sense not only of a higher spiritual Power, but also of morality and ethics. Indeed, all mystics have concurred that the mystic experience is *only* available to one who has largely transcended his materialistic and selfish tendencies. And they are further in accord that the primary purpose of their research is to acquire understanding of the Divine.

This, then, provides us with a fresh perspective: the higher and deeper the level of understanding then the more *spiritual*—not material—does a person become. Conversely, the more material a person's outlook, the more blinkered will be their understanding, even of material phenomena.

Mystics are fully aware of the limited ability of the human mind and the physical senses to understand the deeper mysteries of creation. They also know that the understanding which they derive from mystic experience cannot be conveyed in human language. Nevertheless, certain fundamental principles can to some extent be expressed and, in the many cosmologies of the world, there are hints as to the nature of certain broad principles underlying the processes of manifestation and creation, both in this world and in the more spiritual reality of which mystics speak.

The key to understanding the ramifications of creation is provided by the mystics when they say that the Source of All—the Lord—is one and undivided while His creation arises as patterns of multiplicity and diversity within the ocean of Himself. As regards the phenomena of this world, as well as those of higher realms, the architect of this diversity is said (and depicted in various, often metaphorical, ways) to be a greater or universal Mind. All the phenomena of this world—whether they are considered to be those of physics, chemistry, medicine, biology, evolution, psychology, sociology, or any other branch of science—are ultimately related to the functioning of this greater mind, of which the human mind is only one small part. The world of the senses—which constitutes the field of scientific research—is really a projection or a reflection of a higher reality. The one light of the Source is split and diversified by the greater Mind into the plethora of phenomena which we observe in this world.

It is because man generally ignores the higher principles of mind function—or glimpses them only rather vaguely when he considers ethical and moral issues—that his science is not only so incomplete but has also led to his massive and continuing destruction of the planetary environment. If man could see the higher laws at work governing the play of life in this world, he would automatically rein in the activities of his own mind, behave more in harmony with the divine law, and seek true spirituality within himself.

It is a matter of considerable significance that while we may pride ourselves on the cleverness of our mind in understanding the nature of the universe, we fail to comprehend the actual nature of that very mind. Our understanding, therefore, is related to a point of ignorance: we are quite ignorant of the nature and governing principles of the very means or faculty by which we try to understand the world—our own mind. Hence, the ancient counsel of the Delphic Oracle which mystics have echoed down the centuries, "Man! Know thyself!" This constantly repeated refrain must be telling us something: despite all our supposed scientific understanding of the world, we are ignorant of the nature of our own selves—the self who feels it has acquired so much knowledge.

Now, as the author points out, one of the primary keynotes of modern physics is that the observer is intimately involved with the flow of events in the apparently objective world of the senses. Scientists have demonstrated many times and in many ways that the seeming demarcation between the objective and the subjective is by no means well-defined. And this is also a primary tenet of mystic teachings. From the very outset, therefore, modern physics and mysticism are seen to share common ground and a further study of the relationships between the two turns out to be intriguing, as the author shows.

Physics or science in general is not, of course, one complete and integrated body of fact, theory, and opinion. There still remains plenty of room for interpretation of "facts" and disagreement over theories. It is not essential, therefore, in order to enjoy this book, for a reader to agree either with all of the scientific theories presented, or indeed with all the mystical tenets put forward. History demonstrates that theories in science come and go and what is universally accepted today becomes an idea of the past tomorrow. It would be most naive for anyone to consider that science is boldly and unerringly marching forward towards a body of totally proven, universally accepted and final, physical truths. Much of our scientific analysis of phenomena as well as our technology is based upon certain concepts or ways of seeing things. When the basis of those concepts change, so too do the theories, the analysis, and the technology—sometimes very rapidly, sometimes more slowly.

The author, for example, speaks of the Big Bang theory as the most widely accepted scientific doctrine at the present time concerning the origins of the physical universe. But although the idea has a high level of acceptance amongst scientists, it is by no means proven. A significant minority of well-accredited scientists reject it on perfectly rational grounds, for the theory in itself has some considerable difficulties. Moreover, the idea also presupposes the correctness of the Neo-Darwinian model of evolution (Darwin himself believed in a God who set the ball of life rolling) in which the matter of (at least) this particular planet spontaneously developed so-called "primitive" life forms which evolved into the tremendous diversity of creatures which are seen in the fossil record and which exist today. That consciousness, life, and being have arisen from self-organization of inert matter—whose very existence is considered to have subjective relationships to consciousness and being—is a self-conflicting leap in thought for which there is no supporting data. And although Neo-Darwinism is often taught in schools as if it were reality, at least one of its primary conceptual pillars remains entirely a dubious speculation.

Similarly, the author suggests that physics is *the* fundamental science amongst all sciences. This has been said many times before, but not all scientists or thinking people agree with it. Some have pointed out that the absolutely *incredible* activity, order and organization within living systems (hundreds of thousands of molecular interactions per cell per second, all integrated into one bodily whole) indicate characteristics of matter which are *only* manifested when mind and consciousness—or life—are present. And since physics is a study of sensory phenomena as perceived by these living systems or conscious beings, the question is not so easily answered as to which came first—the conscious observer or the observed. They may have both arisen simultaneously. It can be argued, therefore, that studies of living systems and of the origins of consciousness are at least as primary, if not more so, than those of physics.

Particular scientific theories and points of view, then, may or may not be accurate reflections of the way things really are. But that is not the primary concern of this book. The author wishes to show that a comprehension of nature through the eye of modern physics can lead to an understanding of a higher spiritual reality. To those who question (and even those who do not) the doctrines of scientific materialism, this will provide welcome stimulation for their further thoughts and questioning. This, it seems to me, is the primary intention of the author. And in this task, there is no doubt that he succeeds.

John Davidson
Cambridge, England

Prologue

There is nothing mystical about mysticism.
There is nothing physical about physics.
Both are attempts to know the same reality.

My purpose in writing this book is to dispel the myth that spirituality and material science—exemplified by the new physics—are at odds with each other. This is utterly false. Mystics and physicists each seek to know the truth about Ultimate Reality, which logically must be the same for both. The sides of a mountain are varied and vast, but there is only one peak. Mystics and physicists, it will be argued, use a similar scientific method in attempting to reveal that highest knowledge. If the conclusions reached by spiritual science and material science appear to conflict, it is because these disciplines study different levels of the same reality.

Presently there is an explosion of popular interest in both mysticism and the new physics. Man's search for meaning is never-ending, and the dismal state of the modern world provides added impetus to discover the purpose of our existence. Faced daily either directly or through the media with crime, starvation, war, moral decadence, and natural disaster, a thoughtful person cannot help but question whether all this possibly could be the will of a loving God, or if unfeeling laws of nature guide both the outer world and our inner consciousness.

In modern times, pondering about God has been focused in the sphere of religion, and research into the laws of nature made the responsibility of material science. This was, of course, not always the case. The early Greek philosophers made no such division in their writings. Aristotle, Plato, and Pythagoras viewed reality as a whole, and their philosophical views encompassed both the heavens above and the earth below. They saw the workings of nature as being guided by universal spiritual laws, so the study of one was the study of the other.

This changed with the development of modern science. Recognizing that rigid religious views were standing in the way of knowing the truth about material reality, physics and the other worldly sciences moved to distance themselves from theological speculation. For hundreds of years people continued to believe that the sun revolved around the earth in the face of mounting evidence to the contrary. Why? In large part, because this belief was considered to be more in accord with religious doctrine. When Copernicus demolished the earth-centered universe, religion-centered science began to crumble also.

An uneasy truce between religion and science developed. Preachers sermonized about the will of God, and physicists researched the laws of nature. No one went to church expecting to learn about science, and Ph.D. candidates in physics were not required to cite scripture in defense of their dissertations. Though forays by members of one camp frequently were made into the environs of the other, these were light skirmishes—not fixed battles. The nineteenth century fight in America between fundamentalists and scientists over the teaching of Darwin's theory of evolution was the last major conflict of its kind. Science and religion thereafter kept mostly to themselves.

However, the lines between spirituality and science now are blurring again. And this time it is the material scientists who are taking the boldest steps into the traditional territory of religion. Astronomer John Barrow writes:

> How, when, and why did the Universe come into being? Such ultimate questions have been out of fashion for centuries. Scientists grew wary of them; theologians and philosophers grew weary of them. But suddenly scientists are asking such questions in all seriousness and theologians find their thinking pre-empted and guided by the mathematical speculations of a new generation of scientists[1]

Consider the titles of these books about the new physics, most written by professional scientists:

> *Choosing Reality: A Contemplative View of Physics and the Mind*
> *Does God Play Dice?: The Mathematics of Chaos*
> *Genesis and the Big Bang*
> *God and the New Physics*
> *The Mind of God*
> *Reading the Mind of God*
> *The Tao of Physics*

From the titles alone one can surmise correctly that the authors are not reluctant to draw significant metaphysical conclusions from mostly physical findings. This, I would argue, is not justified and actually unscientific. Does conducting research into the nature of subatomic particles truly qualify one to "read the mind of God?" Is a mathematician warranted in concluding whether or not "God plays dice" by studying events in the physical world? Is it likely that the divine workings of the universe are so transparent as to be laid bare by a few relatively simple scientific experiments or mathematical formulas?

No, it is not. And this is where most authors have fallen short when they have attempted to explore the relationship between mysticism and the new physics. As shall be demonstrated in succeeding chapters, the connections between spiritual science and material science are not plainly evident to the eye of reason. They are as subtle as the essence of God, which is subtle indeed. Those who take scientific findings at face value never will penetrate into the deepest mysteries of Ultimate Reality. To do so requires an understanding of pure mysticism, which is the means of the scientific investigation of non-material planes of existence.

At this point the reader and I already may be beginning to part company. If committed to the traditional view of "science," you may be fearful that the author is about to venture off into a la-la land of crystal-gazing, spoon-bending, spirit-channeling speculation. If committed to the traditional view of "religion," similar alarm bells may be ringing inside you. Please, have no such fear. Rest assured that it is not my intention to disparage any scientific discipline or religious faith, nor to ask the reader to believe in anything which cannot be proved empirically.

The purpose of this book is to interpret findings of the new physics in the light of ancient spirituality—the pure wisdom which can be found at the heart of the teachings of Jesus, Mohammed, Abraham, Buddha, Lao Tsu, Guru Nanak, and every other mystic who attained to great spiritual heights. The core principles of these ageless teachings as they pertain to material science will be laid out and compared with modern findings of the new physics. It will be argued that physicists are hearing an echo of Ultimate Reality, not its source. They are studying the reflection of God's countenance, not His face directly.

Viewed in this light, modern science no longer is seen to be in conflict with religious faith. A seeker after truth can learn about both material and spiritual planes of reality with complete confidence that knowledge of one is consistent with knowledge of the other. For is it not unthinkable that one part of God's creation would be opposed to another part, that the study of one realm would preclude the study of another? In truth, there is only one Law running throughout all of existence, one Voice producing all of the myriad sounds.

When God whispers, creation thunders. But the thunder is a distant echo of that whisper. Recognizing this truth is a first step toward drawing nearer to the source. For as long as we believe that the reality in which we live and breathe is the most real thing, the presence of God remains far away. When an echo is recognized for what it is, those who will not be satisfied with listening to anything but the source begin to move in the direction from which it emanates. But where is that direction? This can be a difficult question to answer even with worldly echoes—and infinitely more so in respect to spiritual ones.

It is my hope that this book will help the reader to distinguish between *reflections* of Ultimate Reality and the divine *source* from which they spring.

Is Ultimate Reality Real, and Who Can Tell?

This book is subtitled, "Echoes of Ultimate Reality in the new physics." So evidently I believe there actually is a transcendent truth that serves as the benchmark against which lesser truths can be compared. If this is not the case, then what basis would there be for concluding that any set of facts is more or less real than any other? If there is no final objective reality beneath this ever-changing physical world, an immovable bedrock that supports the shifting sands of every-day existence, then there is little point to pursue either science or spirituality.

If ultimate truth is more like a wisp of smoke that vanishes when you try to grasp it—instead of a bird that will alight on your hand when you utter the right call—then the chase after absolute knowledge is senseless. Perhaps we should agree with Benjamin Franklin when he said, "But in this world nothing can be said to be certain, except death and taxes." Fortunately, most people do not feel this way. We know deep in our hearts that underneath all of the trivialities that surround us, there is something more. More than sports, television, and cars. More than my spouse and children, my job, and my country. And yes, even more than religion, a philosophy of life, moral values. What this "more" is, we do not know—but that it exists at all, we are sure of.

Such may not seem to be the case, given the obvious widespread superficiality in modern society. When a young person dedicates his life wholeheartedly to the pursuit of idleness, illegal drugs, and casual sex, is he also seeking Ultimate Reality? Is the middle-aged executive who gives little thought to anything but his career and his golf game similarly searching for the meaning of life? I would say that they *are*, for everyone is seeking lasting happiness. That goal is the same for all, though the means by which it is sought are as many as the grains of sand on a beach.

A thirteenth century mystic, Jalaluddin Rumi, will be cited frequently in these pages. This eminent spiritual scientist expressed the human condition thusly: A hungry man will say that he is starving, and call out for bread, milk, cheese, fruit, cake—whatever his empty stomach craves. Yet, says Rumi, "In reality that which attracts is a single thing, but it appears multiple. . . . He [the hungry man] enumerates and names all these things, but the root is one thing: hunger. Do you not see that after he is surfeited with a single thing he says, 'None of these is necessary'? Hence it is clear that there were not ten or a hundred things, there was only one."[1]

Wisdom, then, is in understanding the root of our hunger for happiness and meaning. Having lost touch with the simple truth that what we lack is union with God, we seek solace in substitutes. The physicist seeks the final mathematical description of the forces of nature. The businessman looks for the perfect financial deal. The single man or woman longs for the ideal marriage partner. Our many appetites become confused with our single hunger, and soon those effects come to be seen as more real than their cause. Even the cause—separation from the ground of our being—is forgotten entirely, and we become completely immersed in ultimately futile attempts to assuage our hunger for God by eating the world. That will never satisfy us.

> *All the hopes, desires, loves, and affections that people have for different things— fathers, mothers, friends, heavens, the earth, palaces, sciences, works, food, drink— the saint knows that these are desires for God and all those things are veils. When men leave this world and see the King without these veils, then they will know that all were veils and coverings, that the object of their desire was in reality that One Thing.*
>
> —Rumi[2]

Truth is not a matter of taste

Further, the almost infinite number of ways people use to find temporary happiness can lead to a dangerous moral and intellectual relativism. "Since there obviously are so many approaches to making sense of this world," goes the argument, "isn't it best to assume that they all are equally acceptable? Who has the right to make any judgments?" That is, if truth lies in the eye of the beholder, then no one has the right to say that their vision is more real than anyone else's. If your hunger is fulfilled by green beans, and mine by broccoli, then fine: you are you and I am I, and we can both respect the separate vegetables each of us chooses to put on our plates.

This philosophical view is endemic at present. On college campuses in the United States, for example, there is an increasing reluctance to maintain that one set of ideas objectively is more true than another. Profound teachings of great philosophers that have stood the test of time are replaced in the curriculum by shallow opinions of trendy thinkers whose ideas soon will be forgotten. This unfortunate state of affairs is the result of believing that truth is relative, a matter-of-taste, and ultimately founded in the subjective domain of each person's individual consciousness—rather than in an objective world of shared reality.

Taken to extremes, such a perspective ends in the tragic condition known as *solipsism*, described by science writer Martin Gardner as "the insane belief that only one's self exists. All other parts of the universe, including other people, are unsubstantial figments in the mind of the single person who alone is truly real. It is almost the same as thinking one is God. . . . ."[3] Gardner notes that while pure solipsists are rarely found outside of mental institutions, solipsistic tendencies can be recognized in many aspects of modern life. He decries this, observing that "almost without exception the great philosophers of the past believed in a world independent of human minds. All agreed that the essence of this world is beyond our comprehension."[4]

In other words, even though the nature of that essence of reality is exceedingly difficult—or even impossible—to understand, it was believed that there *is* such an essence. This perspective has been the foundation of both religion and science throughout the ages. However some modern scientists hold that in the subatomic realm, which undergirds all of material existence, there is no deep reality which exists independent of observation. And even what is perceived, they say, somehow is brought into being by the observer. For several reasons these are weak and discordant concepts to found scientific inquiry on.

First, it is transparently obvious that a reality exists about which virtually every person in the world agrees. Even those physicists who intellectually argue that objects in our universe possess no objective reality duck if they see a falling brick about to hit their head. They see the brick; their companion on the sidewalk sees the brick; and the mason working on the scaffolding high above them saw the brick before he dropped it. The very ability to argue that there is no Ultimate Reality apart from appearances is founded on an assumption that other people share enough of your internal reality to understand what you are saying. It is strange that a solipsistically-inclined scientist believes that his world of thoughts is real, but that the world of external objects is not.

Second, the notion that the most real thing you can say about the universe is that there is no more reality to be found in it flies in the face of one of the strongest assumptions of material science: that knowledge always expands and moves forward. One never comes to the end of the path of knowledge. You make progress along that path by following in the footsteps of the scientists who have preceded you, and then taking further strides where they left off. Retrogression is an alien concept in physics. Progress may falter for awhile, or what has been learned may be forgotten temporarily, but knowledge never moves in reverse. Newton uncovered more of the truth than did Aristotle, and Einstein more than Newton. Unveiling of the mysteries of the universe continues apace, moving toward an unreachable goal of final truth.

Science's endless path to knowledge

Unreachable? Does this not conflict with the seemingly boundless optimism of scientists that whatever can be known, one day will be? Science writer Bryan Appleyard points out that actually this is not a belief held by science: "To sustain its effectiveness science insists upon a universally open-ended view of the world that accepts and embraces the permanent possibility of change and progress. At any one time scientific man can only regard his knowledge as provisional because something more effective might come along. He may construct private absolutes of faith and morality, but, in public, he must inhabit a fluid, relative world."[5]

An insidious side-effect of this belief, says Appleyard, is that relativistic leanings creep into many other aspects of our lives, even those far removed from material science. This is because science largely has become the religion of modern man. Advances made by scientists in the twentieth century have led to far-reaching technological changes which generally have made our lives easier, or at least more interesting. While admittedly we now have to cope with radioactive waste and depletion of the ozone layer, we also enjoy laser surgery and cellular phones. Seeing how science has answered so many questions about the material universe in such a short time, there is an understandable tendency to believe that *all* questions of existence similarly will be answered one day.

So for many people, the pronouncements of prestigious scientists have replaced sermons by religious leaders as a guide to the meaning of life. Several years ago I attended a lecture in Portland, Oregon by the renowned physicist Stephen Hawking. After an interesting presentation concerning black holes and other scientific subjects about which Hawking was eminently qualified to speak, a member of the audience rose and asked, "Do you believe there is a God?" Because Hawking has a disease which prevents him from speaking or moving any part of his body other than one hand, he had to laboriously prepare his answer on a portable computer, and then communicate it via a voice synthesizer. All this took a couple of minutes, while the audience waited in palpable anticipation for his answer. Does God exist? Finally we would know.

To Hawking's credit, his short answer was: "That is a big question. It is not for me to say." You could almost hear both the believers and atheists breathe a large sigh of relief. Leaving the question open permitted everyone to remain comfortable with their own conceptions, and was in perfect accord with the pervasive scientific assumption of open-ended progress toward truth—a journey without an end.

Our faith in physical science puts us in a terrible bind. On the one hand, many people either consciously or unconsciously have bought into the belief that materialistic disciplines such as physics, computer theory, and the neurosciences are the surest path to knowledge about the whys and wherefores of the universe. Thus one can go into a bookstore and find such provocative titles as *Consciousness Explained* by Daniel Dennett, a philosopher who interprets research regarding the brain and mental functioning. Dennett's conclusion seems clear:

> The phenomena of human consciousness have been explained in the preceding chapters in terms of the operations of a "virtual machine," a sort of evolved (and evolving) computer program that shapes the activities of the brain. . . . I have argued that you *can* imagine how all that complicated slew of activity in the brain amounts to conscious experience. . . . It turns out that the way to imagine this is to think of the brain as a computer of sorts. . . . Now that the stream of consciousness has been reconceived as the operations of a virtual machine realized in the brain, it is no longer "obvious" that we are succumbing to an illusion when we imagine such a stream occurring in the computer brain of a robot, for instance.[6]

Well, that is depressing news. Our consciousness is the product of a machine. This strongly implies that there is no form of spiritual reality, no life-after-death, no heaven. But wait. Perhaps there is hope after all. On the last page of *Consciousness Explained,* Dennett says: "My explanation of consciousness is far from complete. One might even say that it was just a beginning, but it *is* a beginning, because it breaks the enchanted circle of ideas that made explaining consciousness seem impossible."[7] This is much more palatable, for mystics would agree that thinking about how thinking comes about is a futile dog-chasing-tail activity. I only wish that Dennett had titled his book, "Consciousness Explained a Little Bit," because materialistic science already is overly prone to making claims to knowledge which cannot be backed up with solid experimental evidence.

Nevertheless, this false confidence of physical science that it can answer *all* questions draws many into its web. Spirituality, by contrast, seems so wishy-washy and insubstantial. Science is solid, logical, strong, and dependable. So many people come to depend on it for support: not only for a faster way to cook food by using microwaves, but also to relieve their hunger for meaning and purpose in life. And this is a task which material science is completely unequipped to perform.

Limits of material science

As we have seen, science does not believe in final answers. Disciplines such as physics hold out before us the carrot of knowledge of Ultimate Reality while knowing full well that mankind never will be able to reach it. The promise gets our saliva dripping, which makes it all the more frustrating to find final truth always out of reach. In addition to death and taxes, we can count on this: whenever physics or any other material science opens the door to what appears to be the answer to its last question, what will be found is more questions. Decide for yourself whether this is the rack on which you want to place your hat of faith.

> . . . *someone comes to the seashore.*
> *Seeing nothing but turbulent water, crocodiles, and fish, he says:*
> *"Where are the pearls? Perhaps there are no pearls."*
> *How is one to obtain a pearl merely by looking at the sea?*
> *Even if one measures out the sea cup by cup a hundred times over,*
> *the pearls will not be found.*
> *One must be a diver in order to discover the pearls;*
> *and not every diver will find them, only a fortunate, skillful one.*
> *The sciences and crafts are like measuring the sea in cupfuls;*
> *the way to finding pearls is something else.*
>
> —Rumi[8]

We need to go deeper into the question of why material science is unable to penetrate the mysteries of Ultimate Reality. As I stated previously, the simple fact that every person is searching for life's meaning and purpose in his own way is perhaps the most conclusive argument that a final truth exists to be found. For there is a means to satisfy every other craving in this world: food satisfies hunger; water satisfies thirst; sexual union satisfies lust; companionship satisfies loneliness; money satisfies greed. These hungers, of course, never are satiated permanently.

But if a hunger is crude and fleeting, why should not the fulfillment of it be of a similar kind? Likewise, if we have a deep longing for unchanging truth, why should not this desire be met in equal measure? Mankind's search for final truth is as old as recorded history, and almost certainly much older than that. It would be strange if such a strong desire had been placed in man with no means to fulfill it. Assuming, then, that knowledge of Ultimate Reality *is* attainable, we have to ask whether the methods of material science used by physics, chemistry, biology, and the like will lead to that end. Physicist Joe Rosen wrote about this question in *The Capricious Cosmos*, and his arguments are summarized here. Rosen begins with this definition of science:

> Science is our attempt to understand the reproducible and predictable aspects of nature.[9]

Science defined

By "nature" is meant the material universe with which we can, or can conceivably, interact. So here is a differentiation between the universe, which is everything, and the material universe, which is that portion of everything that has a material character. We can observe and measure it with our physical senses of sight, hearing, touch, smell, and taste. This distinction is important, because it separates material science from spiritual science. Both of these disciplines share a common research philosophy, but reliance on the five senses limits a scientist to investigations of the physical world. If there is anything non-physical in the universe, material science could never know about it. This is a big drawback if one's goal is Ultimate Reality, for a research tool aimed at revealing final truth should not rule out in advance the possibility that the foundation of existence may consist of a non-material substance.

Continuing to examine Rosen's definition of science, Rosen notes that the phrase "Science is our *attempt. . . .* " means that there is no claim for assured success. Material science might be able to learn about everything in the physical world, or it might not. It already has been shown that it will not be able to learn about anything non-physical, so we are left with something like a television that only can be tuned to one channel—and even that image comes in fuzzy! Certainly a limited tool for ultimate truth-seekers. "Science is our attempt to *understand. . . .* " implies that reasons are given for what is observed. It is not enough to simply say, "This is." Science should be able to say, "This is, because. . . . "

Finally, Rosen says that what we are trying to understand are the "reproducible and predictable aspects of nature." Reproducibility means that experiments can be repeated by the same and other investigators. If a scientist claims to make a discovery, but no one else can reproduce that finding, it is consigned to the wastebasket of scientific research. Rosen notes that this "makes science as nearly as possible an objective endeavor of lasting validity"[10] for otherwise researchers would be unable to share their findings and confirm a purported truth about the world. *Predictability* means that the results of new experiments can be predicted from the results of old ones. This is possible because science is not content with merely reporting isolated facts, even if reproducible, but wants to know the broader laws of nature which produce those facts. Experiment and theory go hand-in-hand, one building on top of the other.

Unanswerable ultimate questions

Given this definition of science, and understanding that at the moment we are speaking only about material science and study of the physical world, it is easy to see why science is not capable of dealing with questions of Ultimate Reality. For *ultimate* implies an understanding of everything about the physical universe. Clearly this would include the most basic query:

Why and how was the universe created?
Or, if it was not created, why is it here at all?

These questions are fundamental, but unanswerable by science. Since science is concerned with what is reproducible, and obviously it is not possible to create the universe anew to test a creation theory, then this issue falls outside of the domain of material science. Hence we have *meta*physics and plain physics. According to Rosen, all questions concerning the creation of the universe as a whole necessarily are metaphysical. This does not mean that they are unanswerable—a widely held misconception—only that investigative methods which go beyond those available to physics are needed. These are the contemplative research tools of mysticism, which will be discussed in a succeeding chapter.

Another limitation of material science was stated clearly by physicist Max Planck. Planck is regarded as the father of modern quantum theory, one of the cornerstones of the new physics, and was awarded the Nobel Prize in Physics in 1918. Here is an excerpt from an interview with Planck:

> *Interviewer:* You have often said that the progress of science consists in the discovery of a new mystery the moment one thinks that something fundamental has been solved.
> *Planck:* This is undoubtedly true. Science cannot solve the ultimate mystery of nature. And that is because, in the last analysis, we ourselves are part of nature and, therefore, part of the mystery that we are trying to solve.[11]

Two hundred years before Christ, Archimedes said, "Give me but one firm spot on which to stand, and I will move the earth." The boundaries of scientific investigation have expanded immeasurably since his time, but material science still has been unable to find that firm spot on which to plant its feet. Practitioners of the new physics have journeyed in their research fifteen billion light years away to the edges of the known universe. They have theoretically traversed in time back to a fraction of a second after the moment of the universe's creation. And nowhere in this immeasurably vast domain of time and space have they found a place where their limited human mind is not. For the mind, and the mathematical/technological tools created by the mind, have been both the means of discovery and part of the territory to be explored.

Stand on the floor and suspend yourself in the air with only your own arms. Or look into your right eye with your left eye without using a mirror. When you can do these things, as Planck implied, you stand a good chance of solving the mystery of the universe of which you are a part. Otherwise, the human condition restricts us to studying what lies within the grasp of our consciousness, which normally excludes that consciousness itself. Thus it becomes impossible to grab onto the whole of reality, because what is grabbing cannot be held by itself.

This sounds abstract and confusing, and it is. That is precisely the point. When the mind seriously attempts to look inward at itself, confusion results which precludes further inquiry in that direction. Consider this chain of logic, which summarizes in five lines the ultimately circular approach of modern science:

1. Somehow the physical universe was born.
2. It started as pure energy, which somehow turned into matter.
3. Matter somehow turned into living things.
4. Living things somehow developed consciousness.
5. Consciousness now is trying to know why it and the physical universe were born.

So our condition is akin to a human baby placed by his parents on a space station soon after birth. There are no living beings on the space station, only sophisticated robots. Those machines supply the child with food and drink, and everything else he needs to grow into a conscious adult, including an education. But the robots cannot teach him how he was born. They were not present at his birth, nor programmed with this information, and machines do not reproduce like humans. How then could that child ever unravel the mystery of his birth through logic alone? No matter how well he comes to understand the world of robots and the confines of the space station, the why and how of his existence will be veiled from him. Only with the help of another visitor to his isolated orbiting home can this knowledge be revealed. Thus the question of how you are born can be easily answered by an outside source, but is impossible to answer through your own solitary efforts.

Revolving around the wheel of Gödel's Theorem

A more sophisticated statement of this problem is known as Gödel's Theorem. This theorem is cited by most authors writing about the new physics, but its implications have not been fully recognized by those who believe that material science is a sure route to knowledge of Ultimate Reality. This mathematical proof demonstrates the limitations of any science which attempts to uncover complete truth without moving outside of its own boundaries. Boundaries, in other words, not only serve to *define* the science; they also *confine* it. Inherent in every material science are the seeds of its own destruction, or at least serious limitation.

These two lines illustrate why:

> *The following statement is false.*
> *The preceding statement is true.*

Take a moment to puzzle out the meaning of these sentences, but do not be surprised if you soon get a headache. If we believe the first statement and accept that the following one is false, then the second must be changed to read, *The preceding statement is false.* But if the first statement is false, then we cannot believe it, which puts us back to zero. This is an example of a self-referential system of logic, which is like trying to pull yourself off the ground with your arms, or moving the earth without a place to stand.

Kurt Gödel was a mathematician who proved the theorem which now goes by his name. The mathematics underlying Gödel's Theorem are difficult to understand, but the essence of his proof has been described by physicist Paul Davies: "there will always exist true statements that cannot be proved to be true."[12] Or in the words of science writer John Boslough, "the proof for the validation of any system could not be established from within that system. There must be something outside the theoretical framework—whether the framework was mathematical, verbal or visual— against which a confirming or disconfirming test could be made."[13] Just as you cannot measure the outside dimensions of a house from within the kitchen, so Boslough says that Gödel's Theorem suggests that "no theory of the structure of the universe could be validated from within that structure."[14]

This, then, provides substantive logical support for the theme of this book: that mysticism, or metaphysics, is the only means of validating the findings of the new physics. If I have the blueprints for an entire house, I can tell you all about the kitchen, as well as the overall design of the building. But someone who is confined to one room has no way of knowing how their immediate surroundings relate to the whole structure. Perceiving only that when they flip a switch, a light goes on, or when they turn a faucet, water pours out, they might well draw the conclusion that electricity and water emanate right from the walls of the kitchen. Only a person standing outside would see the power poles and water mains that supply the dwelling from far-off sources.

Gödel's Theorem helps us to understand why physics, and indeed any system of belief which emphasizes consistency over completeness, necessarily will develop protective mechanisms. The system's mechanisms shield the belief system from disturbances which threaten its existence, but unfortunately the blocking of access to the complete truth is a high price to be paid for this protection. Thus self-preservation becomes more important than ultimate knowledge—a bad state of affairs if the purported purpose of the belief system is to find the truth at any cost. Usually the system is not willing to pay the inevitable price to achieve that end: its own death.

Douglas Hofstadter provided a creative metaphor of this dilemma in his book *Gödel, Escher, Bach.*[15] Here is an edited summary of his metaphor. Joe goes out and buys what he thinks is a Perfect Record Player (Hofstadter's book was published in 1979, before compact discs replaced records). That is, Joe thinks that any sound whatever can be reproduced on his machine. His friend, Jane, decides to prove that he's wrong. So she brings him a record called "I Cannot be Played on Record Player 1." Boldly disregarding the title, Joe puts it on his record player and after only a few notes the machine begins vibrating severely and breaks into a large number of pieces.

What happened? Jane had gotten a description of the record player's design from the manufacturer, analyzed its construction, and discovered certain sounds which should set the machine to shaking and eventually falling apart. When played on the record player, they indeed did just this. But Joe is stubborn, and he goes out and buys an even more expensive record player which the salesman promises has higher fidelity than the first. Jane, of course, gets the plan for *that* machine, composes and records a new song called "I Cannot be Played on Record Player 2," and it also causes the player to break into pieces.

Joe's root problem, says Hofstadter, is that if he gets any record player that is sufficiently high-fidelity, a song can be found that "will create just those vibrations which will cause it to break. . . . So it fails to be Perfect. And yet, the only way to get around that trickery, namely for Record Player X to be of lower fidelity, even more directly ensures that it is not Perfect. It seems that every record player is vulnerable to one or the other of these frailties, and hence all record players are defective."[16] Joe and Jane continue on for several more rounds of technological development, and destruction, and eventually Joe is convinced that his quest for perfection in a record player is futile.

He then changes his tactics. Rather than aiming for a record player that can play anything, Joe decides to settle for a machine that can survive. In other words, he opts for consistency—survival—over completeness, or ultimate truth. Thus Joe develops mechanisms to screen out any records that are not his own, since he knows that his records are harmless to the player. He places a secret "stamp of approval" on each of his records, and tests every record before playing so he can smash any that Jane tries to sneak into his collection. In this way Joe is able to enjoy his own style of music, but has been forced to protect his record player from all of the other sounds that it is unable to reproduce without breaking into pieces.

Just like physics and the other material sciences. Bryan Appleyard says that "Science begins by saying it can answer only *this* kind of question and ends by claiming that *these* are the only questions that can be answered. Once the implications and shallowness of this trick are realized, fully realized, science will be humbled and we shall be free to celebrate our selves again."[17] Wouldn't it be wonderful to listen to the majestic sound of all the truth in creation, rather than only that which material science can hear?

As shall be shown in the next chapter, sciences such as physics necessarily restrict their inquiry to the external—or public—aspects of the physical world. Is this all that there is to existence? Mysticism teaches that there are higher regions of consciousness much vaster than this physical plane of reality. Material science has come to the erroneous conclusion that the barriers which it has erected for its own protection are the boundaries of truth.

> *Know real science is seeing the fire directly,*
> *Not mere talk, inferring the fire from the smoke. . . .*
> *All your outcry and pompous claims and bustle*
> *Only say, "I cannot see , hold me excused!"*
>   —Rumi[18]

Divisions of Consciousness

How many times have you heard a speaker say, "There are two kinds of people in this world. . . . "? One of the favorite talk-show comedians in America has a nightly "top 10" list. There are said to be seven deadly sins, not six or eight. For some reason it is most satisfying to pin complex ideas down to a specific number, preferably within the compass of the fingers on our hands. And yet there is something disturbing about any attempt to encompass reality with digits. Is it possible that the universe could be so orderly, or is it our frail mind that seeks to make the complex simple? Samuel Johnson said, "Round numbers are always false." Maybe so, but. . . .

Here comes another round number: there are four ways of being. Yes, four. Not forty, or nine, or 26.3. These four ways of being will tie together concepts which have been addressed to this point, and provide a backdrop for our further investigation of the relation between mysticism and the new physics. But do not ask me to defend my position to the death. After all, the endpoint of my argument will be that abstract concepts never can fully encompass naked reality. So if you have valid reasons for believing that there only is *one* way of being, or a hundred, we are not at odds. Symbols are being dealt with here, not existence as it is. A well-chosen symbol will point to actual truth, but remains decidedly separate from it.

A finger pointing at the moon is not the moon

I hope that you find the conceptual model introduced in this chapter to be interesting and thought-provoking. Just be assured that it is not the final word. There are many ways to cut up an apple: slice it into rounds, dice it into squares, blend it into mush. Every way starts with a whole apple and ends with a divided apple. Similarly, spiritual science teaches that Ultimate Reality is one and yet we exist in the realm of many. The most important thing is the beginning and the end. Unity somehow turns into diversity, wholeness into partness. Slicing and dicing that One can be accomplished *conceptually* in a myriad of ways, but whether or not these concepts bear much resemblance to *reality* is an open question.

Every philosophy, every science, every theology has its own way of carving reality into manageable pieces. People make models of reality because naked existence in itself seemingly is too vast and complex to grasp. Aspects of it are split off from the whole, organized into some sort of logical structure, and called "physics," "sociology," "existentialism," or "mathematics." ~~These divisions are accepted uncritically~~ largely because we have grown up with them, and know no other way of making sense of the world. However, every culture does not divide reality into the chunks favored by technological civilization, and we know that ancient peoples conceptualized the universe in a different manner than is common now.

All this is by way of saying that there are many, many ways of thinking about the world, and here comes another one. As a Zen Buddhist might say, "The finger is pointing at the moon. And also there is the moon." The following analysis will help us to face in the direction of Ultimate Reality. There remains the problem of bridging that gap between the end of our mind's finger and the moon of final truth. This can be done not by thought alone—but by traveling. Everything in this chapter, and indeed this entire book, is merely pointing. "Merely" is a lot, however, when a person is lost and pointing shows the direction to his home.

Unity of consciousness

Let us begin with a circle.

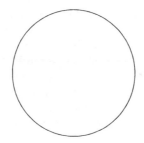

Figure 1. The unity of consciousness

In Figure 1 we see a representation of the mystic conception of the unity of consciousness, or God. Consciousness is not only the space contained within the circle, but in the words of Ken Wilber, "the paper on which the entire diagram is drawn."[1] Wilber calls this all-pervasive substance Spirit, which is how I also generally will refer to it. The term we use is unimportant, for Spirit is far beyond the realm of names. So, feel free to call it what you wish.

If God is viewed as the page on which these words are printed, the space contained within the circle may be seen as our individual consciousness. Personal consciousness, or soul, is of the same essence as universal consciousness, or God—just as this entire page is made of paper, and the area within the circle also is made of paper. The primary difference is that our soul is a drop of the ocean of God, and not the ocean itself. Yet just as a drop separated from the sea remains seawater, so does our personal consciousness enjoy the same qualities as universal consciousness, albeit to a limited extent. If some drops of seawater blown by the wind reach you while standing on the beach, your face feels a slight touch of moisture. But if the ocean unexpectedly throws up a mighty wave as you stand there, you are knocked flat on your back and drenched completely. Both events share the attributes of impact and wetness, but the ocean contains a much mightier force than a few drops.

Our soul enables us to possess a continuing sense of self, because the unity of our personal consciousness remains unbroken throughout life. Think back to your earliest memories as a child. You were playing with toys, learning to tie your shoes, spending hours with a stick and a mud puddle. Now you know so much more, have given up so many childish things, can even find enjoyment in reading a book like this. That child and this person you are now outwardly seem so different—if somehow the two ages of you could be placed side by side at this moment, likely there would be little resemblance between them. However, a constant self-consciousness links your state of being as an adult with your state of being as a child. Not "self-consciousness" in the sense of being nervous or shy, but just being *you*.

Over the years we come to learn many things. Our external surroundings change as we move from place to place, and our internal beliefs vary significantly as the years pass. A liberal becomes a conservative, and later politically apathetic. Almost overnight an atheist can convert to a fervent believer. The love of our life one moment turns into the recipient of divorce papers the next. And throughout all of these changes, which do not cease from the moment of birth to the instant of death, we never stop turning around when someone calls our name. This continuity of consciousness is unexplained by material science, for there is little in the physical substance of our brains which can account for it. No wonder—consciousness stems from the unitary ground of our being: Soul, a drop of Spirit.

Subjective and objective reality

However, the unity of consciousness divides into multiplicity as our attention is directed at the myriad phenomena both "within" and "without" us. We have, for example, a sense of the world being separated into subjective and objective parts:

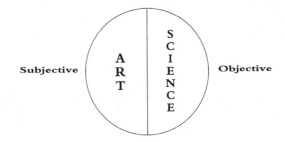

Figure 2. Consciousness divided into subjective and objective parts

Philosophers have written many volumes about the subtle points of subjectivity and objectivity. Yet the essential questions are simple. Is there a reality independent of ourselves? If so, it is objective. Is there a reality that only exists within our own personal consciousness? If so, it is subjective. Saying any more on this subject would entangle us in a bramble of words that add little to our understanding. Common-sense is needed. It appears to me, as I am virtually certain it appears to you, that there is a clear distinction between what has been termed here the objective domain of *science* and the subjective domain of *art*. A scientific truth exists whether or not any particular individual does, while a work of art springs from personal—rather than universal—consciousness. In Bryan Appleyard's words:

> An experiment performed in Chicago will give the same results if it is performed in Tokyo. . . . If science is no more than a local cultural product, this universality is the most inexplicable and bewildering coincidence imaginable. It is the equivalent of discovering a Van Gogh on Mars. It is also a unique state of affairs. Painting or music have utterly different histories in Tokyo and Chicago; religions are mutually contradictory; politics have only in recent decades begun to converge. But there is no Japanese science as distinct from American science. There is only one science and, in time, all cultures bow to its omnipotence and to its refusal to coexist. The only reasonable conclusion appears to be that, for some reason, science is the one form of human knowledge that genuinely does give us access to a "real" world.[2]

Absolutely true. And this holds for both branches of science: material and spiritual. Each seeks to know the truth about objective reality—physical or metaphysical. There is only one reality, divide it up as we will. Still, we cannot ignore the question of what is art if, as Appleyard says, science is the study of the real world. Is all but science unreal? No, certainly not. But both mysticism and physics have a similar perspective about the concept of reality, or truth. Truth is equated with permanence. A higher truth is more lasting than a lower truth. Objective reality has a longer life span than subjective reality. Whatever we are aware of exists, but this is not to say that it is true, or lasting.

A dream usually fades away completely by morning, while the mind which produces dreams wakes up essentially unchanged. Similarly, according to physics the fundamental laws of nature have remained constant during the fifteen billion years or so the material universe has existed, while countless entities governed by these laws have come and gone: ice ages, dinosaurs, stars, people, flowers, books. Thus disciplines which are focused on lasting *laws* of either material or spiritual existence are much more objective than those centered on transitory *phenomena*. Their objectivity is a product of the permanence and universality of their subject matter.

This is why, as Appleyard noted, science is the same the world over, whereas music, painting, philosophy, drama and other arts vary dramatically from nation to nation, year to year, person to person. Walk into an art gallery and stand for a few minutes in front of a painting. Listen to the comments of other visitors. "I love this work. Just look at the mature use of color," says one. Her companion replies, "You're crazy. It's just random splashes of paint. Childish!" You check to make sure that they are looking at the same piece of art. Yes, one painting, two divergent perspectives about it. Admittedly the painting is real, but springing as it did from the personal consciousness of a single human being—its creator—it has no lasting meaning.

On the other hand, enter any basic chemistry classroom in the world and you will hear, "A molecule of water has two hydrogen atoms and one oxygen atom." This statement may be expressed in a variety of languages, but the meaning will be the same in each. And unless a student is seeking to be expelled from the course, no one will argue this fact. There is no place for any difference of opinion. Personal beliefs about the composition of water are irrelevant. It is what it is, the same for everyone in the world.

Symbolic and non-symbolic reality

There is more than meets the eye to this question of objectivity, however. In fact, *more than meets the eye* is precisely the issue. For whatever my physical eye can see, yours can too—assuming we each have normal eyesight. And the same goes for our other senses. Thus we can construct symbols that stand for objects in our shared experience. "Sky" = what is seen to be over our heads. "Bark" = what is heard from the dog when a mailman comes to our door. But what about your or my private experiences which no physical eye can see, nor no physical ear hear? Consider this addition to our model of human consciousness:

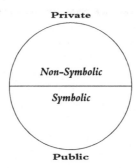

Figure 3. Consciousness divided into non-symbolic and symbolic parts

This way of dividing consciousness draws our attention to an important issue: can everything be represented by a symbol? Again, this is a problem which has occupied some of the best minds in philosophy and science since the dawn of recorded history. Instead of recounting the arguments pro and con on both sides of the question, allow me to take a short-cut to the answer: No, everything cannot be expressed in symbols. Once again I will take a common-sense, rather than academic, approach to defending this conclusion.

Figure 3 reflects the fact that in the everyday state of human consciousness there are private and public domains. The public part is what can be shared with other people. The private part is what remains essentially incommunicable, known only to the person experiencing it. For example, I woke up this morning, looked out the window, saw something white on the ground, and said to my wife, "It snowed last night." She knew exactly what I meant. Through our shared experience and acquaintance with the English language, each of the four words in my spoken sentence conveyed a meaning to her.

However, when I first looked out the window those words were not what entered my mind. I *saw* the landscape and knew immediately what had happened during the night. It had snowed. But in order to share that experience with my wife, I had to convert that private experience into a public one. Symbols were needed to do this, because people do not have access to another person's consciousness.

So even though our private world may appear to be shared, there is a large difference between symbolic and non-symbolic reality. Apart from psychotics, most people share to a remarkable degree a common understanding of the world. When my car comes to a intersection where I have a green light, I confidently assume that other drivers on the bisecting road know that red equals "stop" and green equals "go." If I am a careful driver I will glance in their direction before proceeding. However, it would be impossible to get through daily life if I did not believe that other people shared my perception of reality.

But while much of consciousness is shared with others through symbols—language, art, music, and mathematics—much is not. Emotions, for example, cannot be adequately expressed symbolically. If a mother's child was killed the day before in a tragic accident, what could she say if asked by a well-meaning friend, "How do you feel?" Is there any way that mother could express her pain and sadness in language? "Awful." "Terrible." "Empty." What possible relation could any of these words bear to her emotions that cause tears to stream continually down her cheeks? The look on her face speaks much more eloquently than could any phrase. "Silence," said Thomas Carlyle, "is more eloquent than words."

Dreams are another example of a private experience which generally is non-symbolic. I am referring less to the purported interpretation of dreams, than to their actual content. Some claim that the objects and people in our dreams represent issues which we either are dealing with consciously in real life, or unconsciously. So the monster with which I battle in a dream could stand for my boss—who is refusing to give me a raise—or perhaps my mother, who failed to bring me my bottle on time when I was a wailing newborn. This is a metaphorical view of dreams, where the important thing is not the dream itself, but what it stands for. What, though, about the dream as it actually is?

Most people undoubtedly have had this experience. You rise after a vivid dream in which utterly strange—wonderful—bizarre—confusing—illogical goings-on have taken place, involving people—places—animals—things—worlds-beyond-imagination which have never been seen in conscious reality, and likely never even will reappear in exactly the same form in another dream. Someone asks you at the breakfast table, "So, did you have any interesting dreams last night?" Though still capable of remembering your visions in detail, the thought of trying to put them into words that would bear any sort of resemblance to your inner experience leads you to reply, "Well, yes. But really, I can't describe them to you." Two people may share the same bed, and not their private dreams.

Not everything that is experienced is expressible. Further, it can be argued that never is there an *exact* correspondence between a symbol and reality. Something always is missing from the description. Nevertheless, physics has found that the symbolic language of mathematics is extraordinarily effective in describing the laws of nature, and many physicists go so far as to say that in certain contexts such as quantum physics, the description *is* the reality. Physicist Paul Davies writes that "all fundamental laws are found to be mathematical in form,"[3] and "perhaps the greatest scientific discovery of all time is that nature is written in mathematical code."[4] Unquestionably this is true, at least in regard to the material domain of existence.

Still, not all of existence can be reduced to, or described by, symbols. To believe otherwise is to consider a single tool equivalent to the whole workshop. "To a man with a hammer," goes the familiar cliché, "everything begins to look like a nail." David Lindley writes, "There is a temptingly simple explanation for the fact that science is mathematical in nature: it is because we give the name of science to those areas of intellectual inquiry that yield to mathematical analysis. Art appreciation, for example, requires a genuine effort of intellect, but we do not suppose that the degree to which a painting touches our emotions can be organized into a set of measurements or rules . . . mathematics is the language of science because we reserve the name 'science' for anything that mathematics can handle."[5]

> *The usefulness of words is to cause you to seek and to excite you,*
> *but the object of your search will not be attained through words.*
> *If it were so, there would be no need for strife and self-annihilation.*
> *Words are like seeing something moving at a distance:*
> *you run toward it in order to see the thing itself,*
> *not in order to see it through its movement.*
>
> —Rumi[6]

Four states of being

Now, let us combine the previous figures and see how they result in the four states of being available to human consciousness:

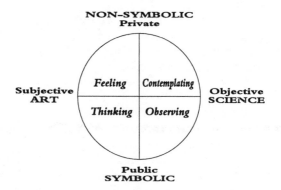

Figure 4. Consciousness divided into the four states of being

Here is another way of describing figure 4, adding an example of a "discipline" that corresponds to each state of being—

This state of being	Is related to this domain of existence	And this discipline
Feeling	NON-SYMBOLIC ART	Imagination
Thinking	SYMBOLIC ART	Philosophy
Observing	SYMBOLIC SCIENCE	Physics
Contemplating	NON-SYMBOLIC SCIENCE	Mysticism

Simple truth

All of this may seem complicated. It is not. Compared with most scholarly examinations of the nature of consciousness, this approach might seem ridiculously simple. Simple it may be, but where is it written in stone that truth must be complex? Since complexity is often a symptom of ignorance, modern scientists use the touchstone of simplicity as a guide in their search for truth. A simpler theory generally is preferred over a more complex theory, in part because it has been found that as one goes deeper and deeper into material reality, things become simpler and simpler. The apparent complexity of the material world is the product of only a few basic laws of nature.

Physicist Paul Davies says, "The entire subnuclear enterprise is founded on the tenacious belief that simplicity lies somewhere at the heart of all natural complexity."[7] And fellow physicist Anthony Zee writes, "Toward the end of the last century, many physicists felt that the mathematical description of physics was getting ever more complicated. Instead, the mathematics involved has become ever more abstract, rather than more complicated. The mind of God appears to be abstract but not complicated."[8] What is good enough for God should be good enough for us. So let us press on with this simple—but admittedly abstract— examination of the four states of being available to our consciousness.

Though the words used to describe these states necessarily are symbolic, *feeling, thinking, observing,* and *contemplating* are as concrete as anything in this world. Actually, more concrete, because consciousness is the means by which we make contact with existence. Whether or not we believe philosophically that a tree falling in a forest makes a sound if nobody is around to hear it, the reality is that unless our ears convey those sound waves to our brain, there is no way that *we* can hear it. Whether God hears it is another question. As human beings, consciousness *is* our reality. What we are aware of is real for us, and what we fail to be aware of, is not. The key issue before us, then, is one raised earlier: how do we distinguish between what is most real and less real in our awareness? Are there, in other words, ~~levels of truth~~ in what we perceive? Or should we accept each of our possible states of being as equally valid?

Let us look at each in turn, although to some extent they are not mutually exclusive. At this very moment, for example, you are simultaneously in three of the states to some degree. Your eyes are *observing* the patterns of ink printed on this page, which in themselves are meaningless—a person who cannot read our language could make no sense of these symbols. But having learned to translate the shapes of letters into words, your mind is *thinking* in concert with your visual observations, turning the words into thoughts and concepts. And there also is a state of *feeling* which pervades your consciousness in a non-rational, non-symbolic manner. In general, at this moment you feel calm, or anxious, or happy, or irritated, or content, or confused, or bored—or any of a myriad of ineffable emotional conditions which can be described only imperfectly. The state of *contemplating* almost certainly is absent from your consciousness, for reasons which soon will become evident.

Feeling

Feeling is what we do when we are in the private and subjective domain of our consciousness. Here are emotions, dreams, imagination, and fantasy. A person can make up whatever he likes. An eight-headed, purple-eyed, diamond-studded dragon with a tail as long as the Nile river never has been spotted on earth, but anyone who wants to create such a creature in his imagination has free rein to do so. This is not to say, however, that everything we feel has no foundation in reality—only that this truth is personal, not universal.

One of the current catch-words in personal development is "create your own reality." A literal-minded person might think that people are being encouraged to design and somehow bring into being their own universe, which apparently would be inhabited by only themselves. Aside from being environmentally wasteful (where would we *put* six billion separate universes, since even a parking space is hard to find nowadays?), and lonely (what if I want a friend in my reality?), this notion is impossible. God creates reality, and what is created lives in it. This is a nice division of labor.

No, the phrase "create your own reality" actually seems to mean either that we can choose how we feel subjectively in the objective reality that has been provided for us, or that we can rearrange to some extent the contents of the furnished apartment—earth—that has been rented to us for a lifetime. But these meanings are so trite as hardly to be worth mentioning. Each of us already creates our own reality every time we feel happy or sad, optimistic or pessimistic, loving or hateful. Watching a horror movie, one person is scared to death and another finds it hilarious. But this is quite different from knowing Ultimate Reality as it is. Being able to steer a car does not mean that you know how the engine works.

Even worse, personal feelings can be mistaken for universal truth. Ignorance is not being able to fix a car when it breaks, but foolishness is believing that you are a master mechanic when you are not. The ignorant can realize their condition and seek illumination, while fools remain content with darkness. Mystics, saints, and sages have spoken about the ineffable nature of God, how the divine presence cannot be expressed in worldly concepts, that the holy Spirit is nearer to us than flesh or bone. So it is not surprising that many people take a cursory look inside themselves, find something ineffable, inexpressible, and intimate, and conclude. . . . "I have found God!" Wrong. They have found their own image of God. The mind is a veil over pure consciousness. When an image is projected onto that veil, it can appear real. The only way to test the validity of such an image is to follow it back to its source, just as a moviegoer can find the projection booth by looking backward from the screen along a beam of light.

> *Every time you imagine something,*
> *You think you lift the curtain*
> *And find the truth.*
> *Your imagination is really your curtain.*
> —Rumi[9]

Language further illustrates the limitations of feelings. When our throat is parched, we usually say, "I am thirsty," not "There is thirst in me." When we are joyful, we say, "I am happy," not "There is happiness in me." Whatever private and subjective state a person is in appears to *be* him, at least until the state changes into something else. So at one moment he feels that he is happy, then sad, then happy again. Because we have come to identify ourselves as one and the same as these subjective states, we tend to say "*I* am such and such." However, the very fact that there is an entity that can say "I" means that my true self is something different from my emotions, imagination, fantasies, or internal sensations. So the domain of feeling—though vital, interesting, and unquestionably a part of being human—is not where the seeker after ultimate objective reality will find what he is looking for.

Thinking

Thinking seemingly would be a better state of being for truth-seekers. Certainly the great philosophers, scientists, and theologians have spent a lot of time there. Walk through the aisles of any large library, and the number of thoughts contained in the pages of all the books can seem almost overwhelming. How could one ever know them all? If final truth is described on one of those pages, can it be found? Yet a more basic question is: So many thoughts have been set loose upon the world. What have they gained us?

I do not mean to sound unduly pessimistic. Like most people, I enjoy thinking, and I admire great thinkers. But anyone who reads the writings of the ancient Greeks cannot help but be struck by how little progress has been made in answering the questions they pondered several thousand years ago. With all the thinking that has gone on since Socrates, Plato and Aristotle walked the earth, it would seem that humans could have answered conclusively at least *one* of those questions. Material science has told us how many elements there are, and the number of planets that circle the sun, and why compass needles point north. Why has not philosophy told us with a similar single voice the nature of the good, or what is truth, or how evil coexists with virtue?

It would not be much of an exaggeration to say that through thinking alone *Homo sapiens*, "wise man," has made no progress at all in arriving at universally accepted answers to such questions. Learned books and journals still debate the same issues which occupied the Greeks twenty-four hundred years ago. Why? Because thinking is an activity which is symbolic—and hence public, meaning it can be shared with others—but still subjective. It is much more of an art than a science, notwithstanding the efforts of logicians to create some sort of calculus of thought. So all of the theories, philosophies, ideas, and concepts which spring incessantly from our minds are a sort of shared imagination. Whereas the domain of *feeling* is private and subjective, *thinking* is public—but no less subjective. It is mankind's collective feeling, so to speak.

This lends thinking a certain seeming validity and solidity which, say, an individual's fantasy about an eight-headed dragon lacks. Thus concepts such as "minimalism" in art, or "idealism" in philosophy, or "capitalism" in economics, come to seem as real as iron, electromagnetism, and chlorophyll. But where can the reality lying behind such concepts be found outside of the minds of those thinking about them? There is a painting. Yes, it exists in objective reality. Where, however, is the quality "minimalist" which an art critic applies to it? Does this concept possess the same degree of reality as the pigments on the canvas, or even the imagination of the artist who created the painting? Physicist Nick Herbert says this about thinking:

> At its core, the process of thinking depends on our ability to tell a good lie and stick with it. Metaphors R Us. To think is to force one thing to "stand for" something that it is not, to substitute simple, tame, knowable, artificial concepts for some piece of the complex, wild, ultimately unknowable

natural world. . . . Each word is a cultural enterprise, a public attempt to dissect the wordless world in one particular way.[10]

Again, if there are solid realities behind such concepts—and mysticism disagrees with Herbert that they are "unknowable"— thinking has proven to be a poor way of finding them. In the realm of the arts, those truths remain as elusive as ever. Only in the sciences has thought proven to be effective as a reflection of reality. And even here, the special form of symbolic thinking known as mathematics has won out over discursive reasoning.

Mathematics, then, occupies a middle ground in our conceptual model between the domains of *thinking* and *observing*. For while numbers unquestionably are symbolic—like words—unlike words they have been found to relate remarkably well to aspects of the material world. The four letters in "tree" do not have any sort of intrinsic relationship to the large entity with a brown trunk and green leaves that stands outside my window. But draw a circle on a piece of paper, and the formula $2\pi r$ describes precisely how the length of the circle's circumference relates to its radius. These mathematical symbols match perfectly with what has been drawn in reality.

Such is not always the case in mathematics, however, so we must be careful not to put too much faith in numeric representations of reality. As was discussed previously, Gödel's Theorem puts bounds on the ability of any system of mathematical theorems to be both complete and consistent. True theorems will exist which cannot be proven to be true from within the system. Things can get even worse than this, though. Computer scientist Joseph Traub and mathematician Henryk Wozniakowski point out that "One important achievement of mathematics over the past 60 years is the idea that mathematical problems may be undecidable, noncomputable or intractable. Kurt Gödel proved the first of these results. . . ."[11]

Those scientists are studying the third handicap: intractability. This means that even though equations exist which describe the dynamics of entities in the physical world—such as the motion of planets and other objects in the solar system—usually it is not possible to compute an exact solution to those equations. Traub and Wozniakowski note that if accuracy to eight decimal places is desired in computing an integral—a calculus term—that has only three variables, a sequential computer that performs ten billion function evaluations a second would take three million years to complete the job. Not surprisingly, physicists come to accept approximations to such equations rather than exact results. This is not good news for the seeker of perfect truth who hopes to rely on the tools of material science.

These speculations of Traub and Wozniakowski present a challenge to a science based on mathematics, or by implication, any discipline that attempts to represent reality in symbols: " . . . information-based complexity might enable us to prove that certain scientific questions can never be answered because the necessary computing resources do not exist in the universe. If so, this condition would set limits on what is scientifically knowable. . . . We can attempt to show that there exist scientific questions such that every mathematical formulation that captures the essence of the question is intractable. We would therefore have science's version of Gödel's theorem."[12]

So even mathematics—the most rigorous form of symbolic thinking—is crippled by the three aforementioned maladies: undecidability means that true statements exist that cannot be proven to be true; intractability puts limits on the ability to compute even proven theorems; and noncomputability of a problem prevents mathematics from addressing it at all. John Ruskin aptly pointed to some of the drawbacks to thinking when he wrote in 1857, "One of the worst diseases to which the human creature is liable is its disease of thinking. If it would only just *look* at a thing instead of thinking what it must be like, or *do* a thing, instead of thinking it cannot be done, we should all get on far better."[13] Modern material science has come to much the same conclusion, which brings us to the third state of being.

Observing

Observing is much different than thinking, though obviously we can think about what we observe. What else, in fact, do we have to think about? Born with an empty page of a mind, the world instantly starts writing on it with sensory impressions: the slap of the doctor's hand on our bottom, the sound of our crying, the lights in the delivery room, the feel of our mother's breast, the scent of her perfume. Observing of one kind or another begins from the moment of birth and continues until death. All of these observations of the material world come through our five senses, which then become grist for the mill of our thoughts and feelings.

Thus in everyday life, the raw observational facts of the objective physical world continuously are translated into subjective meanings. Fact: "There is a piece of cake on the table." Meaning: "I like cake. I am hungry. I want to eat it." Fact: "A large truck is going through the red light at this intersection." Meaning: "If I don't move, it is going to hit me!" Fact: "There is a bouquet of red roses on my desk." Meaning: "My husband must have given them to me. I love him so much." A life of bare facts would be dry and barren without the meanings we attribute to them. This is part of what constitutes our humanness.

Similarly, scientists continually move back and forth between meaningful theories and factual observations, or experiments. Generally they specialize in one area or the other—theoretical physics versus experimental physics, for example—but must remain aware of advances in both areas. Theories that are not grounded in observations remain unproven, while observations not linked to theories remain isolated from the expanding base of scientific truth. Most scientists, though, would say that experimentation is the cornerstone of scientific progress. For this is what grounds theory in objective reality. Thoughts alone can lead us anywhere—toward truth or away from it, nearer to good or to evil. The mind of man can justify anything, and has. Reason alone is a most unreliable guide if our destination is what is true and lasting.

Aristotle, for example, relied almost entirely upon logic and commonsense observations to support his arguments. He was a keen observer, but not a sophisticated experimenter. Contained in his writings on physics is this statement about moving bodies: "We see that bodies which have greater impulse either of weight or of lightness, if they are alike in other respects, move faster over an equal space, and in the ratio which their magnitudes bear to each other."[14] In other words, Aristotle is saying that a heavier object will fall faster than a lighter one, assuming that they are of similar size and shape and are moving through the same medium (such as air). Embedded in the midst of complex and sophisticated arguments about infinity, the void, time and change, this assertion appears to make sense.

Of course, it also is wrong, as Galileo proved. Galileo, according to physicist Robert March, respected the value Aristotle placed on observation, but had greater faith in the Platonic view of the world. For example, says March, Plato argued that while there is such a thing as a perfect triangle, "Yet that wondrous perfect triangle exists only in the human mind, visible only to pure reason. We must look beyond the imperfect world revealed by our fallible senses to uncover a higher, more perfect reality, the only fit object of study."[15]

So Galileo was careful to avoid two pitfalls. As March puts it, "reason, unaided by the senses, can easily be led astray. Passive observation, however, is no better, for nature is too sly an adversary to reveal her most treasured secrets to any fool. . . . She must be tricked into showing her hand by contriving situations that emphasize the hidden reality. This is the essence of experimental science. . . . "[16] Thus through a combination of reason and experiment Galileo demonstrated that in a medium totally devoid of resistance all bodies will fall at the same speed.

Scientific observing, then, differs from everyday observing in several ways. Science controls for extraneous or confounding influences that may affect what is being observed. Drop a feather and a bowling ball from a tall building, and they definitely will *not* hit the ground at the same time, particularly if there is a strong wind. Control for the effect of resistance—by making the objects the same size and shape, or dropping them in a vacuum—and the law discovered by Galileo will manifest itself. Material science also generally enhances man's five senses through technology. A telescope becomes a stronger eye, a microphone a more sensitive ear. Similarly, particle accelerators produce subatomic events which normally are not found in nature, and computers expand the calculating capacity of the human mind.

Perhaps most importantly, science aims not so much to observe surface phenomena, but underlying *laws*. A feather and a bowling ball in themselves are not of particular interest to the scientist studying laws of motion. He could just as well observe a marshmallow and a brick, or a cotton ball and a tuba. Science makes progress by looking beyond the many objects in this universe to the limited number of principles that govern them. Robert Hazen and James Trefil write, "Science is organized around certain central concepts, certain pillars that support the entire structure. There are a limited number of such concepts (or 'laws'), but they account for everything that we see in the world around us. Since there are an infinite number of phenomena and only a few laws, the logical structure of science is analogous to a spider's web. Start anywhere on the web and work inward, and eventually you come to the same core."[17]

While Shakespeare cautioned "mine eye may be deceived," the fact remains that careful observation is the surest route to knowledge of objective physical reality. Certainly it is possible to sense things that are not actually there, such as a mirage in the desert, but the scientific method can keep both the material and spiritual scientist from confusing reality with illusion. The primary difference between these types of investigators, such as the physicist and the mystic, is that the former observes the public world *outside* of his or her own consciousness, while the latter looks *within* to find the root of consciousness itself. Thus in a sense, observing is the antechamber of contemplating.

Contemplating

Contemplating is the final state of being from figure 4 to be discussed. This is fitting, because it is the least known and understood. Many people do not believe that this domain of private, yet objective, reality even exists. It is the equivalent of the *terra incognita*, or hidden territory, which marked the unexplored areas of old maps. However, the ancient cartographers knew the difference between unexplored and nonexistent. Since the earth is a whole, any gap in man's knowledge of it was recognized to be a product of ignorance, rather than of an actual void in the planet. Doubters in the reality of spiritual science fail to make this distinction.

Does it make sense that the unity of consciousness presented at the beginning of this section should, instead, be a pie with the quarter of *contemplating* missing? Did this piece not get baked, or is it just less obvious while the others—*feeling, thinking, observing*—are open to view? Logic and mysticism are not at odds here. If one assumes that there is both an objective and a subjective reality, as well as both a private and a public reality, then each of the combinations of these states of being must exist. This includes a private and objective reality, which is the domain of contemplation. Mysticism teaches that this state of being is the gateway to spiritual worlds beyond this world, divine universes beyond this universe.

This is not fancy, but fact. What makes that teaching of mysticism scientific, rather than dogmatic, is the availability of a means for proving that there are subtle domains of existence in addition to our physical world. This means is contemplation, defined as the complete absorption of consciousness onto a focus of attention. This is at variance with the dictionary definition of contemplation, which is "thoughtful observation." Thoughts have nothing to do with this state of being. In respect to contemplation, they are like water and fire. The two cannot coexist. If you are thinking, you are not contemplating; if you are contemplating, you are not thinking.

> I am fleeing from my own shadow, for the light is hidden from the shadow. . . . His candle is saying, "Where is the moth, that it may burn?" . . . fling yourself into the fire, when His candle is kindled. When you have seen the joy of burning you will no more endure without the flame; even if the water of life came to you, it would not stir you from the flame.
>
> —Rumi[18]

Recall Figure 4: the states of being diagonally across from each other are the most different, and the two states which share one side are the most similar. Thus *thinking* has much more in common with *feeling* and *observing* than with *contemplating*. Thinking is subjective, like feeling, and public, like observing. It has nothing much to do with the objective and private state of contemplation. Not surprisingly, this makes it rather difficult to describe what occurs in this state of being. Mystics advise that contemplation only can be understood through its practice. This, however, is not much different from saying that a person only will understand calculus when they learn calculus.

But contemplating—unlike mathematics—is a discipline which cannot be set down in symbols or learned through books. It is best practiced under the guidance of a master of spiritual science, in the same way as theoretical physics is best taught by someone who already knows this material science. Perhaps an eager physics student could reach the equivalent of a Ph.D. level on his or her own, but this would be exceedingly difficult. Any worldly skill is most easily learned from one who is expert at it, and the same holds true for mystic contemplation.

We can, however, use the above-mentioned approach of "shared sides" to gain some insight into the nature of contemplating. It is like feeling in this way: contemplation is private or non-symbolic, being carried out entirely inside one's consciousness. Thus I do not include contemplation of material objects as providing a genuine example of this state of being. Mystics do, though, tell such stories as the following to illustrate the nature of contemplation.

A woman who desired to learn how to meditate was asked by her spiritual teacher what she loved the most. She answered, "My water buffalo." The teacher instructed her to go sit quietly in another room and contemplate the form of her beloved animal. After several hours the teacher noted that she had not returned and called out to her to rejoin him. "I can't," came the reply, "my horns are too large to fit through the door." The woman already had learned the essence of meditation—one-pointed concentration on an object, or contemplation. In this case her consciousness had merged to such an extent with the focus of her attention that she was unable to separate the two. If during her practice of contemplation any extraneous thought had been present, it would have prevented the absorption of her consciousness from being complete—much as an impurity can keep a chemical solution from crystallizing. This is why contemplating is non-symbolic, akin to feeling.

It is like observing in this way: observation, the scientific variety at least, is directed at objective reality. As was noted in the previous section, various means are used by the material scientist to assure that what is observed actually is "out there" and not "in here"—that is, in the means by which the object is being studied. If through my amateur telescope I see a curious feature on the moon which never before has been observed, before holding a news conference my proper first move would be to wipe the lens with a cloth. That probably would cause both a friendly bug and my astronomical discovery to disappear.

Professional scientists take the same precautions. In 1964, when Arno Penzias and Robert Wilson of Bell Labs found an unexplained signal coming from their giant microwave receiver, they tried various means of getting rid of it—including removing pigeon droppings from the antenna's horn. Science has its moments of humility as well as glory. Eventually Penzias and Wilson became convinced that the signal was real, and now are credited with having discovered the cosmic background radiation of the "Big Bang" which created our universe. But they never would have received their Nobel Prize if pigeon droppings had not been excluded as the cause of the signal.

Since consciousness is both the tool of contemplation and its focus of study, it is doubly important in this state of being to be sure that one is observing reality and not some personal mental excretion. Thus contemplation aims at removing from one's consciousness all but the object of attention. This object will differ according to the goal and level of attainment of the spiritual scientist. But since private—or internal— objective consciousness is the field of study, clearly any thought or image which pertains either to the external material world, or one's personal subjective reality, must be excluded as a basic requisite for further research. This is easier said than done, as the reader may confirm for himself by taking a moment to close his eyes and calmly observing what lies within the domain of his consciousness. Most people find it exceedingly difficult to shut off the chatter of their mind and awareness of external surroundings. If you can do this at will, you are most fortunate.

Meister Eckhart, a thirteenth century Christian mystic, described the essence of contemplative meditation in his writings. Note how well his thoughts are echoed by a Sufi mystic of the same period, Jalaluddin Rumi. First Eckhart:

> Since it is God's nature not to be *like* anyone, we have to come to the state of being *nothing* to enter in to the same nature that He is. . . . But so that nothing may be hidden in God that is not revealed to me, there must appear to me nothing *like*, no image, for no image can reveal to us the Godhead or its essence. . . . The least creaturely image that takes shape in you is as big as God. How is that? It deprives you of the whole of God. As soon as this image comes in, God has to leave with all his Godhead. But when the image goes out, God comes in . . . as long as *anything* is reflected in your mind which is not the eternal Word, or which looks away from the eternal Word, then, good as it may be, it is not the right thing.[19]

> *Purify yourself from the attributes of self,*
> *so that you may see your own pure essence!. . . .*
> *The Absolute Being works in nonexistence—*
> *what but nonexistence is the workshop of the Maker of existence?. . . .*
> *So enter into the workshop, namely nonexistence,*
> *so that you may see both the Worker and the work together!. . . .*
> *What do I know if I exist or not? But this much I do know, oh Beloved:*
> *when I exist I am nonexistent, and when I am nonexistent I exist!*
>
> —Rumi[20]

The Spiritual Science of Mysticism

A story is told of simple-minded fishermen who could not understand why all of the bountiful fish in their lake were at least one inch across. "Without any fingerlings being born, how can we keep on catching so many large fish year after year? This must be a miracle. We truly are blessed." A visitor happened to overhear them talking like this and asked, "Tell me. How large are the holes in the nets you are using?" "Why, it is quite a coincidence," the fishermen replied, "like the fish in our lake, they too are at least one inch wide." Physicist Arthur Eddington used this parable in questioning whether the tools of material science were able to capture all of reality.[1]

Those who fish with a loosely woven net will catch only large fish. If a person wants to be able to land every fish in a lake, he needs to construct a net with a mesh smaller than the smallest prey. Similarly, the seeker after Ultimate Reality wants to assure that nothing at all can elude his knowledge. For "ultimate" means the maximum. Anything less than final and complete truth is not acceptable to the fisher for God. So he wants to be sure that the net being cast out in his search for Ultimate Reality is up to the task. Nothing should be able to slip through it. Pure consciousness is what is required. Anything else is much too crude.

Every science needs observational tools to carry out research into the nature of objective reality. In the material sciences, these can be as basic as the unaided eye and ear, or as sophisticated as the electron microscope and radio telescope. In physics particularly, progress is considered to be dependent on the power and quality of available technology. Hence in the early 1990s the Congress of the United States engaged in a strenuous debate over whether to accede to physicists' requests to fund construction of a Superconducting Supercollider, which would have been the world's largest and most expensive scientific instrument. The power of this particle accelerator is needed to delve deeper into the nature of subatomic particles, and the conditions believed to exist soon after the creation of our universe.

"Give us a better instrument," say physicists, "and we will give you more knowledge." It is difficult to argue with that. However, even the most powerful material observational tool that can be imagined will, of course, result only in the most detailed material information imaginable. Physical tools provide physical data. This is what they are designed to do. But as was noted previously, no observation from inside the confines of the material universe can tell a scientist whether anything exists outside of those confines, or how the universe came to be here at all. Physics and the other material sciences are splendidly equipped to study physical entities, and splendidly unequipped to conduct research into other possible domains of existence.

So mysticism, the investigational discipline of metaphysics, takes up where physics leaves off. Both of these branches of science aim at understanding the same reality. How, in fact, could there be any other than a single reality? The key question is not whether mysticism is a genuine science, but whether there is any reality to be found outside of the material universe. I would think that even the most down-to-earth physicist should be open-minded enough to say, "Well, certainly, it could not hurt to look. I am skeptical that anything exists that is not physical, but science does not make progress by leaving doors closed. It opens them and sees whether anything is on the other side."

Some questions, then, are needed to guide our search. How can the door to a possible non-material reality be opened? And another more basic: Where can the door be found? Even a skeptical question: Probably there isn't any door at all, but how can one be sure of this? Finding the answer to all three questions leads inexorably to a study of human consciousness. The reason is simple. If there is anything objectively real that is not physical, it will not be found "outside" in the material world. Since consciousness is the only means by which human beings can know anything—if I'm not conscious of something, it does not exist for me—then this is the place where one's search for a possible spiritual reality must begin.

Contemplation—the research tool of spiritual science

In a previous section we considered whether the usual activities of consciousness—feeling, thinking, and observing—could lead to knowledge of a non-material objective reality. Feeling and thinking were ruled out because they are subjective; observing with the five senses was ruled out because vision, hearing, taste, sensation, and smelling all are directed toward contact with the physical world. Only one possibility remains: contemplation, which is aimed "inside" toward whatever objective reality may exist apart from the material universe. Contemplation begins with studying the nature of human consciousness itself. Mysticism uses this tool to investigate the hypothesis: "In addition to the *content* of consciousness, there is a conscious reality independent of this content."

The content is all of the impressions of both the external objective world and our internal subjective world. Emotions, fantasies, images, concepts, sensory perceptions, observations, thoughts, dreams—the whole kit and caboodle that we are aware of either when our eyes are open, or closed; awake, or asleep; sober, or drunk. No matter what state of human consciousness we are in, normally it is impossible to escape from this never-ending stream of impressions. It is much as if no matter what channel we turn to on the television of our mind, there *always* is a program on—and the theme *always* is either "The Objective Outer World," or "My Subjective Inner World." Yesterday, today, tomorrow, always the same sorts of programs.

To those who sense that there is more to existence than this, the endlessly repeated show in their consciousness becomes tedious and boring. "Can I tune in something new and fresh?," they ask. The contemplative methods of mysticism provide the means for doing this. But before a new channel can be received—or more accurately, a whole new bandwidth, like going from VHF to UHF on a television—the old channels must be tuned out. Perhaps you have heard two stations coming in at the same time on a radio, or taken a double exposure with a camera. The result is confusion: unintelligible noise or uninterpretable images. In the same way, consciousness works best when it engages in one activity at a time. Contemplation requires concentration and the cessation of all thinking, feeling, and external observing.

To accomplish this, teachers of the higher forms of mysticism generally give their students a word, or words, to repeat mentally. These words have no connection with anything in the material world, and so are much less likely to lead to emotions, thoughts, or images than would a familiar word with many previous associations. Still, the beginning student of contemplation usually finds it difficult to keep his consciousness focused on these words. But with extensive practice, it becomes possible, and eventually even quite pleasant. For mystics universally describe higher consciousness as being marked by a sensation of bliss, peace, and love. This should not be regarded as unscientific, because if bliss, peace, and love are objective qualities found in non-material planes of existence, the spiritual scientist inevitably will encounter them in the course of his research.

Mystic researchers employing the tool of contemplative meditation typically pass through these general stages of discovery. After finding a place to meditate where the outside world does not intrude, shutting one's eyes, and beginning to repeat the words which are to be the sole focus of attention, the result is . . . nothing. What is observed in the interior of one's consciousness apart from the repetition of those words are darkness and silence. Plus images and thoughts which sometimes seem to spring into awareness on their own, and sometimes are created by the student spiritual scientist. All of this is part of the early phase of learning how to focus the beam of consciousness. It usually takes considerable time and effort.

Eventually, however, one's attention begins to remain centered on the words, and even to become so absorbed in them that there is little or no separate awareness of oneself. This is the point at which it becomes clear that there is more to consciousness than its content. There is a "carrier wave" of the images and thoughts which normally fill our minds. This glimpse of pure consciousness occurs only when the contents of consciousness are eliminated. If the surface of a pond is covered completely with leaves, sticks, bottles, trash, and other debris, then an observer cannot see the water underneath. Any movement of the water would produce movement of the debris, but the cause lying beneath the effect would be hidden.

Skim everything off the surface, and the true nature of the pond is revealed. This is the object of contemplative meditation in regards to consciousness. Consciousness is the water, and material images and thoughts are the debris floating on the surface. Remove them and one comes closer to non-material objective reality. However, mysticism teaches that another barrier to spiritual truth remains: the accumulated "dirt" which clouds our consciousness even when gross images have been eliminated. This is a more difficult concept to explain briefly, and is not as essential to the theme of this book. Let us leave it at this: just as it takes some time for the water in a pond to return to its original clarity even after the debris on its surface has been removed, so is there a kind of settling process which must take place in consciousness before the naked purity of Spirit is realized.

It is appropriate to interject a caution here: the pure form of contemplative meditation outlined above *never* should be attempted without the guidance of a highly qualified teacher. It is as foolhardy for a person to believe that through his unaided efforts he can make his consciousness a fit tool for reaching God, as to believe that with no flight training he could sit down in the pilot's seat of a jumbo jet and fly it around the world. In either case, he certainly will crash before his destination is reached.

Mild forms of meditation which aim at mental and physical relaxation certainly can be practiced on one's own. This is akin to being able to float more or less peacefully on the above-mentioned metaphorical pond of consciousness. But be aware that anyone who tries to dive into the depths of that water by himself may encounter "sharks" which are more powerful than he is. These can be either a product of one's own personal consciousness—such as repressed memories and fears—or more objective obstructions. Just as no sensible person would journey into an unknown jungle without a knowledgeable guide, neither should the depths of one's consciousness be plumbed without the companionship of a perfect mystic who knows every inch of the territory to be explored (see "Mysticism and religion" below for an explanation of *perfect mystic*).

Reaching the fountainhead of consciousness

For mysticism teaches that the endpoint of the first stage of contemplation is entry to a higher plane of consciousness, which is as different from our normal waking state as the waking state differs from dreaming. This is attained when the spiritual scientist's attention is focused completely "within" through contemplation on the above-mentioned words or by means of other spiritual practices. Generally human consciousness is scattered throughout one's body, as well as into the world beyond through the five senses. Physician Deepak Chopra writes that "intelligence is present everywhere in our bodies . . . the brain does not just send impulses traveling in straight lines down the axons, or trunks of the neurons [nerve cells], it freely circulates intelligence throughout the body's entire inner space."[2]

Our mind, then, is imbedded deeply in both our body and perceptions of the material world. Contemplative meditation draws the mind inward and upward until there is no awareness of anything outside of the seat of consciousness. This seat is the gateway to the next higher plane of consciousness, just as from street level the first flight of stairs in a building ends in a landing, beyond which one can pass through a door into a whole other floor. Similarly, the first "landing" of consciousness is known in mysticism as the eye-center, beyond which one can enter into a higher dimension of reality that is unknown to material science.

The location of this eye-center is not far from our normal waking state. If at this moment the reader were to close his eyes and observe where the activity of his thinking seems to be taking place, it would be in his forehead—not his feet or arms. This, of course, is where our brain is located. However, the mind and the brain are two separate entities, the brain being an instrument of the mind for functioning on the material plane. In the same way, consciousness and the mind are not one and the same, an important fact which will be addressed in a following section. However, this does not become evident until a higher stage has been reached, so it is valid to equate one's mind and one's consciousness in describing the first "floor" of mystical ascent.

As was noted previously, the process of contemplative meditation sounds esoteric to those who are unfamiliar with it, but actually is simple and scientific. The mind produces thoughts and receives impressions of the external world. To find the source of the mind, the mystic eliminates the distraction of the sensory impressions—since clearly the mind is not *outside* of himself—and through repetition of a single word, or words, follows this emanation of the mind to where it initially springs from consciousness.

Consciousness, like still clean air, is so pure as to be undetectable to the casual observer. Repetition of a word in one's consciousness is akin to blowing smoke rings into the air. The smoke will make evident any movement of the air, just as thoughts reflect the motion of one's mind. However, there usually are so many thoughts in our mind that it becomes like a room *completely* filled with smoke. The haze and incessant swirling makes it impossible to distinguish where all this thinking comes from.

Focus on a single thought, says mysticism, and the confusion begins to be cleared up.

> *When you have seen your own cunning,*
> *follow it back to its origin.*
> *What is below comes from above.*
> *Come on, turn your eyes to the heights.*
> —Rumi[3]

A river flows from the mountains down through hills and valleys. Along its fertile banks a jungle of trees, flowers, weeds, vines, and other vegetation grows in profusion. To find the source of the river, an explorer must clear a trail through the dense undergrowth, get in a boat, and travel upwards in the opposite direction the water is flowing. In the initial stage of contemplative meditation, the river is consciousness; its source is Spirit; the undergrowth is our thoughts and other attachments to the physical world; and the boat is repetition of a single word, or set of words. Contemplation is difficult, because anyone who has paddled a canoe knows what it means to go against the current.

This is why comparatively few people attempt to discover the fountainhead of the stream of consciousness. It is considerably easier to travel downstream in the company of our familiar sensations and thoughts. But this leads us further and further away from God and Ultimate Reality. The mystic boldly and doggedly follows a single emanation of his mind back to its source, which turns out to be a point behind and between the two physical eyes—the eye-center. From this point, or landing at the end of the staircase of the material portion of the human body, there is an entryway to another level of consciousness, or floor of God's creation. Christ says "In my Father's house are many mansions."[4] But is not a mansion larger than a house? How could many mansions fit in a single house? Mystics explain that this refers to the house of the human body, within which many heavenly regions are contained. That is, through the spiritual science of contemplation we can transform our everyday consciousness into a means of entering higher realms of existence.

To know God, become God

Mysticism's message is that every man and woman possesses the means to know God: his or her soul—or highest consciousness—is of the same essence as the all-pervading Spirit which flows from the Supreme Being, and so is able to merge into that final Truth through contemplation. By removing the covers of mind and matter that obscure the reality of our soul consciousness, we can realize God completely, just as a drop of seawater comes to remember what it is like to be the ocean by returning to it. As

Rumi said in a previous quote, this makes it possible to "see both the Worker and the work together." The whys and wherefores of the physical universe are known by reaching the top of creation and looking down. Mysticism thus is able to see from a perspective which places lower laws of nature in their correct relation to higher laws.

If a person knows that they live on the ground floor of a two-story apartment building, and water starts dripping from their ceiling, their first thought is not "My ceiling is raining!" They would run upstairs to the apartment above and tell the resident that his bathtub is overflowing. In the same way, mysticism understands how the "plumbing" of all of the levels of creation fits together. Mystics realize that no matter how long a person on the first floor stares at the drops falling onto his head, their source from the bathtub on the second floor will not come into view. A ceiling stands between the two as a barrier. Rise above the ceiling, and the mystery of where the drops come from is solved.

Material science, on the other hand, laboriously attempts to piece together the fabric of Ultimate Reality by working from the ground up. By learning more and more about the laws that govern this creation, it is hoped that the nature of the Creator will be revealed. Hence, as was noted previously, we find physicists writing books about "the mind of God," assuming that this can be deduced from clues found in material existence. It is much as if an unusually intelligent bird were to carefully study the details of how its birdhouse was constructed, hoping thereby to understand the person who built it. How, pray tell, could such a bird ever grasp the essence of the human mind by counting the number of nails that hold its house together, or by measuring the dimensions of its dwelling? Yet this is the approach used by physicists and other material scientists who assert that by knowing the creation, we can understand the Creator. Does this seem reasonable?

Not to mystics, who say that only by becoming one with God is it possible to know God. In the process, the truth about material reality becomes known as well—just as a person standing on the top of a mountain whose highest summit cannot be perceived from ground level sees the geography of the hills and valleys below. This image can be misleading, however, for it implies that Ultimate Reality is a small point—a pinnacle—with the lower forms of creation spread out below to a much vaster extent. Mysticism teaches that the opposite is true. According to perfect mystics, this physical universe is like a speck of dust floating in the sky of a higher plane of existence. And, in turn, *that* plane is infinitesimal compared to an even higher region.

The terms "specks" and "skies" are, of course, grossly inadequate to describe the nature of spiritual reality. Mystics teach that the regions above and beyond our material universe—though objective—are better thought of as states of consciousness than of concrete domains of existence. This allows them to coexist, so to speak—just as in our consciousness a realm of imagination, a realm of thoughts, and a realm of perceptions exist simultaneously. Whenever we go to sleep and dream we have in a sense journeyed to another level of consciousness. Unfortunately—from a spiritual point of view—this is a lower and more subjective form of consciousness than our waking state. So dreaming is not of much use to us from the standpoint of seeking Ultimate Reality, aside from providing an example of how it is possible to freely move from one level of consciousness to another. From the waking state, mysticism teaches how to go *up*, not down.

Here are some additional observations about the spiritual science of mysticism that are germane to the theme of this book. They are presented in the form of a dialogue between the author and a skeptical questioner.

Mysticism and religion

Why do mystics call Ultimate Reality "God"? This sounds unscientific, if you claim that mysticism is truly a science.

This is only because we are accustomed to thinking of physics, chemistry, and the like as the only genuine sciences. Because "God" is not part of the nomenclature of those disciplines, to some people it rings oddly when used in a scientific context. But if we can accept "quantum" as a term that embodies one of the highest truths of material existence found by physics, why is not "God" acceptable as a representation of the Ultimate Reality found by mystics? That reality is exceedingly loving, powerful, wise, and conscious—so "God" connotes these attributes more accurately than would a less personal word.

There seem to be so many approaches to mysticism. Doesn't this contradict your assertion that there is an objective spiritual science?

Think of it this way: if you went into a second grade classroom in elementary school and saw the teacher giving a physics lesson, it would appear considerably different than a post-graduate physics course for Ph.D.s. The same applies in mysticism, but to a much greater degree. Perfect mystics say that there are *five* planes of reality, or levels of consciousness, above the physical universe—each considerably vaster and more subtle than the one below. Thus just as there are beginning students of physics, plus those who have attained a B.S., M.S., or Ph.D. in the course of study of that science, so also are there many degrees of attainment in spiritual science. Like any worldly subject, the student can learn only as much as the teacher knows.

All right, I'll accept that there are different levels of mystical knowledge. Still, why isn't there a uniform approach to teaching at *each level?*

There is, but it is not obvious. You cannot look in the Yellow Pages under "mysticism" and find a teacher of Ultimate Reality. If you find such an advertisement, you can be assured that the knowledge being offered is elementary. Perfect mystics neither advertise nor accept payment for their teaching. Further, the mystical course of study—which is centered on contemplative meditation—is carried on under the direct supervision of a spiritual "master" (this term is the mystical equivalent of Ph.D.). Not necessarily physical supervision, meaning the teacher and student live in the same place, but inner guidance, which transcends geography. So textbooks are not necessary. The teaching is direct and non-symbolic. Still, there are many books written by perfect mystics or their students, and careful study reveals that the message in each is the same.

I don't agree. The world's religions appear to differ so much. If the message of mystics is the same, why are religions so at odds?

You have hit upon an important point. Mysticism and religion are not at all the same. Mystics nominally may belong to a particular religion, just as they live in a particular country, but their teachings transcend any theology. The particular "son of God," "prophet," or "sage" revered by every great world religion unquestionably was a mystic, since he had a direct perception of God—not merely an intellectual understanding. So deep inside all religions is a core of mystic truth which is common to all. Unfortunately, theological hair-splitting and sectarianism almost always cloud those pure teachings. Perfect mystics reveal the way to know God. They preach the unifying tenets of spiritual science, not the divisive doctrines of religion. So in every religion you find perfect mystics of that faith who are in complete agreement with perfect mystics of every other faith. It is the theologians who quarrel with each other, because they seek to understand the nature of God through subjective thought rather than objective contemplation.

You refer to "perfect" mystics. What does this mean?

"Perfect" means that the mystic has completed his or her course of study: union with God. Since God is perfect, the mystic has become perfect. Many people admittedly find this a difficult notion to accept. Everything we find in the world is imperfect, so we essentially have given up hope of finding perfection. We would gladly settle for "good." Or we believe that in the past there were flawless spiritual teachers, but now they are gone. Not true. Access to perfect truth is not like a vault with a time-locked door that only opens once a century, or once a millennium, or once in the history of mankind. The door to Ultimate Reality opens when God wills it, and when a deserving soul knocks upon it (the knocking being a part of the divine will). So there have been many perfect mystics in the past, and there are perfect mystics living now.

Well, who are they? And can they supply me with next week's winning lottery numbers?

This is why they are not listed in the Yellow Pages. Most people are less interested in returning to God than in bettering the condition of their life here on earth, or in learning enough metaphysical knowledge to impress their friends and influence others. Perfect mystics accept as students only those who have a sincere desire to know the Truth. They have no need to advertise, because anyone with that sort of desire automatically will be drawn to them. Their knowledge and powers far surpass ours, being identical with those of God. It is said that just as a blind man cannot grab hold of someone unless that person with vision desires to be found, the same is true with our groping after divine wisdom. When God wants to hold your hand, He will reach out.

Dogmatism and proof

I find dogmatism distasteful. Are you really saying that the only way of knowing Ultimate Reality—or God—is the method of contemplative meditation outlined earlier?

Admittedly, there seem to be many ways of coming into contact with aspects of higher forms of consciousness. Intuition, for example, is a mental process—or state of being—which can provide valid glimpses into objective material or spiritual reality. This can occur immediately and unexpectedly, as in the familiar "Eureka!" experience, with no disciplined process of meditation involved. Thus a person can have a sudden flash of insight which turns out to be the key to resolving a vexing problem. This commonly occurs both in science and everyday life. After searching the house for a frustrating hour, suddenly you say to yourself "My car keys are in the butter dish!" Heaven knows how they got there, but your intuition turns out to be correct. Still, just as you would not be *sure* that this intuitive flash of understanding was true until you actually lifted the lid of the butter dish, so must less mundane glimpses into the truth about either material or spiritual reality be confirmed through direct experimentation.

Albert Einstein is said to have developed the foundation of his special theory of relativity by visualizing what it would be like to ride on a ray of light. As far as we know, Einstein did not actually ride on such a ray. After all, his own theory says that it is impossible for anything material to go that fast. But the insights he gained from this thought experiment have been confirmed many times. So in this case it turned out that Einstein's visualization accurately reflected objective reality. Experimentation was needed to confirm this, though. In the same way, flashes of insight into the nature of God *must* be followed up by a scientific approach that can confirm or reject them. Otherwise, it never can be known whether the insight is merely personal, or truly universal. If there is no way that one's insight can be reproduced and validated by other spiritual scientists, then it almost certainly is subjective imagination rather than objective reality. The following elaborates on this point.

Science demands proof, proof! I still do not see that you have provided much proof that mysticism reveals the objective private reality which is claimed to exist.

Ken Wilber has answered this criticism:

> It is sometimes said that mystic knowledge is not real knowledge because it is not public knowledge, only "private," and hence it is incapable of consensual validation. That is not quite correct, however . . . a *trained eye* is a *public eye*, or it could not be trained in the first place; and a public eye is a communal or *consensual* eye. Mathematical knowledge is public knowledge to trained mathematicians (but not to nonmathematicians); contemplative knowledge is public knowledge to all sages. Even though contemplative knowledge is ineffable, it is *not* private: it is a shared vision. . . . The knowledge of God is as public to the contemplative eye as is geometry to the mental eye and rainfall to the physical eye. And a trained contemplative eye can *prove* the existence of God with exactly the same certainty and the same public nature as the eye of flesh can prove the existence of rocks.[5]

In other words, Wilber is saying that the findings of mysticism can be confirmed by those who are willing to put in the time and effort needed to learn the methods of contemplative meditation. What begins as a private experience within one's personal consciousness, ends as a shared vision of higher planes of reality. Just as a non-mathematician is not qualified to affirm or deny the validity of a complex mathematical proof, neither can a person ignorant of contemplation affirm or deny the spiritual truths taught by perfect mystics. Mystics, of course, are more than willing to explain how to carry out the experiments by which the validity of those teachings can be tested.

This is what makes mysticism a spiritual *science*. The pity is that so few people study it, preferring instead to turn "thumbs up" or "thumbs down" to mystical concepts without ever entering the laboratory in which this science is practiced: the depths of their own consciousness. If I were to tell a physicist that even though I've never studied the mathematics of Einstein's special theory of relativity, I think the theory is hogwash, he'd be justified in laughing away my skepticism. The same laughter, of course, would be a reasonable response to any physicist who ridicules mysticism without having learned contemplative meditation.

A twentieth century perfect mystic, Sawan Singh, has written: "Men cannot profit from the experience of others in this Spiritual Science in the same manner as they can by the use of scientific instruments and inventions in material sciences. There is no material evidence to convince men of spiritual experience and the experiments of Saints. . . . In order to get the guidance of a Master we have to follow his instructions for going inside our own bodies, which are his laboratories, and thereby make his experience our own. Generally, the inner experience cannot be had collectively. Every soul has to make its own effort and gain its own experience. It is essentially individualistic."[6]

How do I know that the teachings described in this book truly reflect those of perfect mystics (assuming for the moment that they exist)? The author isn't one, is he?

Fortunately it is possible for a person to describe the moves of basketball without being able to slam dunk, or to take a photograph of a painting even though not an artist. My only claim is that I have studied mysticism for twenty-five years, and have been fortunate to have had a perfect mystic as a teacher. The way I look at it is this: if someone stands near a waterfall and has only a thimble to put under the cascading torrent, even that small amount he is able to catch will provide him with a taste of the water. And even though much of that thimbleful may drain out before he passes it on to his friend for a drink, those few drops that are left *still* will taste like the waterfall. I am attempting in this book to give the slightest taste of Ultimate Reality. If that seems at all refreshing, imagine what fully quenching your thirst would be like.

Not a bad metaphor. Yet you haven't answered my question about the validity of the mystic teachings in this book. Why should I believe them?

A certain amount of faith undoubtedly is involved. But this is much the same sort of faith expected of a physics student when he is told about the results of experiments with multi-billion dollar machines that reveal the nature of subatomic particles. The student does not have access to such technology, nor could he operate a particle accelerator if one was placed under his control. His faith in what is being taught stems from several sources. First, the experimental methods by which the results were obtained are open to view. They can be replicated by anyone with the will and ability to do so. Second, the student can go to the library and confirm that many different researchers have arrived at the same conclusions by conducting these experiments. Thus, as was noted earlier, belief in scientific findings is founded in part on the reproducibility of the means used to discover them. Similarly, perfect mystics provide—along with their teachings—the exact methods that can be used to confirm them, as well as evidence that other practitioners of spiritual science have discovered the same truths.

Let's see that evidence. That would help to convince me that what's being presented in this book is more than your own personal opinion.

Many authors do indeed sprinkle their writings about mysticism with quotations from a wide variety of sources to demonstrate that the teachings are common to different times and different places. I considered that approach, but decided against it. After all, my theme is that mysticism and the new physics both are branches of science, one focused on the spiritual realm of existence and the other on the material realm. When I read a book about the new physics, the author does not feel compelled to back up every assertion with quotations from this physicist and that physicist. He may use quotes to elucidate a point he is trying to make, but not to confirm the truth of the statement. The validity of the laws of physics stem from systematic theory and confirming experiments, not because so-and-so says they are true.

After all, why should you believe five quotations from mystics more than one quotation? For if mysticism is a sham science, then everyone who pursues this discipline is deluded and should be pitied rather than believed. In the same vein, if ultimate truth indeed can be found through the contemplative methods of spiritual science, then whoever has reached that end serves as a beacon for other truth-seekers. In a physics classroom all the student requires is one Ph.D. professor who knows his subject. It would be senseless and a waste of time to rush around to other colleges, taking notes from other instructors just to make sure that your professor's teachings are consistent with theirs. If physics indeed is a science, they will be.

In my presentation I have chosen to use quotations from mystics with discretion. In part, this is because of an unfortunate tendency for people to focus on *who* is speaking about spiritual science, rather than on *what* is being said. Such is not the case in material science. As was noted previously, there is not a Japanese physics as opposed to an American physics. Physics transcends the boundaries of nationality, religion, political affiliation, or other personal attributes.

The same is true of mysticism, but many fewer people realize it. Perfect mystics naturally live in a particular place, and may nominally be a member of a recognized religion—but they also will belong to one political party rather than another, and drive one model of car rather than another, and wear one style of clothes rather than another. The same is true, of course, in the case of material scientists. Yet these accidents of birth do not prejudice us toward physicists in the same way as they do toward mystics. This apparently is because many people consider divine truth to be the private property of a particular religion, with anyone not belonging to that exclusive club being forbidden access to perfect knowledge. Thus, goes the sectarian reasoning, only those who profess belief in a certain religious faith can be trusted to speak the truth about God.

Along these lines, let us try a simple test. Here are two quotes about the immanence of God:

(1) "He Himself fills every place by the very fact that He gives being to the things that fill every place."[7]

(2) "Within and without is He the Lord alone: it is He who Fills all places."[8]

These quotations sound very much alike. It would be difficult to agree with one, and not with the other. A reasonable person would consider the message as being more important than the messenger. But be honest. Now that I tell you that the first statement is from Thomas Aquinas, a Christian, and the second is from Guru Ram Das, a Sikh, does this change in some way how you perceive the quotations? If you are a Christian, and I say, "listen to this quote from Guru Ram Das. . ."; would not your ears still be ringing from the words "Guru" rather than listening fully to his message? Similarly, a Sikh might discount any teaching that came from a Christian theologian, even if it were identical to that taught by one of the mystics revered by his own faith.

Rumi—a great spiritual scientist

This is one of the reasons why I have chosen to support the key principles of spiritual science in this book with so many quotations from a single mystic, Jalaluddin Rumi. Rumi is not particularly well known, but those scholars who have devoted themselves to the study and translation of his writings are effusive in their praise: "one of the greatest mystics and poets the world has ever known," "the greatest mystical poet of Islam," "the greatest mystical poetry ever written," "the greatest spiritual masterpiece ever written by a human being," "the supreme mystical poet of all mankind." Strong words, but in my opinion, completely justified. One of his translators, Kabir Helminski, writes of Rumi:

A figure of almost prophetic dimensions, he became for some Muslims almost a second Muhammed, for Christians a second Christ, and for Jews a second Moses. Among those present at his funeral procession were people of different religious traditions, each of whom claimed that Jalaluddin had brought him to a deeper understanding of his own faith. . . . With complete respect for the prophets of Judaeo-Christian-Islamic tradition and with uncommon beauty and insight, he elucidated the mythic inheritance, the shared traditions of his age; yet he was somehow beyond his own culture and time. Although following the details of his Islamic faith, Rumi expounded a religion of Love. Without denying his own Islamic faith he was able to say, "The religion of Love is like no other," and "Gambling yourself away is beyond any religion."[9]

Rumi thus is perfectly suited to speak for all of the mystics who have propounded the truths of spiritual science. He illustrates how mysticism transcends any particular religion, yet is not opposed to the genuine teachings of any faith. Rumi was born in 1207 in what is today Afghanistan. As a youth Rumi became a master of the science and philosophy of his time, following in the footsteps of his father—a well-known teacher of religion.[10]

So up to his middle years Rumi devoted himself to a more or less traditional study of theology, law, and the sciences. However, when he was thirty-seven his life took a completely different course. This was foreseen by his father's close friend who had taken him under his tutelage, and told Rumi: "You are now ready, my son. You have no equal in any of the branches of learning . . . a great friend will come to you, and you will be each other's mirror. He will lead you to the innermost parts of the spiritual world, just as you will lead him."[11] That great friend was known as Shams of Tabriz, a traveling mystic.

As Jonathan Star has noted, up to that time Rumi was an accomplished scholar and fully conversant with the most complex and baffling problems of theology, even becoming the spiritual guide for thousands of disciples. But, says Star, "Rumi was not complete: he lacked a direct experience of the Supreme Reality, that one the Sufis call 'the Beloved.' Intellectually, Rumi knew everything about the mysterious 'wine' of Sufism . . . but he had never *tasted* it himself. One afternoon in 1244 all this changed—Rumi met his Master, Shams-i Tabriz, who gave him that divine taste, that direct experience of God, which transformed his life completely. . . . Several accounts of this event have been recorded, and though each differs in specifics, they all tell of an upheaval in Rumi and his sudden realization that all his book-knowledge was worthless when compared to the experience of the 'unseen world' that only Shams could give him."[12]

Shams-i Tabriz was a spiritual scientist, a perfect mystic, and Rumi became his eager student. Rumi's transformation is similar to that of the great Christian theologian Thomas Aquinas, about whom Thomas Matus has said: ". . . Thomas Aquinas himself was without question a great mystic, a man of profound spiritual experience, but he lived his mystical life on a plane worlds different from his theology. At the end of his life, when the tension between his experience and his theology had become intolerable, he said of his theological work, 'It is all straw!' "[13]

Thus when you read quotations from Rumi's poetry and prose, understand that his words flow from direct perception of Ultimate Reality, and that his mockery of intellectualization stems from his realization of the vast difference between this limited sort of knowing and genuine spiritual illumination.

> *You are certainty and direct vision,*
> *so laugh at opinion and imitation!*
> *You are all contemplation,*
> *laughing at transmitted sayings and news.*[14]

> *Whoever finds the way to vision in the spiritual retreat will never seek support from the sciences. Since he has become the companion of the spirit's beauty, he is tired of hearsay and knowledge. Vision dominates over knowledge.*[15]
> —Rumi

Rumi is not one to mince words. If the emperor is not wearing any clothes, he is not afraid to point out nakedness. But even though the following quotations about the limitations of material science and the intellect are hard-hitting, the intent lying behind them is entirely loving. Like all perfect mystics, Rumi wants us to *wake up!* The words he uttered—which were written down by his students, and thereby saved for posterity—are from a man whose consciousness has fully awakened. Observing that most everyone else is still asleep, he is calling out to us, *loudly.* Although becoming awake can be difficult, Rumi's call is true compassion.

> *That iniquitous man knows hundreds of superfluous matters in the sciences, but he does not know his own spirit. He knows the properties of every substance, but in explaining his own substance he is like an ass.*[16]
> —Rumi

> *Know that the intellect's cleverness*
> *all belongs to the vestibule.*
> *Even if it possesses the knowledge of Plato,*
> *it is still outside the palace.*[17]
> —Rumi

The Material Science of the New Physics

Physics is considered by many—particularly physicists themselves—to be the king of the mountain of the material sciences. In this view, all of the other sciences are derivatives of the fundamental laws of nature which are the province of physics. Chemistry, for example, generally is acknowledged to be theoretically reducible to the principles of quantum physics, which govern the activities of individual atoms and subatomic particles. However, in practice quantum physics is incapable of dealing with the vast number of particles involved in chemical reactions, so chemistry—which takes a "broader brush" perspective by looking at atoms in combinations—remains an essential branch of science.

So while there are strong arguments in favor of considering physics to be the foundation of material science, as a practical matter this discipline is ill-equipped to study many aspects of physical reality. The earth's crust is made up of atoms and governed by the laws of physics, but geology is the science of choice for understanding the dynamics of earthquakes. While the raw material that constitutes the stage of physical existence may indeed be reducible to the small number of fundamental particles known to physics, the number of ways in which these particles can act and interact essentially is infinite, and beyond our comprehension.

Look around and observe the amazing variety of material things within view. All are subject to the fundamental laws of physics—as are you—but cannot be reduced to those laws. No physicist can tell you what the exact shape and size of your house plant will be a year from now, or even if it still will be alive. In fact, as was noted previously, no physicist can predict with absolute accuracy the orbits of the planets in our solar system. Even though governed by known laws of motion and gravitation, how those simple laws operate in the complex actuality of our universe is beyond the present ability of man to compute precisely. Thus if physics is limited to understanding only the most basic features of the physical world, imagine how ill-suited is this science for unraveling the mysteries of spiritual domains of existence. This fact admittedly is recognized by many physicists, who have an appropriate humility regarding their discipline's ability to know Ultimate Reality. Yet the very language of physics can convey a different message.

Take, for example, the phrase "Theory of Everything." This is the Holy Grail of theoretical physics, the equation which contains within it the explanation for everything in the physical universe. Astronomer John Barrow writes: "Science is predicated upon the belief that the Universe is algorithmically compressible [simplifiable] and the modern search for a Theory of Everything is the ultimate expression of that belief, a belief that there is an abbreviated representation of the logic behind the Universe's properties that can be written down in finite form by human beings."[1]

This notion of a Theory of Everything will be discussed in more detail later. For now, I simply want to point out that this term *exists*, and that its very existence speaks volumes about the material science of physics. To most people the idea that the complete meaning of the universe could be put into writing is arrogant and absurd. To a perfect mystic, doubly so, for he has actually contemplated that final truth and knows that it is infinite, ineffable, and inexpressible. This is why mystics such as Rumi sometimes sound so harsh when referring to the physical sciences. They are not impugning the ability of physics, chemistry, geology and like disciplines to uncover the material laws of nature. But when these laws are represented as being the only laws, or the highest laws, perfect mystics feel the necessity to dissent. Physics as we know it today is a means well-suited for laying bare only the topsoil of physical reality, and it should not claim any more than that.

Physics and mysticism: cards from the same deck

Along these lines, it is unfortunate that the current connotations of the words "physics" and "mysticism" imply that these sciences are separate and distinct means of knowing Ultimate Reality. As was noted earlier, such was not always the case. Physicist Fritjof Capra says, "The roots of physics, as of all Western science, are to be found in the first period of Greek philosophy in the sixth century B.C., in a culture where science, philosophy and religion were not separated. The sages of the Milesian school in Ionia were not concerned with such distinctions. Their aim was to discover the essential nature, or real constitution, of things which they called 'physis.' The term 'physics' is derived from this Greek word and meant therefore, originally, the endeavor of seeing the essential nature of all things. This, of course, is also the central aim of all mystics. . . . "[2]

Perhaps the Greeks were right, and science needs to return to a more unified perspective. Such is the theme of this book: a set of playing cards which has been divided into two piles by color—red and black—needs to be shuffled into one deck again. One pile is mysticism, and the other the new physics. In the eye of God and from the standpoint of Ultimate Reality, there always has been only one deck, one truth. Over the centuries mankind has chosen to separate knowledge of reality into realms of spirituality and of materiality, religion and science, mysticism and physics. Perhaps the time is coming when this dichotomy will end. It serves no purpose and confuses truth-seekers who are led to feel that they must choose between the path of material science and the way of spiritual science, as if it were possible to walk only on one or the other.

Actually, there is *one* path. Just one. There is scientific truth about the objective reality of existence, and there is unscientific conjecture. Every physicist who believes this walks side by side with every mystic who shares that view. Certainly there are closed-minded people in both camps who choose to believe that their discipline is the only genuine science. If they are happier in choosing to remain in the company of only true believers similar to themselves, who can dissuade them? But it is sad that in a world already so divided by sex, and race, and nationality, and religion, and political persuasion, that spiritual and material scientists would distance themselves from brethren with whom they share such a common perspective.

Fritjof Capra, who is well acquainted both with physics and mysticism, noted their similarities in *The Tao of Physics:*

> The parallel between scientific experiments and mystical experiences may seem surprising in view of the very different nature of these acts of observation. Physicists perform experiments involving an elaborate teamwork and a highly sophisticated technology, whereas mystics obtain their knowledge purely through introspection, without any machinery, in the privacy of meditation. Scientific experiments, furthermore, seem to be repeatable any time and by anybody, whereas mystical experiences seem to be reserved for a few individuals at special occasions. A closer examination shows, however, that the differences between the two kinds of observation lie only in their approach and not in their reliability or complexity.
>
> Anybody who wants to repeat an experiment in modern subatomic physics has to undergo many years of training. Only then will he or she be able to ask nature a specific question through the experiment and to understand the answer. Similarly, a deep mystical experience requires, generally, many years of training under an experienced master and, as in the scientific training, the dedicated time does not alone guarantee success. If the student is successful, however, he or she will be able to "repeat the experiment." The repeatability of the experience is, in fact, essential to every mystical training and is the very aim of the mystics' spiritual instruction.
>
> A mystical experience, therefore, is not any more unique than a modern experiment in physics. On the other hand, it is not less sophisticated either, although its sophistication is of a very different kind. The complexity and efficiency of the physicist's technical apparatus is matched, if not surpassed, by that of the mystic's consciousness—both physical and spiritual—in deep meditation. The scientists and the mystics, then, have developed highly sophisticated methods of observing nature which are inaccessible to the layperson.[3]

Let us remember physicist Joe Rosen's previously-discussed definition of science: "Science is our attempt to understand the reproducible and predictable aspects of nature." Even though Rosen chose to limit the meaning of *nature* to the material universe, his definition holds for either a physical or spiritual science. In the quote above Capra has emphasized that repeatability—equivalent to reproducibility—is as important in mystical practice as in physics experiments. If a purported finding cannot be reproduced under the same experimental conditions, then it must not be accepted as truth.

This, by the way, serves to separate scientific mysticism from what might be termed "happenstance" mysticism. This includes a wide range of phenomena such as spontaneous out-of-body and near-death experiences. As physician Raymond Moody, Jr. and many others have concluded, near-death experiences appear to be evidence that a person's consciousness is separable from his physical body and is not dependent upon the brain for existence. Mysticism is in agreement with this. However, even though Moody has documented that many commonalties exist among the observations of people who have had near-death experiences, it is questionable whether this truly qualifies as a finding of spiritual science.

For while a necessary condition for such an experience is, obviously, being close to death, this is not a sufficient condition for reproducing the generally uplifting and inspiring visions reported by some who have almost died. "Some" is not all, for many (or most) of those who pass through death's door and then back again do not remember anything about it. Moody says that in regard to this issue, he "can't detect *any* difference between those who do and those who don't have such experiences during their 'death' in their religious background or personality, in the circumstances or cause of 'death,' or in any other factor."[4] Thus even though research into near-death experiences is important—especially to human beings, all of whom are to die someday (which accounts for the interest in Moody's work)—it lacks a solid scientific theoretical foundation.

Properties of scientific theories

Physicist Joe Rosen notes, "We want to *explain* laws of nature, to know the reasons for them; we want to *understand* the reproducible and predictable aspects of nature, not just describe and predict their phenomena. That is science. Scientists' term for an explanation of a law of nature is *theory*."[5] Physics has many theories, of course, as do the other sciences. It is not particularly difficult to come up with a theory. People do this all the time, in everyday life as well as in scientific research. "Cats are better pets than dogs," for example, is as much a theory as "Light always travels in straight lines." The main difference between them is that the Best Pet theory clearly is subjective and may be proven or disproven by each pet owner for himself, while the Straight Light theory is objective and its validity or invalidity is evident to every experimenter.

There is no way to put a subjective theory to an objective test. The United States Declaration of Independence from British rule contains these sentiments: "We hold these truths to be self-evident, that all men are created equal, that they are endowed by their Creator with certain unalienable Rights, that among these are Life, Liberty and the Pursuit of Happiness." No explanation for the validity of those truths is given, as they are considered to be "self-evident." Political statements, like other subjective perceptions, do not need to be justified—though their importance is undeniable. But in terms of our model of the four states of being, they exist in the domain of art rather than science.

The Declaration of Independence recognizes that people are free to pursue happiness in most any way they please—though society does restrict this right when one person's pursuit unduly interferes with other peoples', as in the case of a pickpocket who finds fulfillment in stealing wallets. There are no generally sanctioned Criteria for an Acceptable Philosophy of Life to which everybody must conform. In the case of science, on the other hand, there are such criteria that define the properties of a valid theory. This is one of the characteristics that separates objective science from subjective art, and general truth from personal perception.

Following the scientific method, then, requires criteria by which a theory can be evaluated—for observation or data alone are insufficient for this task. Rosen says that there are ten properties of an acceptable theory, something like a Ten Commandments for valid science: *logical implication, causation, truth, falsifiability, generality, fundamentality, naturality, simplicity, unification,* and *beauty*. Let us look at each, with the intent of using these properties as benchmarks for determining the extent to which the spiritual science of mysticism and the material science of physics are capable of discovering the ultimate theory of reality.

The emphasis is on "ultimate." For I do not question the ability of physics to discover relative truths about the physical world. Its capacity to lay bare the essence of highest existence is another question. Just for fun, I'll score each property of a valid theory as if it were a round in a boxing match: win, lose, or draw for the friendly combatants of mysticism and physics. (Understand that this does not conflict with my desire to have these disciplines viewed as brethren. Two brothers can be great comrades, but one still may be a more appropriate leader than the other. Perfect equality is a wonderful goal. Yet if a person knows how to find his way out of a dense jungle, and his companion does not, he who is lost would be well-advised to follow in the footsteps of the one who knows the path homeward.)

Logical implication

The most important property, Rosen says, is that whatever is explaining logically must imply that which is being explained. This is known as *logical implication*. I could develop a theory, for example, that says whichever side of bed I got out of in the morning determines the degree of happiness I will experience during the rest of the day. At first glance this does not sound like a totally ridiculous supposition, because when someone is grouchy people often say, "You must have gotten out of the wrong side of bed!" However, if I want my theory to find its way into the annals of science I had better come up with some logical linkages between human mood swings and the nature of beds. It would not be enough to conduct experiments to test this purported relationship, because a correlation between two things does not equate to an explanation of their relationship.

Every time I put my key into the ignition switch of my car, for example, it makes a slight sound as the edges of the key slide into the hole of the switch. A second or two later, in the usual order of things, the car's engine starts. Calculate the statistical correlation between the sound made by the key, and the successful ignition of my car's engine, and it would be a perfect match. Perhaps, then, it is the sound that starts the car. But a more scientific investigation of causative factors in car-starting would consider whether the intermediate activity of turning the key might also play some important part. And, of course, it would turn out that not only is this also correlated perfectly with engine ignition, but is a much more satisfying logical explanation of that effect.

Physics would seem to have the edge in this area because it relies heavily on mathematics, the most logical of symbolic languages. The principles of mysticism, though not opposed to logic, have their root in domains beyond mind and matter. Thus they cannot be expressed precisely in either words or numbers. What symbol could encompass a non-symbolic reality? How is it possible for a finite mind to comprehend such questions as why God created the material world? Mysticism holds that to fully understand the linkages between God, Spirit, and Creation, it is necessary to reach the level of consciousness from which the creative process emanated. Then the laws of spiritual science will be perceived as clearly as the laws of material science. So let us consider this round to be a *draw* between mysticism and physics.

Causation

Rosen adds that "another property of an acceptable theory is that what is explaining should be perceived not merely as logically implying what is being explained but as actually causing it."[6] Interestingly, he notes that the word "perceived" means that whether the causal mechanism exists in objective reality is not so important as that the theory contain such a causal linkage.

As an example, Rosen says that a theory might state laws of motion that are valid for all bodies and not just for planets. Laws that describe the motion of the solar system as a special case could be derived mathematically from the more general laws. Yet additionally one would need an actual Sun, and planets being attracted to the Sun, to understand the causative implications of the mathematical theory. This is a rather obscure point for those who are not schooled in the philosophy of science, but it does have implications for our examination of the relationship between physics and mysticism.

In that context, this criterion for a successful theory implies that causal linkages between various domains of existence do not have to be demonstrable in *each* domain. Thus if an all-pervading conscious energy known as Spirit is theorized to be the cause of effects in the material universe, it is not necessary for the existence of Spirit to be verifiable by physical science. After all, since by definition Spirit is a spiritual entity, it is impossible for it to be perceived with the material observational tools of science. Yet because mysticism posits logical causative linkages between Spirit and its effects in the physical universe, this qualifies as a scientific theory—even though it is only by reaching a spiritual plane of existence that this theory can be experimentally verified as fact.

A metaphor that illustrates this point, albeit crudely, is Galileo's theory regarding the rate of descent of freely falling objects. As was noted previously, the physical law that causes objects with different masses to fall at the same rate in a vacuum cannot be proven experimentally if you stand on the roof of a tall building and drop a feather and a bowling ball into the teeth of a howling gale. The bowling ball will reach the ground in a few seconds and the feather may be blown for hours before falling to earth.

As is usually the case in science, the myriad and complex *phenomena* of existence veil the simple and few fundamental *laws* which create the conditions of the everyday world. Just as in a hurricane the laws undergirding the motion of falling bodies are difficult to discern, so are the subtle spiritual laws of consciousness hidden in a turbulent mind full of never-ending thoughts and images of materiality. However, Galileo's laws of motion operate equally in a vacuum and in a gale, and spiritual laws affect both the mystic and the skeptic. A bowling ball has no awareness of the laws of motion, but when dropped it falls to the ground nonetheless.

Having studied the approaches of both spiritual science and material science in some depth, I am confident in awarding a *win* to mysticism in this round. As will be discussed more fully in a following section, physics not only has failed to explain the causes of certain fundamental aspects of nature—but actually tends to view these failures as *accomplishments*. The inability of physicists to determine simultaneously the position and momentum of a subatomic particle such as an electron has been called the "uncertainty principle," and elevated to the status of an elemental law of nature. This is one of the few times in the history of science when ignorance of the cause of some event has been celebrated as an extension of man's knowledge of the universe.

Randomness, in other words, is being elevated by some material scientists to almost the position of a deity. Assumed "causelessness" of events at the subatomic level even has been theorized to be the root of human free will. From the point of view of mysticism, it is strange that an inability to know the cause of something can be construed as validating the power of human consciousness. So an interesting turnabout of sorts is taking place in the new physics: rather than seeking causes for physical events—up to and including the creation of the universe—randomness and causelessness are seen by many physicists as fundamental principles of existence.

Mysticism, on the other hand, agrees with this response of Albert Einstein to those who refuse to look beyond the uncertainty of quantum physics: "You believe in a God who plays dice, and I in complete law and order. . . . "[7] So on this round of causation, a *win* has to be given to mysticism, which holds that laws of cause and effect govern all aspects of the physical universe.

Truth and falsifiability

Calling a theory true means that it is not contradicted by experimental findings. If theoretical laws of motion are held to predict where billiard balls end up on the table after being struck by the cue ball, then the actual position of the balls should be in accord with theory—taking into account unavoidable experimental error. *Falsifiability* is closely related to truth, because if a theory can be proven to be true, it also must be capable of being proven false. This is what separates science from dogmatism.

Up to a certain age children are willing to accept this parental response to "why?": "Because I told you so!" However, as we grow older this retort becomes less acceptable as a validation of the adult claim to truth. Reasons lying behind assertions come to be expected. "Eat your vegetables". "Why?" "Because they will help keep you healthy." This can be frustrating to parents, but the child simply is taking more of a scientific view of life. Rather than viewing his mother and father as Godlike, possessing infallible knowledge and authority, he wants to have proof that their pronouncements make sense.

Rosen notes that "to be falsifiable, a theory must predict something in addition to what it was originally intended to explain. . . . "[8] In other words, that prediction is tested against experimental results that were not taken into account when the theory was devised. Physics, as was discussed earlier, makes no claim to be able to predict anything lying outside of the material laws of nature. Viewed as a whole, physics is intended to explain the physical universe, nothing more.

Mysticism, on the other hand, claims complete knowledge of *all* domains of existence above this one. In a certain sense, then, by Rosen's criterion the highest mystical theories are not falsifiable—if nothing can escape the vision of God-consciousness, it is not possible (and nonsensical) to predict what lies beyond God. When one reaches the summit of a mountain, the map you have been following does not show anything higher than that peak. The map cannot be faulted for not predicting what lies ahead. It has gotten you to your final goal.

So for the property of truth let us call the bout a *draw*. Both physics and mysticism have developed experimental methods which are able to confirm the validity of their theories. That the theories are about separate domains of reality is irrelevant. Regarding the property of falsifiability, physics theories admittedly are more capable of being falsified because they aspire to less. A theory concerning only a small part of the physical universe easily is falsified when understanding expands about the whole in which the part is embedded.

Until the twentieth century astronomers thought that the stars in our Milky Way galaxy constituted the entire universe. It came as quite a shock when what was first thought to be a fuzzy image of a "star" turned out to be an entire other galaxy. Almost overnight every fundamental theory about the nature of the universe had to be reexamined. What once was thought to be the whole universe turned out to be a tiny fraction of it. Now the universe is estimated to contain at least 100 billion galaxies, not just *one* as was thought up to 1932. So the openness of physics theories to being falsified in part is due to their being so limited, and thus possessing so much falseness to be found. This does not deserve a win, so let us also call it a *draw* on falsifiability.

Generality and fundamentality

Rosen links another pair of properties of an acceptable theory: *generality* and *fundamentality*. He writes, "what is explaining should be more general than what is being explained and should also be more fundamental than the latter."[9] The more general a theory is, the more it encompasses. Trying to understand how galaxies form demands a broader perspective than studying the formation of individual stars. Galaxies admittedly are a huge collection of stars—billions upon billions—but need to be examined in their own right. For most cosmologists believe that stars are born out of a diffuse cloud of galactic raw material, not that stars come into birth on their own and then migrate together to create a galaxy.

For a theory to be fundamental, its components must be more basic than the phenomena being explained. This concept often can be difficult to tie down, as in the familiar example of the chicken and the egg. Which is more basic? One produces the other. Hence modern science increasingly finds the notion of "systems" to be more realistic than the idea of components working in isolation. Is the heart more fundamental to human life than the brain, or the lungs more than the intestines? Take any one of these organs away, and life soon will end unless an artificial organ is substituted. Still, it often is possible to find genuinely fundamental parts of an integrated system. An automobile can be driven if one of the cylinders in its engine fails to work, but not if the gasoline in its tank runs dry. Hence fuel is more fundamental for a car to run.

Mysticism, of course, is dedicated to the pursuit of both generality and fundamentality. It seeks to know the broadest truth about existence, and to understand the most basic force which sustains all of creation. Physics, it must be admitted, has chosen to limit its scope of inquiry to the material world and, within that sphere, to less than fundamental problems. Thus we have seen that physics is not capable of addressing the question of how the physical universe came into being—even though this does not stop physicists from speculating about this conundrum. When material scientists attempt to answer such truly basic questions, they necessarily end up resorting either to trivial reasoning (such as "the universe is here because otherwise we would not be asking why it is here"), or to assumed unexplained causes (as in "the universe resulted from a quantum fluctuation," leaving aside what created the laws of quantum physics in the first place).

So the referee—me—has no hesitation in awarding a *win* to mysticism in both rounds, generality and fundamentality. Physics may be able to stand shoulder to shoulder with spiritual science in other areas, but not here. No discipline other than mysticism possesses such a broad vision, or seeks to answer such basic questions of existence through a scientific method of inquiry.

Naturality

According to Rosen, this is an attribute of a good theory because "a theory should explain one aspect of nature by another and should not look outside of nature for its explanations. For example, 'the apple falls by the will of God' is not accepted as a theory among scientists, since godhood, by its very definition, lies outside nature."[10] Uh-oh. This looks like a clear-cut defeat for mysticism coming up, and the referee—though feigning neutrality in this bout—has a decided fondness for that contestant. But before awarding a win to physics, let's look more closely at the rules of the game to see if they are fair.

Remember from our original definition of science that Rosen considers nature to be "the material universe with which we can, or can conceivably, interact." So it is understandable that if physical science limits its inquiry to material existence, both the questions it asks and the answers it finds will be within that domain. There is nothing wrong with this. Physics is entirely justified in excluding metaphysics from its arsenal of investigational tools if it desires to be a purely physical science. Mysticism has acted similarly, since spiritual science pays little attention to explaining the details of physical phenomena.

The perspective of mysticism, though, is much like this: if a prisoner has been able to get out of jail, either by escaping or serving out his sentence, the layout of his cell ceases to be of much interest to him. Intoxicated with newfound freedom, now he has a whole world to explore. Mountains, cities, the seashore, jungles, deserts, Disneyland. When he could not move beyond the confines of a small cell, every detail of his surroundings was important: the closeness of the bars on his single window, the glimpses of countryside visible past the guard tower, the arrangement and composition of the bricks that wall him in. But who, once free, would choose to return to prison and pursue a further investigation of his cell's characteristics? Freedom means not being confined anymore, the condition of perfect mystics—who also are much more concerned with liberating inmates of the physical universe than with making them knowledgeable about the conditions of their imprisonment.

Recall from our discussion of Gödel's Theorem that materialistic belief systems tend to either erect barriers that shelter them from "alien" external influences, or to self-destruct as a result of being open to truth from any source. At present, physics is tending toward the first option, though it will be shown in a following section that many physicists recognize a need to eliminate this self-imposed rule that only material phenomena can be used to explain the material universe. In a sense, physics already has biased answers to the big sweepstakes question, "What is Ultimate Reality?," by the very rules it uses to screen out certain entries. In effect, the fine print of the entry form says, "Any answer concerning the true nature of nature which unnaturally looks outside of nature for explanation will not be accepted."

Aside from being bad grammar, this is an unscientific attitude. One would think that a more important criterion than *naturality* for an acceptable final theory about the universe is that the theory be *true*. For the boundary of nature is ever-shifting, while final truth is eternal. Neil McAleer points out that "In just 15 years (1917-32) the size of the known Universe expanded 1 trillion times" as a result of astronomical discoveries.[11] Given this humbling fact, is it likely that material science *now* has found the limits of existence? Physics and cosmology would be better off if they heeded these teachings of Lao Tzu, the Taoist mystic: "The supernatural is just a part of nature, like the natural. The subtle truth emphasizes neither and includes both. . . . The universe is already a harmonious oneness; just realize it."[12]

I am reminded of a supposedly true story about an Amish farmer. The Amish are an orthodox Anabaptist sect who do not believe in using much of our modern technology. For instance, they choose horses and buggies over automobiles. This farmer had a large fenced field, and tended to his crops in the hot Pennsylvania summer without the aid of a tractor. Possessing both a parched throat and strong religious faith, he approached his non-Amish neighbor with a proposition: "I'll pay you to string an electrical line from your house to the edge of my property," said the clever Amish farmer, "That way I can put a refrigerator there and be able to reach through my fence for a cold drink on a hot day."

This shows that it is possible to remain true to the tenets of a belief system, while benefiting from a wider view of the world. Electricity exists, whether or not a person chooses to make it an integral part of his life. God and Spirit exist, whether or not one believes in them. This will be evident to physics only when the hand of science reaches through the fence of materiality that has been built around that discipline. When it does so, mysticism promises that what will be grasped is most refreshing:

> *Think of the soul as source*
> *and created things as springs.*
> *While the source exists,*
> *the springs continually flow.*
> *Empty your head of grief*
> *and drink from the stream.*
> *Don't think of it failing—*
> *This water is endless.*
>
> —Rumi[13]

While physics may choose to limit itself to material explanations of material phenomena, this is an arbitrary interpretation of the scientific method which need not apply to spiritual science. So the purported property of naturality for an acceptable theory is invalid for our purposes, and a *draw* is declared for this round. Perhaps this quotation from Rumi will make the reasoning behind this decision clearer. Here he argues that practitioners of the material sciences are as mystical as mystics in their own way. A pot, the saying goes, should not call the kettle black.

The astronomer says, "You claim there is something other than these spheres and the terrestrial globe that I see. As far as I am concerned only these things exist. If

there is something else, show me where it is!" His demand is invalid from the
outset. He says, "Show me where it is," but that thing has no place. So come,
show me where and from whence is your objection. It is not upon the tongue, in the
mouth, or within the breast. Dissect all of these, piece by piece and atom by atom.
You will find nothing of this objection and thought in these places. So we realize
that your thought has no place. Since you do not know the place of your own
thought, how should you know the place of the Creator of thought?
                                                                      —Rumi[14]

Simplicity and unification

Regarding these attributes of a sound scientific theory, Rosen says that
"what is explaining should be simpler than what is being explained and
should also be more unifying than the latter."[15] Thus as physics traces
chains of cause and effect backward from gross physical phenomena to the
much subtler atomic and subatomic forces, reality appears both simpler and
more united. This is what makes the basic principles of the new physics
easier to comprehend than those of the old physics. As this science expands
its sphere of knowledge, the deepest laws of nature turn out to be simple
and few.

The Big Bang, which is how scientists refer to the moment of our
universe's creation some ten to fifteen billion years ago, provides a perfect
example. Referring to how the unified energy of the Big Bang came to
produce a tremendous diversity of phenomena, physicist James Trefil says,
"If I had to pick out a single overall characteristic of the evolution of the
universe, it would be the development of complexity from simplicity. The
universe seems to get simpler as we move backward in time. An atom, for
example, is a relatively complex structure—a conglomeration of protons,
neutrons, and electrons. A quark [which makes up protons and neutrons],
on the other hand, is a simple structure—indivisible and unstructured."[16]

Nobel prize winner Steven Weinberg goes so far as to say that the
convergent pattern of scientific explanation is perhaps the deepest thing that
science has learned about the universe. If this seems almost self-evident, it is
only because unification and simplicity are so deeply embedded in the very
nature of things that it is difficult to conceive of the world being otherwise.
However, it is possible to imagine a universe where the nature of reality
appears more confusing and complex the deeper one delves into it. There,
physicists transplanted from our kinder and gentler creation would lead
miserable lives, likely tormented into insanity by their increasing realization
that it will be impossible to ever make sense of existence.

Thankfully, such is not the case in *our* universe. I say, "thankfully,"
because most people prefer simplicity and unity to complexity and
separateness. This is what makes lying on a sunny tropical beach, clad solely
in a bathing suit and thinking only of when to go back into the warm

ocean, so enjoyable. We feel ourselves melting into the soft sand, and floating as one with the gentle waves. In a sense, these pleasant sensations are remembrances of a time when our soul—the essence of consciousness—was not so deeply entwined in the confusion of this material world. Mysticism is in complete agreement with physics that explanations of existence will become simpler and more unified as they move closer to the ground of being. As physicist Paul Davies notes, this search for truth can take on almost religious connotations for even hard-headed material scientists:

> The physicist approaches his subject with something near to reverence, compelled by his belief in the mathematical beauty and simplicity of nature, and convinced that by digging deeper into the bowels of matter, unity will emerge.[17]

If I change only a few words, these sentiments can be made to apply equally well to mystics:

> The mystic approaches his subject with reverence, compelled by his belief in the beauty and simplicity of creation, and convinced that by digging deeper into the depths of consciousness, unity will emerge.

So we have to call it a *draw* on this round. Physics and mysticism are devoted similarly to the principles of simplicity and unification. If anything, mysticism could be viewed as stronger in this area, because contemplative meditation—the research tool of spiritual science—itself is simple and unified. Both the methods and goal of mysticism, then, are commensurate and point in the same direction. Physics shares, in Weinberg's terms, the goal of coming to the endpoint of the "convergent pattern of scientific explanation." But any serious student of this material science would have to agree that its predominantly mathematical tools are in no way as simple and unified as what is being sought.

That is, physicists theorize that at the instant of the Big Bang energy—or nascent matter—was in the simplest state imaginable. To understand why the Big Bang took place, and how that energy evolved over time, physicists utilize some of the most complex mathematical models imaginable. More correctly, they are unimaginably complicated to the layperson. Regardless, the mathematics of quantum cosmology certainly is not the simplest way to describe the creation of a universe. The direct explanations found in holy religious books are much more in the spirit of the pure unity from which our universe was born:

Sikh *Adi Granth:*	One Word, and the whole Universe throbbed into being.[18]
Christian *Bible:*	In the beginning was the Word, and the Word was with God, and the Word was God.[19]
Islamic *Koran:*	Creator of the heavens and the earth from nothingness, He has only to say when He wills a thing: "Be," and it is.[20]
Jewish *Old Testament:*	And God said, Let there be light: and there was light.[21]
Taoist *Tao Te Ching:*	The nameless is the beginning of heaven and earth. The named is the mother of ten thousand things.[22]
Hindu *Upanishads:*	Accordingly, with that Word, with that Self, he brought forth this whole universe, everything that exists.[23]

Are these remarkably similar descriptions of creation only myths, or echoes of scientific fact? Mysticism holds that these simple words are much closer to objective truth than the complex equations of modern cosmologists. This assertion will be discussed further in another section. For now, we must examine one more criterion of a valid theory.

Beauty

Because *beauty* is a core principle that guides both material and spiritual scientists in their search for ultimate truth, it is fitting that this is the final quality to be addressed. Yet as Rosen notes, this quality is in the eyes of the beholder. Like other non-symbolic realities, it exists yet cannot be adequately described. Just as the difference between a lovely painting and an unsightly one is clear to an art critic, so can scientific theories be distinguished using this criterion. "A scientist," says Rosen, "will always prefer a beautiful theory to an ugly one, other things being equal, and even often at the expense of some other desirable property of acceptable theories."[24] What makes for a beautiful theory? He believes that the previously-mentioned qualities of simplicity, unification, and generality are central to a perception of beauty.

Evidence from everyday life supports this view. At one time or another everyone has had the satisfying experience as finally having "all the pieces fit together" concerning a vexing problem. This solution may not have been arrived at through a logical process, or indeed by any means that could be described to another person, but somehow you simply *know* that the decision you have reached is correct. It is simple, obvious, and clear what to do now. The actual issue could range from the type of car to purchase, or who to marry, or what to eat for dinner tonight. The sensation of "Yes, this is right" that comes over one at such a time is much the same as what a scientist feels in the company of a beautiful theory. The opposite reaction of "No, this is wrong" arises in the presence of ugly solutions. Anthony Zee includes this story about Albert Einstein in his book, *Fearful Symmetry* (appropriately subtitled, "The Search for Beauty in Modern Physics"):

> What I remember most clearly was that when I put down a suggestion that seemed to me cogent and reasonable, Einstein did not in the least contest this, but he only said, "Oh, how ugly." As soon as an equation seemed to him to be ugly, he really rather lost interest in it and could not understand why somebody else was willing to spend much time on it. He was quite convinced that beauty was a guiding principle in the search for important results in theoretical physics.
>
> —H. Bondi[25]

It must be kept in mind, however, that in either material or spiritual science this intuitive perception must be tempered by reason and experiment. After all, we have seen that there are nine other properties of an acceptable theory besides beauty. Too often, people searching for truth in life jump aboard bandwagons which prove to be unable to take them where they want to go. The music they hear at first is enchanting, but soon sour notes predominate. Faith must coexist with an attitude of, "This has the ring of truth. Yet I must test its soundness." Blind faith in either religious doctrine or scientific orthodoxy is dangerous. A beautiful facade often hides an ugly interior. Beauty, then, is a useful guide in the search for truth as long as it is recognized to be better at pointing the way, than of making the journey.

While I find nothing more beautiful than the principles of mysticism. I recognize that the physical laws of nature possess their own splendor which evoke the same sort of admiration in material scientists. So a *draw* on this property seems called for. For I hope that this discussion has made it clear that mystics and physicists have a similar reverence for the essence of reality. Physicist Zee writes:

> Physicists from Einstein on have been awed by the profound fact that, as we examine Nature on deeper and deeper levels, She appears ever more beautiful. Why should that be? We could have found ourselves living in an intrinsically ugly universe, a "chaotic world," as Einstein put it, "in no way graspable through thinking." Musing along these lines often awakens feelings in physicists best described as religious. In judging a physical

theory purporting to describe the universe, Einstein would ask himself if he would have made the universe in that particular way, were he God. This faith in an underlying design has sustained fundamental physicists.[26]

Scorecard, please

All in all, then, I came up with a draw between physics and mysticism on seven of the ten properties for an acceptable scientific theory, with three of the rounds called a win for mysticism. If this actually was a contest between opposing teams, we might expect to hear cheers for the winner. But such is not the case here. The game of truth produces only winners, whether the play is carried out on the field of material or spiritual science. My only purpose in carrying out this exercise was to demonstrate that even when using criteria designed to be applied to the rigorous discipline of physics, mysticism at least holds its own as a branch of science.

A mystically-inclined reader who already is convinced of this could, of course, accuse me of whipping a horse which already is across the finish line. By the same token, an inveterate materialist may consider that I have been flogging a dead carcass. The previous section, and indeed this entire book, thus is directed more toward those who are receptive to the notion that spiritual realms exist in addition to material reality—but who are skeptical of relying solely on religious faith as the foundation for this belief. "Proof, proof, give me proof," they plead.

The investigative methods of mysticism *can* provide the spiritual scientist with direct evidence that God and Spirit are as real—no, more real—as gravity and electromagnetism. This science uses the same criteria as physics to distinguish between the genuine and the false, between partial truth and ultimate truth. Any seeming dichotomies between these disciplines stem primarily from the separate domains of existence being studied (spiritual vs. material) and the dissimilar observational tools being utilized (pure consciousness vs. reason and physical perception)—and *not* from any difference of opinion regarding the importance of laying bare the nature of reality by following the scientific method.

A geologist studying the earth's crust will practice a form of science distinct from a cosmologist investigating the characteristics of distant galaxies. In the same way, a mystic exploring higher domains of consciousness will conduct his research in a manner different from a physicist delving into the nature of subatomic particles. So it is absurd to say that one of these investigators is not practicing "science" only because his observational tools and sphere of interest differ from the others. By that token, there could be just *one* genuine science—and geology, cosmology, physics, chemistry, biology, and the like would have to fight it out between themselves as to which would claim that distinction. If that seems nonsensical to you, then the idea of excluding mysticism from the club of scientific disciplines should appear equally arbitrary.

In the quotation below Rumi is reassuring material scientists of his time that they should have no fear that attending his school of mysticism will interfere with their research concerning physical reality. Quite the opposite—spiritual science will give life to their other investigations:

> *Those persons who have made or are in the course of making their studies think that if they constantly attend here they will forget and abandon all that they have learned. On the contrary, when they come here their sciences all acquire a soul. For all sciences are like images; when they acquire a soul, it is as though a lifeless body has received a soul. All knowledge has its origin beyond, transferring from the world without letters and sounds into the world of letters and sounds. In yonder world, speaking is without letters and sounds.*
>
> —Rumi[27]

Theory of Everything, or God?

Having, I hope, further dispelled the myth that physics and mysticism are unrelated approaches to revealing the nature of Ultimate Reality, we can take a look at how each of these sciences generally describes this final truth. For here also, seeming differences between material and spiritual science turn out to be more illusory than real. The endpoint of physics is referred to as the "Theory of Everything." Mysticism seeks to know God. Are these disciplines searching for different truths, or are varying terms being used for the same goal?

Astronomer John Barrow, author of *Theories of Everything—The Quest for Ultimate Explanation*, writes: "Today, the real goal of the search for a Theory of Everything is not just to understand the structure of all the forms of matter that we find around us but to understand why there is any matter at all, to attempt to show that both the existence and the particular structure of the physical Universe can be understood, to discover whether, in Einstein's words, 'God could have made the universe in a different way; that is, whether the necessity of logical simplicity leaves any freedom at all.'"[1]

This is quite a statement. If it is to be believed, material science now has even higher metaphysical aspirations than does mysticism. For if a Theory of Everything seeks not only to know why and how the universe was created, but also whether it could have been made in any different way, this amounts to being able to second-guess God—to knowing more than He does about the process of creation. Better to be a mystic than a physicist if you are not prepared to take on such ultra-divine responsibilities. Lekh Raj Puri, in *Mysticism, the Spiritual Path*, offers this considerably humbler description of the goal of mysticism: " . . . a spiritual trance in which the soul beholds the light of God, and merges its individuality in the universality of Omnipresent Spirit."[2]

In other words, mysticism aims to know God from the inside as it were, and having reached that sublime state of consciousness be able to understand directly the whys and wherefores of existence. Material science, by contrast, seeks to study creation from the outside. By delving into the intricacies of the process by which the physical universe was born, and subsequently evolved, the hope is that somehow the mind of the Creator can be known. As was mentioned previously, mysticism says that this is as likely as a bird being able to understand the consciousness of the person who built its birdhouse, merely by studying how that dwelling is constructed.

Truth is beyond words

Yet if we leave aside for a moment the question of whether material science is up to the task it has set out to accomplish, we are left with a remarkable similarity between the endpoint in physics of finding a "Theory of Everything," and the goal of mysticism to know "God." This can be made clearer by taking another look at the ten properties of an acceptable scientific theory. Let us state each of these criteria *in the extreme*, which would be necessary if an ultimate theoretical framework were to be evaluated. Hence, such a theory—or intertwined set of theories—would need to be:

1. The most *logical implication* of everything in existence.
2. The *cause* behind all other causes found in creation.
3. *True*, not being contradicted by experimental findings of any kind.
4. *Falsifiable,* or capable of predicting correctly all other facts of existence.
5. More *general* than everything being explained.
6. The most *fundamental* explanation of existence possible.
7. As *natural* as needed to explain all of nature.
8. The *simplest* theory imaginable.
9. Able to *unify* every aspect of creation.
10. The most *beautiful* explanation of existence.

If one looks behind the scientific phrasing, these essentially are qualities of a divine being. So let us not be distracted if two terms point to the same reality. A physicist has every right to say that a "Theory of Everything" is the cause behind all other causes, the unity linking all separateness, the truth of all truths. Similarly, a mystic may prefer to use the word "God" for that which has these attributes. Since both are pursuing the same quarry, why not join together in the hunt? Rumi tells the story about four people who did not know they all wanted the same thing: grapes.

> *A man gives one coin to be spent among four people.*
> *The Persian says, "I want* angur.*"*
> *The Arab says, "*Inab, *you rascal!"*
> *The Turk says, "Shut up all of you.*
> *We'll have* istafil.*"*
> *They begin pushing each other about,*
> *then hitting each other with their fists, no stopping it.*
> *If a many-languaged master had been there,*

He could have made peace and told them:
I can give each of you what you want
with this one coin. Trust me, keep quiet,
and you four enemies will agree.
I know a silent, inner meaning
that makes of your four words one wine.[3]

Metaphysics in physics

If a Theory of Everything ever is discovered by physics, it would both
encompass and transcend the known laws of nature. Thus it necessarily
would possess the characteristics of those laws—whatever attributes the
separate laws of nature enjoy, the unifying Theory of Everything would
display as well, and in abundance. Physicist Paul Davies describes those four
generally agreed-upon properties.[4]

First, the laws of nature are *universal*: they apply unfailingly in every
corner of the physical creation. Second, the laws are *absolute*: they do not
depend on who is observing nature, or the actual state of the world; the
laws govern the universe, not the other way around. Third, the laws are
eternal: timeless, unaffected by the dynamics of ever-changing phenomena.
Finally, they are *omnipotent*: nothing escapes the grasp of the laws of nature;
and they are all-powerful. Plus, in Davies' judgment, they are *omniscient* in a
sense, because a physical system does not have to provide the laws with any
information for them to do their job of governing that system.

Where does this leave us? Physics tells us that the laws of nature, and by
inference a Theory of Everything, are universal, absolute, eternal,
omnipotent, and omniscient. This should remind us of some other entity
which is said by mysticism and religion to have these same attributes. Again,
do not the terms "God" and "Theory of Everything" appear to be referring
to the same ultimate truth? "Curiously," says Davies, "the laws have been
invested with many of the qualities that were formally attributed to the God
from which they were supposed to have come."[5] So even though most
physicists would claim to have left metaphysics behind in their scientific
pursuit of knowledge, the characteristics of the laws of nature which have
been discovered by physics bear a remarkable resemblance to the attributes
of God.

It seems neither unreasonable nor unscientific to suggest, "Perhaps
those laws look like God because He is their father." Yet this premise
appears to be unacceptable to most material scientists, even though it
happens to be true. I am not sure why. Clearly many of those scientists
claim to be religious and at least intellectually believe in the existence of
God. Still, there is a general reluctance to recognize openly that metaphysics
has a place in physics. Not a complete reluctance, however, as will be
demonstrated below. But on the whole, metaphysical assumptions creep

into physics almost unnoticed through the back door, rather than being put forward in an explicit manner. In the words of B. Alan Wallace:

> Let us, for the moment, define religious doctrine as a system of belief concerning such issues as the nature of the Ultimate Being, the essential meaning of human life, the origin of the universe, the destiny of the individual following death, and the means to salvation. Viewed in such terms, it becomes immediately apparent that contemporary science is thoroughly suffused with religious doctrine. That doctrine may vary anywhere from agnosticism to unequivocal atheism. In this sense, twentieth-century physics is as profoundly influenced by metaphysics and religion as it was during the time of Newton.[6]

Take the simple assertion of a materialistic physicist that "God did not create this universe." Certainly everyone is welcome to put forth their own Theory of Everything, and this statement could be a component of such a theory. But the scientific method requires that a theory be defended. What evidence can be put forward that it is true? Can that physicist prove conclusively that some other force or being *was* responsible for creation, or that the universe has existed eternally? No, he cannot. Hypotheses regarding these questions abound in the new physics, but none are even close to being proven. So a negative assertion, "God did not create this universe," is being put forward as truth with no positive evidence to back it up.

In the thirteenth century Jalaluddin Rumi discredited this approach, and his reasoning is equally valid today. Though Rumi used simple language and examples from everyday life to make his point, this should not detract our attention from the sound logic he used. He says:

> *Statements based on affirmatives are easier to make than statements based on negatives. When you make a negative proof, it is like saying that So-and-So has not done something. It is difficult, however, to know any such thing. It necessitates having been with that person from the beginning of his life to the end, day and night, during his sleep and waking hours, in order to say that he has absolutely never done such a thing. . . . For this reason testimony based on a negative statement is not admissible because it is not within the realm of possibility. On the other hand, testimony based on an affirmative is within the realm of possibility and quite simple. One need only say, "I was with him for a moment, and during that one moment he said thus and so and did thus and so." Such testimony is acceptable because it is within the realm of human capability[7]*

Thoughts have no end; love does

Some might question, of course, whether it is humanly possible to be present at the creation of the physical universe so as to be able to "testify" about what caused that event. Firmly embedded as we are in the meshes of time and space, this seems to be a fantasy. However, mysticism unequivocally states that not only is this possible, but many have done so while still existing in the human body—just like you and I. The nature of Ultimate Reality can be known here and now. It is not necessary to physically die in order to become one with God, and privy to divine wisdom.

Through contemplative meditation, perfect mystics have been able to reach a level of consciousness that is far beyond time and space as we know it. Sawan Singh has said that in this domain "the whole creation looks like bubbles forming and disappearing in the Spiritual Ocean."[8] This direct perception allows the highly advanced spiritual scientist to assert with complete confidence that the universe was created by the Spirit of God. Every major religion and every perfect mystic agrees on this point. So using scientific phrasing, we can say that whenever serious research has been carried out concerning the question, "Did God create the universe?"—the answer has been in the affirmative. A negative answer has been given only by those who have not conducted proper research on the issue, which requires diligent effort in learning and practicing contemplative meditation.

Given the difficulty of proving the non-existence of spiritual domains, and the preponderance of experiential evidence that refutes atheism, it seemingly would take considerable effort to develop a Theory of Everything which is purely materialistic. After all, this would require proving several profound assertions, including but not limited to: one, that everything in our universe has a material nature, or at least a material origin; two, that there is nothing non-material outside of the physical universe which could be influencing matter in this domain; three, that the human mind is not itself subject to the laws of nature governed by the Theory of Everything, which allows man to go "one up" on that Theory and lay bare its secrets.

Let us consider that last assertion for a moment. If physicists are able to discover an ultimate Theory of Everything, this implies either that everything in the universe is governed by the theory except the investigative methods of physics, or that the theory controls the process of its own discovery along with everything else. Along these lines, physicist Stephen Hawking writes that "if there really is a complete unified theory, it would also presumably determine our actions. And so the theory itself would determine the outcome of our search for it! And why should it determine that we come to the right conclusions from the evidence? Might it not equally well determine that we draw the wrong conclusion? Or no conclusion at all?"[9]

Hawking's answer to this problem is not entirely satisfying. He says that Darwin's principle of natural selection tells us that variations in genetics and upbringing will produce some individuals in a species who are better able than others to draw right conclusions about the world around them and act accordingly. Being better able to survive and reproduce, eventually their superior abilities of behavior and thought will come to dominate. Thus, in a sense, the theory of evolution supposedly assures that the Theory of Everything one day will be found by humans who have been fine-tuned to understand their surroundings. But this argument still leaves us either with having to assume that natural selection is a law that is separate from the Theory of Everything (which means that it does not really explain *everything*), or that the Theory of Everything wants to be known—and so includes within its scope a law that permits the human mind to understand the law that permits the human mind to understand the law that permits. . . .

I stopped just in time. We were heading back into the self-referential confines of Gödel's Theorem again. Remember, this is that hall of mirrors where the mind of man either is reflected back onto itself as far as the eye can see—in which case no escape is possible—or both the mirrors and mind are broken, thereby allowing one to escape the illusion at the cost of purported "rationality." Mystics unfailingly choose the latter option. Rumi advises:

> *Sacrifice your intellect for the love of the Friend;*
> *in any case, all intellects come from His side.*
> *The true possessors of intellects have sent their intellects to that side;*
> *the fool has remained on this side, where the Beloved cannot be found.*
> *If your intellect departs from your head in bewilderment,*
> *every hair on your head will become a head and an intellect.*[10]

Physicists generally are still inclined toward the first choice: try to use human reasoning to understand what has created human reasoning. This then requires that we try to use human reasoning to understand what has created our ability to try to use human reasoning to understand what has created human reasoning. And so on, and so on. Thought piled upon thought, concept heaped on concept. Strangely enough, those with a materialistic view of the world fail to find any solid place to stand, while mystics who have left body and mind behind rest on the immovable foundation of God.

Ultimate Reality: One or Many? Matter, Mind, or Spirit?

At the risk of further muddying the already turbid conceptual waters we have entered, this seems to be the place to look more closely at the meaning of "materialism." After all, throughout this book I have used the terms *material, materialistic,* and the like—often in contrast to *spiritual* and *non-materialistic.* Defining these words more precisely will provide us with further insight into the dilemma faced by physics as it strives to arrive at a Theory of Everything. For what that "everything" consists of obviously is a key question. The answer not only is the central concern of both physics and mysticism, but even moving toward an answer requires that the question be addressed in some depth. What is sought determines to a large extent what shall be found.

More precisely, what is seeking is the same as what will be found. For consciousness is part of the phenomena being studied, as well as the means of investigation. This understandably creates complications which are much less of a problem when consciousness is directed outwards toward the apparently material world. Whatever answer we come to concerning the basic "stuff" the universe is made of, naturally has implications for *ourselves-* —we being part of the universe. So the debate in this area tends to be carried on with a bit more feeling than does, say, research concerning the nature of basic subatomic particles.

As physicist Nick Herbert noted in his book *Elemental Mind,* there are two main divisions of thought in what he termed the "fledgling science of consciousness" (I would disagree, of course, with that characterization if it fails to recognize mysticism as a full-grown spiritual science). Either mind and matter are separate, or there is only one fundamental substance. The first position is dualism, the second monism. Within dualism, one can hold that mind is a byproduct of matter, matter is controlled by mind, or the two mutually influence each other.

I will not discuss dualism further, because most physicists—and all perfect mystics—strongly favor the competing theory of consciousness, monism. It is not difficult to understand why this is so. Material science, as was noted in a previous section, strives for "simplicity" and "beauty" in theories. Clearly any theory—and particularly a Theory of Everything— which reduces the complexity of phenomena to a single fundamental substance will be simpler than one which posits *two* pillars of existence. Monism also is a more beautiful theory than dualism.

This appears to be a subjective statement, but physics can provide objective support for it. Physicist Anthony Zee says that, in line with beliefs of the ancient Greeks, beauty in the domain of physics is related to "symmetry"—a precise mathematical concept involving the concept of invariance.[11] A circle, for example, is more symmetrical than a square, because rotations around its center leave a circle invariant, while a square is left unchanged only by rotations through angles of 90 degrees, 180 degrees, 270 degrees, and 360 degrees. Simply put, you can turn a circle any which way around its middle and it always will look the same. But move a square just a little bit, and you can tell that its position has been changed. Since physicists consider the laws of nature to be eternal, then final truth should be highly—even perfectly—symmetrical, unchanging, and beautiful.

Mystics state the reason for their belief in monism more briefly: God is One. This appears to be a tautology—unity explains existence because existence is unity. Yet if existence truly is composed of a single substance, *any* explanation for why it is as it is would be a tautology. In other words out of that Oneness have come the phenomena to be explained, the mind which is doing the explaining, and the explanations themselves. Merge the phenomena, the mind, and the explanations back into their source, and there is nothing left but One. Robert Bly offers this version of a poem by a perfect mystic of the fifteenth century, Kabir:

> I said to the wanting-creature inside me:
> What is this river you want to cross?
> There are no travelers on the river-road, and no road.
> Do you see anyone moving about on that bank, or resting?
> There is no river at all, and no boat, and no boatman.
> There is no towrope either, and no one to pull it.
> There is no ground, no sky, no time, no bank, no ford!
> And there is no body, and no mind! . . .
> Kabir says this: just throw away all thoughts of imaginary things,
> and stand firm in that which you are.[12]

Since physics and mysticism agree that the essence of Ultimate Reality is monistic—a unity—where is the conflict between these sciences? In the nature of that unity. Nick Herbert goes on to point out that a monistic materialist believes that matter is all that there is. Mind, and our illusory "consciousness," merely is one of the attributes of matter. He quotes William Uttal, a modern materialist, as saying that "Mind is to the nervous system as rotation is to the wheel."[13] Thus most brain researchers believe that the motion of the mind is produced mechanically when nerve cells reach a sufficient degree of complexity. This reasoning is what allows some computer scientists to hold that robots will become conscious when their hardware and software are sufficiently sophisticated.

That philosophy is called "emergent materialism," since the idea is that consciousness emerges from matter only when a system is highly complex. Herbert says: "Because of its simplicity, concreteness, falsifiability, and general concordance with the present fashion of scientific thinking, emergent materialism has itself emerged as the dominant mind/matter philosophy of the scientific community."[14] In the context of this book that statement would be better phrased as, "of the scientific community *which studies physical existence.*" For mysticism holds a different conclusion about the fundamental substance of creation. The other side of materialism is idealism, or the belief that mind underlies matter. Mysticism goes beyond the traditional philosophical perspective of idealism, however. It agrees that mind is more basic than matter, but considers Spirit as more basic than mind. This is not just an academic distinction, or a matter of words. From the standpoint of mysticism, Spirit is the essence of consciousness and the soul, not mind. Just as most physicists erroneously consider the mind to be a product of matter, those who believe that mind is the root cause of matter are similarly mistaken. Spirit is the Ultimate Reality that lies behind both mind and matter.

Since the nature of Spirit will be the focus of succeeding chapters, it will not be addressed more fully here. For the moment it is sufficient to recognize that both physics and mysticism are monistic sciences, the former holding that matter is the substrate of existence, and the latter that Spirit is. This would seem to result in an unbridgable gulf between these sciences. Yet if we take a closer look at the materialistic world view of physics, the division between physics and mysticism narrows considerably.

For, in truth, reality is an undivided whole: matter emerges from mind, mind from Spirit, and Spirit from God in a continuous flowing stream of all-pervading conscious energy. Knowing neither the source of this stream, nor the boundaries of the channel within which it flows, material science considers the stagnant pool of the physical universe to be the entire watercourse. As we shall see, mysticism teaches that matter is Spirit far reduced in energy, or vibratory activity. Far from being the essence of reality, it is the crudest manifestation of God's creative action. To put it plainly, believers in emergent materialism have got the truth backwards. And it is not even necessary to look beyond the findings of modern physics itself—the linchpin of materialistic science—for evidence to support this assertion.

Matter's Magic Act—Now You See It, Now You Don't

While some of what will be discussed in this section may appear mysterious and illogical, be assured that the following overview of particle physics reflects mainstream material science—though admittedly simplified and popularized. Many details have been omitted because my object in this book is not to make the reader an expert in the new physics, nor am I qualified to do so. Some excellent books on various aspects of the new physics are listed in the bibliography. These contain indepth descriptions of findings which I will merely touch on.

Keep in mind that my not-so-secret agenda is to convince the reader of this assertion: if your goal is knowledge of Ultimate Reality, you will not attain it via the means of physics, or any other material science. But if you are content with an echo of final truth, these disciplines will serve you well. The laws of physics reach to the very edge of spiritual domains of existence—where the reverberations of Spirit can be perceived by those able to toss off the blinders of a materialistic world view.

Let us see, then, how far we are able to trace the origin of matter. Consider the page which holds these words. It certainly feels relatively substantial. Though paper can be bent, folding does not destroy it. But to make this page disappear for good, you could burn it. This would produce quite a bit of heat, some ashes, and smoke. The piece of paper would no longer exist, but these other entities—heat, ashes, smoke—would have taken its place. If you placed the ashes in a blast furnace, they too would be converted into heat by the even higher temperature.

Matter equals energy

This reflects a fundamental law of physics: matter and energy are equivalent. If people know only one scientific formula, it often is "$E = mc^2$," Einstein's famous equation which states that energy is equivalent to mass times the speed of light squared. As Hazen and Trefil point out, "Because the speed of light, c, is such a large number, the conversion of a little bit of mass can produce a lot of energy. By the same token, it requires a great deal of energy to produce even a small particle. A block of cement small enough to fit under your kitchen table could run the entire United States for over a year if it were converted completely to energy."[1]

So there is a tremendous amount of energy locked up in matter. In the words of physicist Heinz Pagels, "All the mass you see around you is a form of bound energy."[2] But how exactly does this transformation of energy into matter take place? This is one of the great mysteries which remains to be unraveled by science. Physics cannot provide the final answer to this question, but it has been able to make significant progress in laying bare the secrets of matter. Take the period at the end of this sentence. Within that tiny space a million atoms could be placed end to end. Atoms, of course, are one of the building blocks—or levels—of matter.

Matter mostly is immaterial

Pagels notes that five distinct levels of matter have been identified by physicists: molecules, atoms, nuclei, hadrons, and quarks. In a telling phrase, he says that "In order to understand any level it was necessary to penetrate deeper."[3] And the deeper into matter one delves, the simpler its structure becomes. Pagels says that the eight dozen or so atoms (there are ninety-four naturally occurring elements, plus others which can be created in a laboratory) are a simplification over millions of molecular compounds— combinations of atoms. In like manner, those eight dozen atoms are made up of just two entities: nuclei and electrons.

In my high school physics class the teacher likened atoms to miniature solar systems, with electrons playing the part of orbiting planets and the nucleus the role of the sun. I remember him holding up a model which was supposed to give us an idea of what the subatomic world looked like. It had little solid balls, the electrons, circling around another ball, the nucleus. As is often true in education, I have had to unlearn this lesson. Actually subatomic particles are neither as firm and perceptible as this model implied, nor could the scale of the atomic world be represented accurately by a handheld prop.

The actual situation is like this: if a hydrogen atom could be magnified until its nucleus was the size of a ping-pong ball, then its electron—as small as the tiniest speck of dust—would be whirling about some three hundred meters away. In other words, picture setting a ping-pong ball on the edge of the first of three football fields laid end-to-end, and putting a dust speck on the far side of the third field. This is all the matter that makes up the hydrogen atom: between the nucleus and the electron is a void—no football fields, of course, or anything else that can be perceived. And most of an atom's mass is in the central nucleus, not the electron. If the whole human body were compressed to nuclear density, each of us would take up no more space than a pinhead.[4] (So the next time someone yells at you, "You're a pinhead!," just smile at him and agree.) Take out all of the empty space in the atoms of which you are composed, and that is what size you would be. In fact, you would effectively disappear entirely in a "black hole," where matter is compressed to a much higher degree. Nick Herbert says that "a black hole as massive as earth would be only about 2 cm in diameter—the size of a large grape."[5] Think of it. Squash matter down to its most elemental nature, and our entire planet is reduced to the dimension of a grape. Taken to extremes, then, a materialistic philosophy ends up with not much matter to base itself on. What gives our world its apparent solidity is the speed with which subatomic particles move.

Electrons zip around at six hundred miles a second. This is extremely fast even for something moving across the face of the earth. At that rate you could travel all the way across the United States in five seconds. But imagine confining an entity that moves so quickly within a space one-millionth the width of a period at the end of a sentence. According to Fritjof Capra, this is what gives matter its solid aspect and makes the atom appear as a solid sphere.[6] Without energy, then, matter would not exist as we know it. In the words of physicist Edward Harrison:

> . . . atoms vibrate with rhythmic modes and are excited into states of various energies. As the modes of excitation evolve, waves of light are emitted and absorbed. The waves travel in space, and on arrival at the retina of the eye, other atoms absorb the incident waves. This vibrational give-and-take of atomic energy, in which atoms act individually and collectively, accounts for our visible world.[7]

Quantum theory

Interestingly enough, though, the very energy which enables atoms to be the foundation of the material world at one time seemed to make it impossible for them to even exist. Classical physics—in contrast to the new physics—could not account for the obvious stability of atoms. The laws of physics known early in the twentieth century predicted that electrons should radiate electromagnetic energy away as they orbit. And since they orbit so rapidly, their fall into the nucleus should take place in about a billionth of a second. Since the material universe is about ten billion years old, and theory predicted that it should have existed less than a second, physicists recognized that a deeper law of nature remained to be discovered.

And indeed, it soon was: *quantum mechanics.* Through the efforts of Max Planck, Albert Einstein, Louie de Broglie, Neils Bohr, Werner Heisenberg, Paul Dirac, Erwin Schrödinger, John von Neumann, and other pioneering physicists of the 1920s and 1930s, many of the mysteries of the atom were resolved through quantum theory. It was found that in the atomic and subatomic world—where the laws of quantum mechanics are most evident—energy comes only in discrete packets, or quanta. Energy is taken in or given out in a series of "quantum jumps." Since the quanta cannot be divided, there is no such thing as a fraction of a quantum.

According to physicist F. David Peat, with this discovery "the atomic picture suddenly came into focus. An electron, Bohr argued, cannot lose energy by spiraling inward. The only way it can change its state is by giving up a whole quantum of energy and jumping to a lower orbit. But if this lower orbit is already occupied by another electron, then there is nowhere for the first electron to go. It has to stay where it is, going around and around in its orbit forever. Adding Planck and Einstein's quantum to Rutherford's planetary model produced atoms that were totally stable, demonstrating, for example, why rocks—which are, of course, made out of atoms—have survived over geological time scales."[8]

Of course, there is much more to quantum theory than this. Other aspects of this fascinating subject will be addressed in later sections. For now, I want only to raise the issue of quantum "thinglessness," a term coined by Nick Herbert which points to several strange qualities of the quantum world. One is that in the course of a quantum jump, such as an electron moving from one orbit to another, the entity that is jumping supposedly gets from "here" to "there" without passing through "in between." This is difficult to picture, and indeed quantum theory says that it is impossible to visualize the nature of quantum reality—which to the eye of reason appears irrational and paradoxical (much like mysticism).

There are wide differences of opinion among physicists about what causes this "thinglessness." In other words, how can an entity which looks solid and real at one point in time and space suddenly pop up looking similarly solid and real at another point in time and space, while apparently being thingless in between? This is a complex question that has been admirably addressed by Nick Herbert in his book, *Quantum Reality*. He discusses a number of approaches to this problem, among which is the "Copenhagen interpretation" favored by most physicists (Copenhagen is where Bohr and Heisenberg first developed their theories of quantum mechanics).

This mainline view, which emphasizes the primacy of *measurement*, is summarized by Herbert: "Copenhagenists do not deny the existence of electrons but only the notion that these entities possess dynamic attributes of their own. Although an electron is always measured to have a particular value of momentum, it is a mistake, according to Bohr, to imagine that before the measurement it possessed some definite momentum. The Copenhagenists believe that when an electron is not being measured, it has no definite dynamic attributes."[9] Thus the essence of this perception of the quantum realm is: there is no deep reality at all to it. What is observed is, at heart, what is observed. Though objective, the quantum world is considered to be objectless.

Any further discussion of quantum theory would lead us too far astray, so do not be concerned if this brief introduction has left you confused. The confusion lies primarily in the subject matter, and does not much lessen with increased knowledge. This is what Werner Heisenberg, a brilliant physicist, had to say about the early days of quantum theory:

> I remember discussions with Bohr which went through many hours till very late at night and ended almost in despair, and when at the end of the discussion I went alone for a walk in the neighboring park I repeated again and again the question: "Can nature possibly be as absurd as it seemed to us in these atomic experiments?"[10]

For our purposes, it is sufficient to keep this in mind: much of the seeming absurdity of quantum physics comes from the realization that apparently material objects like electrons actually have more of a wraith-like nature. In quantum leaps they move from here to there without traversing the intervening space. In acts of measurement their dynamic attributes, such as position and momentum, are brought from potentiality into actuality. The subatomic realm thus does not have the same flavor to it as does our every-day world. When I go to bed at night, in the morning I have every right to expect that the coffee maker will be in the same spot on the kitchen counter. It is not going to take a quantum jump into the garage. Similarly, our dog definitely is sleeping on the sofa whether or not someone is looking at her. Her position in space, unlike the Copenhagen view of an electron, exists even before a measurement is made.

Physicists, therefore, have found that while the observable world appears solidly material, its subatomic foundation is much less substantial. The Copenhagen interpretation admittedly leaves unanswered the question of what kind of reality lies at the heart of the quantum realm. Regardless, there is complete agreement among practitioners of the new physics that whatever that reality may consist of, it differs markedly from the common-sense world of observable phenomena. This is not particularly surprising, of course, because little can be observed by the physical senses in the deepest interior of atoms. Microscopes using advanced technologies can provide tantalizing glimpses of the universe within each atom, but not more.

On the way to a hoped-for Theory of Everything, physicists are grappling with what are termed Grand Unified Theories—which attempt to unify three of the four known forces of nature: electromagnetism, the weak nuclear force, and the strong nuclear force (a Theory of Everything adds in the fourth force, gravity). Physicists believe that these forces were united soon after the Big Bang, when the universe was very young. The expansion and consequent cooling of the physical universe caused the forces of nature to "freeze out" of the original single superforce. Because the universe was extremely small when it was very young—much as an embryo consists of only a few cells soon after it has been conceived—

delving deeper and deeper into the subatomic world is equivalent to getting closer and closer to the moment of the Big Bang.

Worlds within the atom

How deep physicists are going, at least in theory, is illustrated by this quote from Paul Davies: "The world of GUTs [Grand Unified Theories] and proton decay is millions of millions of times smaller than the world of quarks and gluons probed by our accelerating machines so far. It is like a universe within the proton, as inaccessible to us in its smallness as are the extragalactic spaces in their remoteness."[11] Protons and neutrons, however, are small enough in their own right, compared to the relative vastness of the atom. These subatomic particles are *hadrons*, the next level of atomic structure. This brings us to the edge of the mysterious realm inside the atomic nucleus alluded to by Davies.

There are supposedly an infinite number of hadrons, but most are highly unstable and disintegrate in a tiny fraction of a second. The stable proton and neutron are the best-known and commonest hadrons. They are held together in the nucleus of an atom by the strong force, and until the 1960s were considered to be the "bottom rung" of atomic structure. This is understandable, given how incredibly small protons and neutrons are. Remember that a million atoms can fit within this dot: .

Now consider that the nucleus occupies only a trillionth (or one-thousand billionth) of the atom's volume, and the hadrons buzz around inside of *that* at a speed of about forty thousand miles a second. According to Fritjof Capra, the nucleus is best pictured "as tiny drops of extremely dense liquid which is boiling and bubbling most fiercely."[12] Gary Zukav says that in order to see the nucleus, an atom would have to be as high as a fourteen story building, and then that nucleus only would be about the size of a grain of salt.[13] Thus, the protons and neutrons would be specks of dust within the grain of salt that is resting in the middle of the sphere as large as the fourteen story building.

Amazingly, though, the new physics holds that there is yet another level to material reality: that of the *quarks*. Named after a line in James Joyces's *Finnegan's Wake*, quarks never have been seen. Still, most physicists are confident that they exist because the quark theory does such a good job of explaining hadrons. Heinz Pagels says, "We can now appreciate the great simplification that the quark model represented—the problem of an infinite set of hadrons was reduced to a problem in the dynamics and interactions of just three quarks."[14]

Quarks are considered to be point-like particles without further structure. All ordinary matter is composed of "up" and "down" quarks. A proton is believed to contain two up quarks and one down quark, while a neutron has two down and one up. Because each quark has a fractional (2/3 for "ups," 1/3 for "downs") charge, nature cleverly combines quarks to produce the positively charged proton (2/3 + 2/3 - 1/3) and neutral neutron (-1/3 + -1/3 + 2/3).

It is worth pausing at this point to reflect on the beauty and simplicity of both the quark theory, and the entire structure of the atomic realm—the intricate nature of which I have barely touched upon in these few paragraphs. All of that intricacy is the product of just three particles: the up and down quarks uniting in triplets, plus electrons. The dynamics of those three particles are mediated by only four forces: electromagnetism, the weak nuclear force, the strong nuclear force, and gravity. And physicists are making strides in understanding how all of these particles and forces, as few as they are in comparison to the myriad phenomena in the universe, are the product of a single unified force which briefly was in evidence at the moment of the Big Bang—and continues to be the ultimate foundation of the subatomic world.

Physics is not physical

But discussion of this superforce must be deferred to a later chapter. My primary goal in presenting this overview of particle physics is to underscore a point made on the first page of this book: *physics is not physical.* Many lay persons believe that this discipline is founded on direct observations of purely material entities. So when a physicist expounds a materialistic philosophy, it is all too easy to assume that his views are a direct outgrowth of scientific research. In reality, we have seen that the material nature of existence quickly fades away in the subatomic realm. The deeper reaches of that world within the atom are far beyond the grasp of the five senses, so physicists increasingly find it necessary to resort to pure reasoning and mathematics to make further progress in understanding so-called "material" reality.

When an entity cannot be directly observed, or even indirectly sensed in any manner, how is it possible to continue assuming that it is material? What, in fact, does *matter* even mean when physics has shown that this term is equivalent to *energy*, and that the four forces of nature—which are the root of all forms of energy—are considered to be the product of a single superforce? Since the Big Bang caused our universe to expand from an infinitesimal point to the immense size it is now, moving into the smaller and smaller world of subatomic particles and forces is equivalent to growing closer to the moment of creation. However, the traditional observational methods of physics are useless for delving into either the most minute divisions of space or the earliest moments of time.

In these spheres of investigation physicists have found that mathematical models must be substituted for actual observations of material objects or forces. This can be confusing to the layman. As Paul Davies and John Gribbin point out, "scientists often use the word 'discovery' to refer to some purely theoretical advance."[15] For example, they say that Stephen Hawking often is credited as having "discovered" that black holes emit heat radiation. In various popular books this finding is discussed at some length, including the implications that this discovery has for other areas of physics. Yet in Davies' and Gribbin's words, "nobody has yet seen a black hole, much less detected any heat radiation from one."[16]

The rationale for assuming that mathematical models are a valid reflection of imperceptible aspects of physical existence is that they work so well in explaining perceptible phenomena. Physicists have proven beyond a shadow of a doubt that many of the material laws of nature can be precisely reflected by mathematical equations. Man's capacity to calculate the exact solutions to those equations often limits his ability to accurately predict the behavior of phenomena, but this does not detract from the beauty of how relatively simple mathematics is able to mirror the complex dynamics of physical systems.

This has led some mathematicians and physicists to agree with Sir James Jeans that "the universe appears to have been designed by a pure mathematician."[17] Spiritual science certainly has no quarrel with that statement, for perfect mystics tell us that laws of cause and effect—which to some extent can be modeled by equations—govern the physical universe. Jeans, who was a noted mathematician, physicist, and astronomer, had a properly humble attitude regarding material science:

> The essential fact is simply that *all* the pictures which science now draws of nature, and which alone seem capable of according with observational fact, are *mathematical* pictures. Most scientists would agree that they are nothing more than pictures—fictions, if you like, if by fiction you mean that science is not yet in contact with Ultimate Reality. Many would hold that, from the broad philosophical standpoint, the outstanding achievement of twentieth-century physics is not the theory of relativity with its welding together of space and time, or the theory of quanta with its present apparent negation of the laws of causation, or the dissection of the atom with the resultant discovery that things are not what they seem; it is the general recognition that we are not yet in contact with Ultimate Reality. To speak in terms of Plato's well-known simile, we are still imprisoned in our cave, with our backs to the light, and can only watch the shadows on the wall.[18]

Physics, then, needs to come to terms with this central question: when is mathematics a suitable tool for laying bare the laws of existence, and when is it not? As Sir James Jeans pointed out, there is no argument that certain observations of the physical world are best explained by mathematical "pictures." However, Jeans and the spiritual science of mysticism agree that Ultimate Reality lies far beyond any symbolic schema—including that of modern mathematics. The intricate equations of the new physics are reflecting an echo of final truth, the shadows of Plato's cave. To hear clearly the sound of God's essence, and to see directly His refulgent light, it is necessary to turn away from echoes and shadows.

> *You must especially avoid the reasoning of the lower senses*
> *concerning that Revelation not bound by any limits.*
> *Although the ear of your senses is capable of hearing words,*
> *your ear that perceives the Unseen is deaf.*
> —Rumi[19]

Yet the mainstream of physics is not taking this course. Pure mathematics is the tool of choice for investigating "ultimate" questions such as: How did the universe come to be as it is today? What is the fundamental constituent of matter and energy? Was there a beginning to the physical universe, or has it existed eternally? While astronomer David Lindley acknowledges the value of mathematics in certain spheres of inquiry, he strikes a cautionary note:

> On the other hand, the inexorable progress of physics from the world we can see and touch into a world made accessible only by huge and expensive experimental equipment, and on into a world illuminated by the intellect alone, is a genuine cause for alarm. Even within the community of particle physicists there are those who think that the trend toward increasing abstraction is turning theoretical physics into recreational mathematics, endlessly amusing to those who can master the techniques and join the game, but ultimately meaningless because the objects of the mathematical manipulations are forever beyond the access of experiment and measurement.[20]

A matter of taste

A concrete example will make this critique clearer. In Stephen Hawkings' book, *A Brief History of Time*, he discusses a theory of "quantum cosmology." The details need not concern us here. Basically Hawking is attempting to explain the origin of the universe through certain extensions of quantum theory. As was noted previously, this makes sense to physicists because when the universe was very young, it also was much smaller than an atom—and so supposedly was governed by the laws of quantum mechanics. Hawking's hypothesis rests in part on a notion of imaginary time: "That is to say, for the purposes of the calculation one must measure time using imaginary numbers [which give negative numbers when multiplied by themselves], rather than real ones."[21]

Well and good, there is no harm in exploring new concepts—even if it means throwing out our familiar sense of time with an imaginary replacement. But Hawking is walking along the narrow edge of a slippery argument here. On the one hand, he apparently wants us to regard his nascent theory as a legitimate possible explanation of how the physical universe came into existence. That is, all this is not "recreational mathematics," as Lindley puts it. Yet at the same time, Hawking is not willing to claim that his theory bears any relation to objective reality. After surmising that perhaps imaginary time actually is more basic than what we call real time, he negates that supposition with a significant "but":

> But according to the approach I described in Chapter 1, a scientific theory is just a mathematical model we make to describe our observations: it exists only in our minds. So it is meaningless to ask: Which is real, "real" or "imaginary" time? It is simply a matter of which is the more useful description.[22]

This is an interesting statement, coming as it does from one of the most highly respected physicists in the world today. I was under the impression that material science sought to know the truth about reality, not about our own minds. Hawking implies that this is not the case, at least when physics ventures into areas of investigation where direct observations by the physical senses cannot be made. This reminds me of what happened after a university lecture on quantum physics that I attended a few years ago.

A distinguished physicist had been invited to give a talk in the evening that was open to the general public, as well as the college community. You do not need to know what subject he was speaking about to appreciate this story. It was a complex issue, and this enthusiastic scientist had not one, but two, overhead projectors set up to illustrate the points he was making. For over an hour he dashed back and forth between the projectors, taking off one slide and putting on another slide in rapid succession. I found myself able to follow the general thrust of his arguments, though some of the details were lost in the shuffle of the overhead transparencies. Finally, with a flourish he put the last slide on the screen and told us the inescapable conclusion: "This means that such-and-such is true."

I sat there in the audience for a few seconds, both dazed from the flurry of arguments that had been presented, and—I admit—relieved that the onslaught was over. Then it dawned on me that even after all of the rational arguments this noted physicist had put forth, I was not convinced that his explanation of the quantum phenomena being addressed made the most sense. For I had just finished reading a book that offered a different way of looking at a similar question involving quantum theory. Gearing up my courage, I put my hand in the air and became the first to ask a question of the visiting scholar. "I found your presentation interesting," I said, "but isn't it correct that an equally valid explanation of the phenomena you discussed is this-and-that, rather than such-and-such?" To my amazement, he thought only for a moment and then replied, "Well, yes. Which explanation you choose is really a matter of taste."

A matter of taste! I had just sat through over an hour of a challenging but mind-numbing lecture, and the conclusion was a *matter of taste*? This experience brought home to me how far removed the new physics was from objective physical reality. There was no touchstone that could serve to distinguish whether the perspective of a highly trained professional physicist, or the view of an interested layman, was more correct.

Is this any way to run a science? Or, as David Lindley puts it, "what is the use of a theory that looks attractive but contains no additional power of prediction, and makes no statements that can be tested? Does physics then become a branch of aesthetics?"[23] In other words, if theories cannot be proven to be true or false through experimental means, perhaps one day the sole criterion for their acceptance will be how "beautiful" they are. Competing physics theories will be trotted out for judging like exquisitely groomed poodles at a dog show, rather than rigorously tested through controlled research.

A thought of a thing is not that thing

We have come, then, to an interesting conclusion. As regards learning the truth about Ultimate Reality, physics is more esoteric than mysticism, and mysticism is more rigorous than physics. As will be demonstrated more clearly in following chapters, the spiritual science of mysticism sets forth specific principles concerning the nature of final truth which can be confirmed in the laboratory of contemplative meditation. What is discovered there are not abstract symbolic equations, but the pure conscious essence of Ultimate Reality. Mysticism makes a clear distinction between thoughts about objective truth, and that truth as it is in actuality. Willem Drees, a physicist and theologian, draws a similar line between mathematics and reality in the course of critiquing Hawking's probabilistic theory of quantum cosmology:

> The probability of throwing a head when tossing a coin is one half, but a fifty percent chance of getting an actual head is there if and only if someone tosses a coin, if and only if one of the possible outcomes is realized. A mathematical idea of getting a universe from nothing does not give a physical universe, but only the idea of a physical universe—assuming that there is a difference between the Universe and a mathematical idea

about the Universe. There has to be some input of "physical reality." Perhaps that is an aspect of the nothingness, but that makes it into a physical entity and not nothing at all.[24]

Interestingly, in the final pages of *A Brief History of Time* Hawking himself appears to agree with this reasoning. He writes, "Even if there is only one possible unified theory, it is just a set of rules and equations. What is it that breathes fire into the equations and makes a universe for them to describe? The usual approach of science of constructing a mathematical model cannot answer the questions of why there should be a universe for the model to describe."[25] That, of course, is the province of spiritual science, which contains—but is not contained by—the domain of material science. Mysticism steps into the breach to address questions which require research tools that are not in the arsenal of modern physics.

While discussing the four states of being, I mentioned that historically physics has emphasized observing objective physical reality. This is a *symbolic* domain of existence, where words, numbers, equations, and the like can be used to describe one's observations. This is what makes material science a public enterprise, as those symbols can be shared with others, whereas private sense perceptions cannot. Hence an astronomer observing galaxies through the eyepiece of an optical telescope is able to take a photograph of what he sees, or to convert those images into digits that can be manipulated by a computer. In so doing, observations of physical existence which began as private perceptions can end as public symbolic representations of that reality.

This approach served physics well until twentieth-century discoveries in such areas as relativity, quantum physics, and Big Bang cosmology led to the *new physics* supplanting the old. As we have seen, two unique characteristics of the new physics are:

(1) It deals with questions concerning the edges of physical existence, such as the ultimate origin of the universe, and the most fundamental constituent of matter and energy.

(2) Due to (1), it has passed beyond investigations of observable phenomena into research that relies on abstract mathematics and pure reasoning to test the validity of its theories.

So the mainstream—or at least a strong current—of the new physics has turned from the symbolic *science* of observing to the symbolic *art* of thinking (please refer to Figure 4 again if this point is unclear). Having come to the limits of what can be observed with the five senses, most physicists have decided that the only recourse is to fall back into the domain of thinking. Here, at least, it is felt that symbolic representations of Ultimate Reality, of a Theory of Everything, can be developed—even if that final truth itself remains forever beyond man's capacity to perceive directly. But as Stephen Hawking admitted in the quote above, those symbols lack the essence that "breathes fire" into words and equations. We are left with a model of a universe that answers every question except the most important one: how did that model become a reality?

Physics begins to embrace metaphysics

Many physicists are disturbed by this evident trend away from objective science to subjective mathematical speculation. This chapter closes with some quotations from thoughtful material scientists who argue that a turn toward metaphysics is needed for reasons which I hope have been made evident. The rapid progress of the new physics has brought this discipline to the edge of sense perception and rational thought. If there is an objective reality which lies beyond the five senses and mental cognition, it will not be found by observing and thinking—the traditional research tools of physics.

The choice for material science is clear: either acknowledge that the fourth state of being—*contemplation*—is a valid means of learning about realms of existence which transcend mind and matter, or continue using the investigative tools at hand. The spiritual science of mysticism is not neutral on this question, but unequivocally states that it is impossible to think one's way to knowledge of Ultimate Reality. Physicists, of course, are most welcome to try. But the Perennial Philosophy of mystics, which will be described more fully in the remaining chapters, is founded on a simple direct realization: God-consciousness is based on love, which requires merging one's personal individuality into divine totality. Sensory perceiving and mental thinking are as useful in God-realization as is a pitchfork in bailing out a sinking ship. A tool has to be appropriate to the task.

Soon we will move to an examination of the specific principles of spiritual science which are reflected, dimly or poorly, in findings of the new physics. Before doing so I want to share without comment some thoughts from several authors about the relation between physics and metaphysics. These writers certainly do not represent the views of all—or even most—material scientists. Still, it is encouraging that the search for Ultimate Reality, the Theory of Everything, God, or however we wish to refer to that Final Truth, has begun to be regarded as an endeavor which should unite—not divide—truth-seekers of all practices, persuasions, and predilections.

Physicist Joe Rosen:
Cosmology reaches beyond the domain of science and is basically metaphysics . . . the objective reality science has led us to believe in is partially hidden from us, is a veiled, clouded, fogged reality. Science allows us a few clear glimpses of parts of it as well as provocative hints about more of it . . . science, in its study of nature, cannot fulfill our demand for

objectivity. The quantum aspect of nature involves observers too much for that. So any objective reality must be "farther" from us than nature, than perceived reality. It must transcend them. . . . As long as objective reality, which science guides us to believe in, itself lies beyond the domain of science, let us not shut ourselves off from the possibility that we might possess other channels to it.[26]

Physicist Paul Davies:

Is there a route to knowledge—even "ultimate knowledge"—that lies outside the road of rational scientific inquiry and logical reasoning? Many people claim there is. It is called mysticism . . . many of the world's finest thinkers, including some notable scientists such as Einstein, Pauli, Schrödinger, Heisenberg, Eddington, and Jeans, have also espoused mysticism. . . . Mysticism is no substitute for scientific inquiry and logical reasoning as long as this approach can be consistently applied. It is only in dealing with ultimate questions that science and logic may fail us. I am not saying that science and logic are likely to provide the wrong answers, but they may be incapable of addressing the sort of "why" (as opposed to "how") questions we want to ask.

    . . . in the end a rational explanation for the world in the sense of a closed and complete system of logical truths is almost certainly impossible. We are barred from ultimate knowledge, from ultimate explanation, by the very rules of reasoning that prompt us to seek such an explanation in the first place. If we wish to progress beyond, we have to embrace a different concept of "understanding" from that of rational explanation. Possibly the mystical path is a way to such an understanding. I have never had a mystical experience myself, but I keep an open mind about the value of such experiences. Maybe they provide the only route beyond the limits to which science and philosophy can take us, the only possible path to the Ultimate.[27]

Astronomer John D. Barrow:

Not every feature of the world is either listable or computable. . . . These attributes that have neither the property of listability nor that of computability—the "prospective" features of the world—are those which we cannot recognize or generate by a series of sequence of logical steps. They witness to the need for ingenuity and novelty; for they cannot be encompassed by any finite collection of rules or laws. . . . The prospective properties of things cannot be trammelled up within any logical Theory of Everything. No non-poetic account of reality can be complete. The scope of Theories of Everything is infinite but bounded; they are necessary parts of a full understanding of things but they are far from sufficient to unravel the subtleties of a Universe like ours.[28]

Science writer Bryan Appleyard:
It may be true that quantum mechanics points to a deeper, spiritual realm—but the knowledge of that truth must come from outside and be independent of the quantum, otherwise it remains dependent on the whims of science. We must, in effect, know the truth before we can discover it in the quantum.[29]

Astronomer David Lindley:
The physicists must hope instead that they can complete physics in the manner the ancient Greeks imagined, by means of thought alone, by rational analysis unaided by empirical testing. . . . This theory of everything will be, in precise terms, a myth. A myth is a story that makes sense within its own terms, offers explanations for everything we can see around us, but can be neither tested nor disproved. . . . This theory of everything, this myth, will indeed spell the end of physics. It will be the end not because physics has at last been able to explain everything in the universe, but because physics has reached the end of all the things it has the power to explain.[30]

> *We're quite addicted to subtle discussions;*
> *we're very fond of solving problems.*
> *So that we may tie knots and then undo them,*
> *we constantly make rules for posing the difficulty*
> *and for answering the questions it raises.*
> *We're like a bird which loosens a snare*
> *and then ties it tighter again*
> *in order to perfect its skill.*
> *It deprives itself of open country;*
> *it leaves behind the meadowland,*
> *while its life is spent dealing with knots.*
> *Even then the snare is not mastered,*
> *but its wings are broken again and again.*
> *Don't struggle with knots,*
> *so your wings won't be broken.*
> *Don't risk ruining your feathers*
> *to display your proud efforts.*
>
> —Rumi[31]

Universities of Ultimate Reality

In the preceding couplets, the perfect mystic Jalaluddin Rumi has described the condition of the new physics: lured by the enticements of reason and mathematics, conceptual snares are loosened and tightened, over and over again. A solution to one problem immediately brings into focus another problem. The sought-for Theory of Everything remains forever hidden behind the impenetrable barrier of Gödel's Theorem: in any system of logic, there always will be true statements which cannot be proven to be true. Thus the essence of Ultimate Reality never can be realized through the investigative methods of material science.

Physics can learn a great deal about Limited Reality, however, and unquestionably has made great progress in laying bare many of the subtle laws of the physical universe. So long as physics remains within the sphere of its expertise, it finds few insurmountable obstacles in the path of knowledge. Given sufficient time, technology, and tenacity, over the centuries physicists have proven themselves able to answer seemingly intractable questions. Only when the door of *Ultimate Reality* has been knocked upon, has it been impossible for material science to push it open even a crack. And as we have seen from the words of those scientists quoted at the end of the last chapter, even fervent believers in the efficacy of the new physics are coming to recognize that the truth of a Theory of Everything never will be confirmed by this science.

Perfect mystics have perfect answers

Since material science has been unable to reach any verifiable conclusions regarding such ultimate questions of existence, common sense would argue that deference should go to those who have. If theologians and physicists alike are constrained by the limits of reason and sense perception to engage in nothing more than an "is/is not" argument about the existence of God, then recourse should be made to the teachings of the only persons who have conducted a truly scientific investigation of this question—the perfect mystics. Through the research tool of contemplative meditation that is the hallmark of spiritual science, definitive answers have been determined to the very questions which religion and material science have been debating for centuries.

The preceding statement likely will come as a surprise to many people. For the prevailing view is that beliefs about the existence, or non-existence, of God are just that—beliefs. As such, they fall into the subjective domain of philosophy and theology, where thoughts are traded back and forth between theists and atheists in a never-ending speculative bazaar. Certainly many people choose to buy—and hold onto—one metaphysical position or another, but the coin of purchase generally is considered to be subjective faith, not objective science. Thus physicist Steven Weinberg found it possible to make this statement:

> . . . the great majority of the adherents to the world's religions are relying not on religious experience of their own but on revelations that were supposedly experienced by others. It might be thought that this is not so different from the theoretical physicist relying on the experiments of others, but there is a very important distinction. The insights of thousands of individual physicists have converged to a satisfying (though incomplete) common understanding of physical reality. In contrast, the statements about God or anything else that have been derived from religious revelation point in radically different directions. After thousands of years of theological analysis, we are no closer now to a common understanding of the lessons of religious revelation.[1]

Religion is not spiritual science

Actually, I have little quarrel with this conclusion. But how does this fit with my assertion that spiritual science has arrived at conclusive answers about the nature of Ultimate Reality? Once again, we must keep in mind the significant differences between *mysticism* and *religion*. The former is a science; the latter is occupied almost entirely with intellectual philosophizing and emotional proselytizing. In terms of our model of consciousness, religion as it is practiced today is focused either on the symbolic and subjective domain of thinking (theology, dogma), or the non-symbolic and subjective domain of imagination (faith, conversion). Faith does not require reasons for the existence of God, and thinking supplies only inadequate reasons. So it is not surprising that, as Weinberg correctly pointed out, the world's religions have failed to come to agreement on any but the most general conclusions.

And these agreed-upon premises primarily are in the realm of ethics, not metaphysics or cosmology. Every major religion is in agreement that killing other people is bad, and loving them is good; that it is better to give than to receive; and similar tenets. But agreement is lacking when it comes to more "ultimate" questions such as: Is God single or many? Do heaven and hell exist? Have there been numerous prophets or sons of God, or only one? Do the physical laws of nature operate independently of their Creator, or under His direction?

Weinberg notes that in modern times religious conservatives differ about what they believe to be true—which at least makes for stimulating discussion—but their liberal counterparts often fail to take any stand at all: "Many religious liberals today seem to believe that different people can believe in different mutually exclusive things without any of them being wrong, as long as their beliefs 'work for them.' This one believes in reincarnation, that one in heaven and hell; a third believes in the extinction of the soul at death, but no one can be said to be wrong as long as everyone gets a satisfying spiritual rush from what they believe."[2] Given this evident mish-mash of religious beliefs, it is easy to understand why material scientists put little stock in the idea of a genuine spiritual *science*.

Still, these are the plain facts: the spiritual science of pure mysticism *does* exist; it is being practiced at this very moment by millions of people in every corner of the world; and its principles are as precise, objective, and demonstrable as those of physics or any other material discipline. Further, the same mystical truths have been known since the dawn of recorded history—if not before—and can be discerned at the core of every major religion. Unfortunately, these truths are easily forgotten or misinterpreted by those who lack the perfect mystic's direct knowledge of Ultimate Reality. This is a central reason why the world's religions are so much at odds. The founder(s) of each religion preached the truth about God as it was known to them through direct perception. Then, in most cases, those teachings were written down and elaborated upon by followers of that prophet, sage, or saint who generally had not attained to his spiritual height.

Everyone knows how a rumor changes form as it passes from one person to another. Each tends to embellish what he has been told, and puts his own subjective interpretation on that item of information. It makes little difference if it is fact or fancy—after being handled by only a few people, there often is just a passing resemblance between the original communication and its final message. Imagine, then, the increased difficulty in maintaining the message's accuracy if it were translated from language to language over centuries, or millennia. And what if some of those entrusted with passing on the information had a vested interest in emphasizing some points, and concealing others? In such circumstances, we should not be surprised that the original message ended up being altered, but rather be amazed if the first version was even close to the last.

An additional difficulty—and a large one—arises when mystical knowledge is being transmitted. For we have seen that the truths of spiritual science, born from contemplative meditation, are objective yet non-symbolic. This means that just as the subjective experience of "sadness" cannot be expressed accurately to another person by any word, equation, or other symbol, neither can the perfect mystic's objective perception of Ultimate Reality—unless that person reaches the same level of consciousness enjoyed by the perfect mystic. Far from being an indictment of spiritual science, this fact is a logical implication of the central finding of pure mysticism: God and Spirit, the highest truths, are far beyond the domains of mind and matter. How, then, could the divine essence be represented by any symbol pertaining to the material universe?

> *To every image of your own imagination you say, "Oh, my spirit, my world!"*
> *Were these images to disappear, you yourself would be the spirit and the world.*
> —*Rumi*[3]

To prove there is sleep, wake up!

This quotation helps in understanding the nature of proof, or validation, in spiritual science. Clearly it is not possible to prove the existence of a non-symbolic objective reality, which is beyond the domain of mind and matter, through any means available in this physical universe. If such a proof were possible, it would invalidate the mystical truth purportedly being proven. For example, since mysticism asserts that the all-pervading conscious energy of Spirit cannot be perceived by the physical senses or understood through mental reasoning, any evidence that could be provided of its existence through those avenues would refute that very assertion.

Similarly, a mathematics professor could cover an entire blackboard with the iron-clad proof of a theorem, but if his students have fallen asleep in their chairs, that evidence will remain unknown to them. To my mind, the logic of this argument is clear, but many physicists and other material scientists are unrelenting in their demands for concrete proof of God or Spirit. Since the divine essence of Ultimate Reality only can be known on a higher plane of consciousness, such proof is an impossibility.

It is as if a man fell asleep and was trapped in a dream from which he could not awake. While still in that state, an entity appears before him and says, "Remember, my friend, you are only dreaming. Do what I say and you will wake up. Then the relative truth of this dream will be clear to you." Would it not be absurd for that man to reply, "I don't believe you. Give me proof that there is something other than the consciousness I possess now." For all that entity could tell him would be, "The evidence for wakefulness lies in waking up. Whatever I could tell you now about being awake *still would be part of dreaming*. There is no purpose to that, because I have come into your dream not to add to illusion, but to break the spell of your slumber."

Thus from the viewpoint of perfect mystics, the popular preoccupation with "psychic phenomena" is meaningless. Yes, mysticism teaches, it is possible to perform so-called miracles once one rises a little way above the domain of everyday consciousness, but what purpose does this serve? Does bending a spoon bear any relation to creating a universe? Does guessing correctly the suit of a playing card equate in any way to knowing God? Such parlor games never are encouraged by perfect mystics, who teach their students to ignore all echoes of Ultimate Reality in order to find the source as directly as possible. In the *Hua Hu Ching*, oral teachings of the Taoist sage Lao Tzu, we find:

> The clairvoyant may see forms which are elsewhere,
> but he cannot see the formless.
> The telepathic may communicate directly with the
> mind of another, but he cannot communicate with
> one who has achieved no-mind.
> The telekinetic may move an object without touching
> it, but he cannot move the intangible.
> Such abilities have meaning only in the realm of duality.
> Therefore, they are meaningless.
> Within the Great Oneness, though there is no such
> thing as clairvoyance, telepathy, or telekinesis,
> all things are seen, all things understood,
> all things forever in their proper places.[4]

Mystics agree, theologians don't

Yet notwithstanding the ineffable nature of spiritual science, much evidence exists in support of the contention that the findings of mysticism are the same wherever and whenever this discipline is practiced in its pure form. This was the message of Aldous Huxley's classic book, *The Perennial Philosophy*. With impressive scholasticism, wit, and attunement to mystical teachings, Huxley demonstrated that (in his words):

> . . . the metaphysic that recognizes a divine Reality substantial to the world of things and lives and minds; the psychology that finds in the soul something similar to, or even identical with, divine Reality; the ethic that places man's final end in the knowledge of the immanent and transcendent Ground of all being—the thing is immemorial and universal. Rudiments of the Perennial Philosophy may be found among the traditionary lore of primitive peoples in every region of the world, and in its fully developed forms it has a place in every one of the higher religions. A version of this Highest Common Factor in all preceding and subsequent theologies was first committed to writing more than twenty-five centuries ago, and since that time the inexhaustible theme has been treated again and again, from the standpoint of every religious tradition and in all the principal languages of Asia and Europe.[5]

So in a deeper sense, Steven Weinberg could not have been more wrong when he implied that there is no "common understanding" of spirituality. This commonality is not easily perceived when one looks through the blurry eye of theology or philosophy, but is crystal clear when viewed through the sharp focus of mysticism. Huxley observes that his anthology contains few writings from professional men of letters, and hardly anything from professional philosophers. For he says that the nature of the reality described by the Perennial Philosophy "is such that it cannot be directly and immediately apprehended except by those who have chosen to fulfill certain conditions, making themselves loving, pure in heart, and poor in spirit."[6]

In accord with the teachings of all perfect mystics, and the theme of this book, Huxley says that there is not much outward evidence that the consciousness "of the average sensual man" contains anything that resembles the Ultimate Reality which the Perennial Philosophy points to. However, just as the true nature of matter is revealed only by making subtle physical experiments, so are the potentialities of mind and soul discovered only by "making psychological and moral experiments." And, Huxley says, there is scant evidence that most of those who write about spiritual matters have done anything more than read what others have written on the subject. Lacking direct mystical knowledge, their views about God are subjective and unreliable. On the other hand:

> . . . in every age there have been some men and women who chose to fulfill the conditions upon which alone, as a matter of brute empirical fact, such immediate knowledge can be had; and of these a few have left accounts of the Reality they were thus enabled to apprehend. . . . To such first-hand exponents of the Perennial Philosophy those who knew them have generally given the name of "saint" or "prophet," "sage" or "enlightened one." If one is not oneself a sage or saint, the best thing one can do, in the field of metaphysics, is to study the works of those who were, and who, because they had modified their merely human form of being were capable of a more than merely human kind and amount of knowledge.[7]

While I agree wholeheartedly with these sentiments, Huxley could have made his point even more emphatically: the seeker of Ultimate Reality should not be content with merely "studying the works" of saints, or perfect mystics, but should endeavor to attain their same level of consciousness. Only in this manner can a person know exactly what a perfect mystic knows, and perceive God's essence as completely and directly as does that saint. The reading of sacred writings is not of much use in accomplishing this divine task, because in so doing we never leave the domain of thought and emotion. As Rumi stated previously, for the student of mysticism words primarily serve to "cause you to seek and excite you, but the object of your search will not be attained through words."

Attaining the goal of oneness with God—Ultimate Reality, Final Truth—is accomplished through contemplation of the Lord's Spirit, which is audible as sound and visible as light on higher planes of consciousness. Spirit is the link between God and our soul. In fact, it is the all-pervading essence of everything in existence, material or non-material. Because Spirit is what unites every part of creation into a seamless whole, it often is said that "God is Love." Spirit, in the words of the perfect mystic Sawan Singh, is "the substance of which love is the attribute."[8] So this is the path by which seekers of either perfect *truth* or of perfect *love* will reach their goal. The road, the means of transportation, and the final destination are the same for all, no matter what limited words we use to describe that journey toward oneness with God. Lekh Raj Puri says:

> In that supremely transcendent stage, bliss, knowledge, truth, love, existence, reality, spirituality, being—all these mean the same thing, for there is only One. It is Oneness through and through, call it what we may. . . . It is "One," which appears as "many" to our deluded eyes; but in itself, it is always and for ever the Indivisible One.[9]

How perfect mystics teach

This makes it possible to understand one of the central differences between a perfect mystic, the highest rank of spiritual scientists, and a highly competent Ph.D. in the material science of physics. The latter is concerned with learning, adding to, and teaching symbolic knowledge concerning laws of the physical universe—though not all physicists are both researchers and instructors (and neither are perfect mystics). As such, it is not necessary for either he or his students to alter their consciousness, since the activity of material science is conducted almost entirely in the usual waking states of thinking and observing (with perhaps a few excursions into sleep during a boring lecture.) Hence the link between those who know the methods and findings of physics, and those who want to learn them, is *symbols*: words, numbers, diagrams, photographs, and the like that are formed into textbooks, lectures, research papers, equations, and such.

Certainly a human being is desirable, and perhaps even essential from a practical perspective, as a mediator and communicator of those symbols. As was noted previously, it might well be possible for a student to attain to a doctorate-level proficiency in physics by studying entirely on his own, but this would be most difficult to accomplish. Still, it is feasible. This is due to physics being a *symbolic* science, which enables knowledge of the material universe to be transmitted to the minds of students without their having to contact directly the source of that information. These symbols simply are added to the content of everyday consciousness, while the form of consciousness remains unchanged. So to understand the findings of subatomic physics, it neither is necessary to personally carry out all the experiments involving particle accelerators that have contributed to that knowledge base, nor to merge one's consciousness with the quantum realm of electrons, neutrons, and protons.

The situation is quite different in mysticism. Here both the method of attaining knowledge (contemplative meditation) and the final goal (truth of God, or Ultimate Reality) are *non-symbolic*. Knowledge is obtained by expanding the essence of one's consciousness, rather than by cramming additional concepts into a limited sphere of understanding So the prospective student of spiritual science is faced with problems not found in material science: How does one learn to enter a state of consciousness which is unknown to all but a few? And how can one even find a qualified teacher of the highest spiritual science when no outward signs of accreditation exist? No diplomas on the walls, no certifications from a professional organization, no authorship's of research papers. The situation may seem almost hopeless. For without the guidance of a perfect mystic, there can be no perfect knowledge of God.

Exploring the reasons for this last statement would take us too far afield, but a few hints may be given. First, the evolved consciousness of a perfect mystic can in no way be equated with our personal, subjective minds. His consciousness is one and the same as that of God. When Jesus was accused of being a blasphemer because "Thou, being a man, makest thyself God," he replied, "Is it not written in your law, 'Ye are gods?. . . .' Though ye believe not me, believe the works: that ye may know, and believe, that the Father is in me, and I in him."[10] Mysticism teaches that union with God is possible, not just for a few rare souls, but for anyone who deeply and sincerely yearns for that end—and is willing to undertake the disciplines needed to attain to God-consciousness.

So a perfect mystic stands as both the means, and the goal, for a student of spiritual science. The means, because he instructs and guides. The goal, because he already has reached the state of consciousness to which the student aspires. Thus whatever the perfect mystic *is*, the student one day *will be*. At first glance, this may not appear to be so different from the case of a physics professor, and those he is instructing. Yet it is not at all the same. Those in the classroom of a Ph.D. hope to reach his level of knowledge, and perhaps even surpass it, but not to merge their consciousness with their professor's consciousness. That instructor may be an alcoholic, a womanizer, a gambler, and an insufferable egotist. But as long as he knows and can teach his subject matter, those students will get what they need from him. For they are not out to become him, just to absorb the symbolic understanding of physics he communicates.

A perfect mystic, on the other hand, has united his consciousness with God.

> *Whoever possesses a partial intellect is in need of instruction, but the Universal Intellect is the originator of all things. Those who have joined the partial intellect to the Universal Intellect so that the two have become one are the prophets and saints.*
> —*Rumi*[11]

If the student of spiritual science aims to attain to the same divine state, at some point the consciousness of God, the perfect mystic, and the student all will become one. Jesus said, "and there shall be one fold, and one shepherd."[12] This, of course, not only is a matter of faith, but of solid logic. If the essence of Ultimate Reality indeed is monistic, or a single substance, then those who are able to merge themselves into that substance will also be one. If this does not occur, then either they have not attained to that final truth, or the nature of God actually is dualistic. The prevailing view in physics, mysticism, and the major world religions is that monism, rather than dualism, is the true philosophical perspective.

Second, a perfect mystic is essential for the seeker of Ultimate Reality because that end is too far removed from our current state of consciousness. If we have only to cross the street to get to where we want to go, we can walk with our own two feet—or crawl, for that matter. But if our destination is halfway around the world, with vast oceans in between, we would be advised to buy an airplane ticket and fly with a pilot who had made that journey many times before. As shall be discussed in a following section, the spiritual regions of consciousness which must be crossed in order to reach the kingdom of God are inconceivably vast.

One does not simply "cross the street" of our everyday waking state and find himself in the lap of God. I am aware that this claim has been made by many who have had near-death experiences, or a relatively minor elevation of consciousness, but the teaching of mysticism is plain: these states of being are far, far away from Ultimate Reality and the perfect unity of God. The very fact that myriad "spiritual" phenomena are described by those who have left this material plane, and returned to tell us about what lies beyond, is proof enough that Oneness has not been reached. How can there be separate entities in absolute unity?

> *There are a hundred thousand ranks. The more a person becomes purified, the higher he is taken. . . . What can I say about the stations of those who have attained union, except that they are infinite, while the stations of the travelers have a limit? The limit of the travelers is union. But what could be the limit of those in*

union?—that is, that union which cannot be marred by separation. No ripe grape
ever again becomes green, and no mature fruit ever again becomes raw.

—Rumi[13]

The spiritual science of mysticism says that our condition here in this crude physical world is something like this: if God is the ocean, Spirit is the waves, and our soul is a drop. The physical universe is like an ice floe floating in the midst of the ocean of God. It is composed of the same substance as God, Spirit, and soul—just as the ocean, waves, drops and ice all are water. Our soul has been "washed up" onto the floe and finds itself far from the ocean. Though still a drop, it is encased in the ice of matter and mind, and is utterly incapable of making its way back to its source through its own efforts. Even the ocean of God is of no use, because there is no direct connection between the perfect unity of that sea of Love and the frozen separateness of that isolated drop of soul. Only when a wave rises out of the ocean, breaks upon the ice floe, and sweeps the ice-encrusted drop back into the sea, can the soul be united with God.

Perfect mystics are a wave of God, Spirit made flesh, the Way home.

God does not speak to everyone, just as the kings of this world do not speak to
every weaver. They appoint ministers and representatives so that through them
people may find the way to them. In the same way God has singled out certain
servants so that everyone who seeks Him may find Him within them. All the
prophets have come for this reason. Only they are the Way.

Since you are not a prophet, follow the Way! Then one day you may come out of
this pit and reach a high station. Since you are not a sultan, be a subject! Since you
are not the captain, take not yourself the helm!. . . . Since you have not become
God's tongue, be an ear!

—Rumi[14]

Finding a School of Ultimate Reality

Assuming that a person has decided to take Rumi's advice and seek the guidance of a perfect mystic (a wise decision, since Rumi says that "Whoever enters the Way without a guide will take a hundred years to travel a two-day journey"[15]) a central problem remains: where does one obtain this instruction in spiritual science? This is a large question, and difficult to discuss adequately here. Once again, I will try to provide only a few suggestions which may be of help to the reader. I realize, of course, that most of those who read this book will not choose to pursue the advanced course of Ultimate Reality. It certainly is not to everyone's liking, and isn't a requirement for all in this school of life. Those rare souls who are drawn to pure mysticism have concluded that no other discipline is able to provide the complete knowledge that they seek, the union with God which they yearn for.

Fortunately, this conclusion enables the scientifically-minded seeker after *final truth* to pay little attention to virtually all of the meditation and personal growth approaches which are so plentiful today. One question eliminates most from further consideration: "Does this discipline claim to be able to provide the student with direct knowledge of an objective Ultimate Reality through a precise demonstrable method?" In almost every case, the evident answer is "no." Generally, these non-mystical approaches aim at improving a person's situation in life (through more happiness, more money, more loving relationships, and so on); or they promise increased understanding of metaphysics but not complete union with God; or purported spiritual truths are communicated by someone who is unable, or unwilling, to describe the exact means by which those truths can be realized (and thereby confirmed) by oneself.

If a philosophy, religion, meditation practice, or self-improvement approach does not even claim to be able to lead one to knowledge of Ultimate Reality, a seeker after final truth can be certain that this is not the path for him. The next problem, then, is to evaluate the validity of those paths that do make such a claim. In this category we reasonably could place most of the world's religions, and a fair number of mystical disciplines. I will leave physics and the other material sciences out of this group of those who claim to possess a path to Ultimate Reality, because even though they may assert a potential ability in this area, no one believes that an actual Theory of Everything is close to being confirmed.

Further, we must respectfully discard most of the aforementioned religions and mystical disciplines from our increasingly short list of possibly valid spiritual sciences. I say "respectfully," of course, because their adherent's faith in the existence and goodness of God is sincere and deep-rooted. However, from the standpoint of science these faith-based religions and disciplines always lack a crucial characteristic: an "experimental design" which describes exactly what a person needs to do in order to know Ultimate Reality completely, to unite one's consciousness with God fully. I do not mean to belittle the importance of faith, because every scientist must believe that he is on the correct road to truth or his research will be half-hearted and listless. But when faith in the reality of a destination is not combined with concrete actions to move toward that end, one is left with little more than good intentions.

Many religious people will reply to this argument with words such as: "No, you are wrong. Our devotion and good actions in this life will lead to union with God after we die. Thus we indeed are moving along a path that leads to perfect truth. But we believe that it is not possible to come to the end of that journey while existing in our human form." Well, there is not much that can be said in response to these sentiments. A person certainly is entitled to hope for salvation and ultimate knowledge after death. However, mysticism promises that union with God is possible here and now, in this very life. This is what makes it a spiritual *science*.

Suppose that a physicist wrote a letter to the committee that decides on the recipients of Nobel prizes in his field, saying: "This year's physics award should go to me, because I have discovered all of the secrets of the material universe. Through my efforts, everything has been explained. While the evidence for this claim only can be provided to the committee members after you die, I want my Nobel prize *now*. Just have faith in me." That letter, we must assume, would find its way quickly into the wastebasket. (If not, I need to begin composing my own correspondence to send to Sweden. There is a blank spot on my mantel where a Nobel prize would fit nicely.)

Kabir, a fifteenth-century perfect mystic, wrote:

> *O friend, rely only on that*
> *Which you get while living. . . .*
> *Those who say, 'After death*
> *Your soul will merge in God,'*
> *Only give you false assurances;*
> *Whatever you gain now*
> *Will be with you hereafter. . . .*
> *While living, if a man fails to cut*
> *The deadly noose of delusion*
> *But looks forward to liberation after death,*
> *He is like the thirsty man*
> *Who thirsty remains*
> *Even after visions*
> *Of water in his dreams.*[16]

Another criterion which a seeker after final truth should keep in mind when evaluating potential mystical disciplines is their "track record." This term usually is not used in such a context, and certainly it is not proper to equate a perfect mystic with, say, the C.E.O. of a corporation. The ability to unite souls with God is of a much different order than the capacity to improve a company's profit margin. Still, it generally is as important to consider the history of a "university" of spiritual science in which one is considering enrolling, as it would be in the case of a worldly institution of higher learning.

In either case, a chronology—though by no means able to provide conclusive evidence of legitimacy—can aid in evaluating the value of the course of study. Harvard University was founded in 1636 and has produced many thousands of satisfied graduates over the centuries. Faced with a choice between attending Harvard, and a college which is just beginning its first year of operation, a serious student would be wise to enroll in the institution with a longer track record. This does not guarantee that he will receive a good education, but the very fact that Harvard has not only survived, but prospered, for hundreds of years is an assurance that its product—knowledge—actually can be obtained by those who seek it. Those who deliver shoddy goods are unable to stay in business for so long.

Similarly, mystic schools of God-realization which have a substantial history possess a credibility which "start-ups" lack. In discussing how to assess the legitimacy, authenticity, and authority of spiritual paths, Ken Wilber notes that *lineage* "is one of the greatest safeguards against fraudulent legitimacy."[17] One reason is that the truths of the highest form of spiritual science have been known since the dawn of history. Anyone who claims to be discovering something new about God or higher planes of consciousness actually is revealing their ignorance of such matters. Genuine perfect mystics are humble, and never outwardly put forth a claim to any special knowledge—though inwardly they possess truth in abundance. All credit is given to their own spiritual teacher, the perfect mystic who brought them to his own level of consciousness and complete union with God.

Accreditation is all-important

Another reason to consider a teacher's lineage is that only a living perfect mystic can certify that a student of mysticism has completed his course of study, and has graduated to final knowledge of Ultimate Reality. In other words, only a saint can "accredit" another. Here sainthood is equated with a divine state of being, and not with a life of good works—as is the case in some religions. This is not a distinction which can be conferred by a committee of non-saintly people. A perfect mystic, or saint, is one who has reached an objective state of God-consciousness. That state only can be recognized by another perfect mystic with the same spiritual attainment—just as it takes a panel of Ph.D.s in physics to determine whether a doctoral candidate has completed the requirements for that degree, and so is qualified to teach others and conduct independent research.

It is as meaningless for a person beginning to learn spiritual science to say, "God is One, so I am God," as it would be for a neophyte physics student to say, "The essence of material reality is mathematical, so I know mathematics." Saying so does not make it so. Explication of a hoped-for goal cannot be equated with its attainment. (My daughter, when she was a little girl, would play "doctor" with her stuffed animals. But in no way would I trust her to do open-heart surgery on me. So the gullibility of those who uncritically accept the claims of spiritual teachers never fails to amaze me.)

No one would turn their automobile over for a tune-up to a shabby-looking guy who stands on a street corner with only a screwdriver in his hand, and a hand-lettered cardboard sign that says, "I am a mechanic. Leave car at curb with keys in ignition." Yet almost anyone who advertises himself as a purported teacher of spirituality has no difficulty finding at least some trusting students. Are we less concerned with the care of our soul than with the upkeep of our automobile? Perfect mystics advise that even if an entire lifetime is spent finding a genuine University of Ultimate Reality, this time has not been wasted, but well-spent. It is much better to take only a few measured steps in the right direction, than frantically run the wrong way.

One of the best means of evaluating a prospective teacher of spiritual science is to spend some time in his company, if possible. The teacher, of course, must be alive in order to do this. Those who believe in the efficacy of mystical instruction from an "ascended master" or "departed son of God" are depriving themselves of the benefit of having a living spiritual guide. Charan Singh, a modern perfect mystic, says:

> You have met so many doctors, so many advocates, engineers, and none of them have become doctors, lawyers or engineers simply by going to the laboratory or by reading books. . . . For this worldly knowledge, worldly gain, we have to go to teachers, to learn from them. And spirituality is the most difficult subject. For this we need a Master, a teacher in spirituality. He tells us why we have to meet the Lord and where the Lord is; how to find the Lord; what is between us and the Lord that keeps us away from Him; and how to remove these coverings, these barriers between us and the Lord. You may call him Master, teacher, elder brother, friend. These are just names given to recognize somebody in this world.[18]

There are, however, both perfect masters and imperfect masters of spiritual science. Hence, "look before you leap," is as essential in this area as in other aspects of life. Part of this preparation is confirming that a potential teacher of Ultimate Reality possesses, at the least, the finest positive qualities of a human being. Because a perfect mystic has merged his consciousness with God, whatever personal attributes he manifests will be divine. While a living saint outwardly has the form of a human being, inwardly his soul shines with the Lord's light.

> *Know that from head to foot the shaykh* [perfect mystic] *is nothing but God's Attributes, even if you see him in human form. In your eyes he is like foam, but he describes himself as the Ocean; in the eyes of men he is standing still, but every instant he is traveling. You still find it difficult to grasp the shaykh's state, even though he displays a thousand of God's greatest signs—how dull you are!*
> —Rumi[19]

This radiance *must* be reflected in every of his actions, and each of his words. Thus if a teacher of spiritual science has any evident personal weaknesses, this is a sign that even though his level of consciousness may be above average, it is not one with highest truth. Lust, anger, greed, attachment, egotism, and associated character defects will not be found in a perfect mystic.

A pure consciousness reaches pure truth

Earlier we examined in some detail the ten criteria of a valid scientific theory, finding that each pertained equally well to the material science of physics and the spiritual science of mysticism. There is, however, a central difference between these disciplines in the manner these criteria come to be applied. In physics, the validation of a theory occurs from the *outside*, so to speak. A physicist places a potential truth about material reality upon the examining table of observation, experiment, and rational discourse. Using the scientific method and those criteria for a valid theory, he seeks to determine whether the object of his examination is a "live and kicking" truth of existence, or only a dead carcass that lacks the vital characteristics of valid knowledge.

A student of mysticism, on the other hand, seeks to turn his consciousness into the very form of the highest mystical truth—Spirit. This enables the validation of Ultimate Reality to take place from *inside* consciousness itself. Physics abstracts symbolic knowledge from the raw reality of physical existence, and then puts those symbols to the test. A mystic confirms the validity of what is perceived in meditation by having formed his consciousness into a pure, sharp, and unbiased means of contemplating spiritual truth. In other words, that consciousness possesses the same qualities which previously were used to describe a valid scientific theory, which includes being: *simple, unified, fundamental, natural* and *beautiful.* A perfect mystic, of course, manifests these qualities perfectly.

This transformation of consciousness into a valid and trustworthy means of investigating spiritual domains of existence is another reason why mysticism is a science. Since non-symbolic planes of reality are being explored, it is not possible in spiritual science to stand back—as would a physicist—and attempt to validate one's findings through systematic tests and rational criteria. As any student of meditation knows, engaging in such mental gymnastics precludes entry into the objective, non-symbolic domain of consciousness known as contemplation. It already has been noted that thinking and contemplating are as fire and water: completely incompatible with each other.

So consciousness itself must become both the means of investigation, and the method of validation. To put it simply, a pure consciousness reaches pure truth. What is contemplated by a perfect mystic is known to be Ultimate Reality, God's essence, not because he has intellectually contrasted his mystical perceptions against the criteria of scientific truth. His consciousness has become Truth, and so manifests final knowledge in absolute clarity and completeness. There is nothing to contrast, analyze, or compare in that state of divine unity. It is only when attempts are made to describe direct mystical realization in symbolic terms, in words and concepts, that analytical reason comes to bear on spiritual truths. This, of course, is just what is being done in this book, which is why whatever you read in these pages should be considered no more than a weak reflection of the brilliant sun of Ultimate Reality.

> *That is the Ocean of Oneness, wherein is no mate or consort.*
> *Its pearls and its fish are none other than its waves.*
> *Oh absurd! Absurd! That any should ascribe partners to Him!*
> *Far be it from that Ocean and Its undefiled waves!*
> *There is no partnership and complication in the Ocean.*
> *But what can I say to him who sees double? Nothing! Nothing.*
> *Since we are paired with double-seers, oh idolater,*
> *It is necessary to talk as if we ascribe partners to Him.*
> *That Oneness is on the other side of descriptions and states.*
> *Nothing but duality enters speech's playing-field.*
> —Rumi[20]

Questions to ask before enrolling

In summary, here are some suggestions to guide your search for a path of spiritual science that leads to complete knowledge of God and Ultimate Reality:

1. Does this discipline even claim to know final truth, and the means of reaching it?

2. If so, does that means consist of a precise, demonstrable scientific method which holds out the possibility of knowing final truth *in this* lifetime?

3. Does this discipline have a solid lineage or history? Has the teacher been "accredited" by being part of a succession of perfect mystics, or does this person claim to have attained God-consciousness through his own unaided efforts?

4. Are any human weaknesses or character defects evident in this teacher?

And I should add a final question: is there a monetary charge for instruction in the spiritual science? For an unfailing characteristic of perfect mystics is that they accept no compensation for uniting souls with God. If any person does so, you can be assured that he is a not a teacher of the highest order. Saints are givers, not takers. There is no financial barrier to knowledge of Ultimate Reality. The course of study involves no tuition, other than complete commitment to the discipline of meditation and the ethical guidelines taught by the perfect mystic. Believe me, that price alone is more than all but a few are willing to pay. If final truth could be bought with any amount of money, its cost would be cheap indeed. Instead, the coin of purchase is an absolutely pure and calm consciousness. Charan Singh writes:

> There is a secret concerning seeing God in meditation. If you keep your heart pure, you will see God. Christ says, "Blessed are the pure in heart, for they shall see God." If your heart is ruled by lust and greed, you will not see God, just as you cannot see your reflection in dirty water. The trees and shrubs near a pool, and the light of the sky above it, are all clearly reflected in its water. But if a breeze stirs the water, the reflection vanishes. It is the same with our vision of God. If there is no dirt in our heart and no anger and hatred agitates it, we can see the vision of God in our hearts.[21]

Why so many metaphors?

Hearing such sentiments, the scientifically-minded reader may be inclined to discount both the message and the medium of expression. "Why engage in all these metaphors? Dirty water, pools, breezes, reflections. Why cannot mystics speak plainly and precisely? These words are not science, but religious poetry." This reaction is understandable; the words of the mystical path frustrate many who seek to comprehend spiritual science. As has been discussed previously, material sciences often are able to establish a close correspondence between mathematical and verbal symbols, and objective physical reality. This is not possible in the non-symbolic—yet objective—domain of mysticism. Thus to provide us with even a hint of truths that can be known directly only on higher planes of consciousness, perfect mystics are forced to use indirect means of expression: metaphors, analogies, poetry, allusions, seeming paradoxes.

Rumi says that once a mystic was teaching, and "in the midst of his words a fool said, 'We need words without any analogies.'" The mystic replied, "Come without analogy! Then you will hear words without analogies." Rumi adds: "After all, you are an analogy of your self. You are not this. Your bodily person is your shadow."[22] Rumi also noted both the necessity and inherent limitation of analogies:

> *Thy Attributes cannot be understood by the vulgar without analogy, yet analogy increases the mistaken idea of Thy similarity with the creatures. But if a lover should ascribe a form to Thee in his yearning, the Sea of Incomparability will not be*

tainted. If poets compare the crescent moon to a horseshoe, their nonsense will not detract from the moon's moonliness.[23]

It is important to keep in mind the special character of analogies and metaphors in the spiritual science of mysticism. We are accustomed to hearing such phrases as "the crescent moon is like a horseshoe," and realizing that this is merely an artistic use of words that does not bear any resemblance to objective reality. The moon has nothing in common with horseshoes, apart from sharing some of the same atomic structure. But when a mystic says, "our consciousness is like a dirty pool of water," his words need to be attended to as other than subjective poetry. Such a statement is an echo of objective reality on a higher domain of existence. Pools, water, and dirt as we know them do not exist in that domain, and the phenomena there are not perceptible *here*.

The only way of describing spiritual truths to those who, as the Bible says, "have eyes, and see not; which have ears, and hear not"[24] is to use images familiar to physical eyes and ears, and the rest of the five senses. Consciousness, God, Spirit—these have no form and no substance. Metaphors and analogies help to bridge the gap between our symbolic physical world and non-symbolic spiritual planes of existence. More precisely, they are signposts that point the way to the bridge of contemplative meditation, and offer guidance as to how one can travel across that span in the easiest and safest manner. Actual transport of one's consciousness into higher spheres is, however, completely separate from all thoughts and images, metaphors and analogies. Lekh Raj Puri writes:

> Mystic knowledge is a transcendent experience—a subtle condition of spiritual transport and ecstasy, an ethereal state of the soul in its naked refulgence, a merging of the drop of individual existence into the limitless ocean of Universal Life, an identification of one's finite Being with the all pervading Eternal Infinite. . . . Mystic transport is beyond all words and expressions . . . [yet] To say that a thing is beyond human intellect, that it cannot become the content of human thought, is not equivalent to saying that it is nothing, that it does not exist. . . . Our sense-organs and intellect render us useful service in this world; but before the transcendent wisdom of mystic transport, they are useless.[25]

Importance of purity and stillness

In the quotation above, Charan Singh was describing two conditions that must be met if the student of spiritual science is to carry out successfully his experiment of God-realization in the laboratory of consciousness. These conditions are: *purity* and *stillness*. Far from being abstract metaphors, they are concrete attributes of consciousness which are obvious to every person who takes a few moments to observe the content and dynamics of his mind. What do you think about all day long? How does your consciousness manifest these thoughts? Mysticism teaches that when one not only knows the answers to these questions, but has achieved complete control over the functioning of his mind, he has moved significantly closer to knowledge of Ultimate Reality.

An apt analogy can be found in the 1993 mission of the American space shuttle Endeavor to repair the Hubble Space Telescope. News reports noted that the telescope was supposed to be the most powerful astronomical observatory ever built. Unfortunately, soon after this instrument was placed in orbit it was learned that the telescope's primary mirror was defective and photographs transmitted to earth were blurry. The astronauts on board Endeavor were able to correct various problems with the Hubble observatory, including two which confront every student of mysticism who begins to practice contemplative meditation.

The first problem was that the primary mirror had been ground incorrectly, and so was unable to focus light precisely enough to produce sharp images. Even though the outer edge of the mirror was too flat by only one-fiftieth the width of a human hair, this was sufficient to render the ninety-four-inch main mirror useless for its intended purpose of observing distant astronomical phenomena. If such a minor flaw makes it impossible to properly focus comparatively crude beams of material light, imagine what purity of consciousness is needed to perceive clearly the infinitely more subtle radiance of God.

Second, solar arrays that power the Hubble Space Telescope and its gyroscopes (which help point it in the right direction) were expanding and contracting as the arrays passed in and out of the Earth's shadow during each ninety-minute orbit. According to an article in the magazine *Astronomy*, "the motion of the array causes the telescope to jitter and this, in turn, blurs the images."[26] Thus the very means of energizing the observatory and controlling its direction was preventing it from carrying out its mission. A similar problem generally afflicts a beginning student of mysticism. The same mind which is drawn to know God, and thus impels the student to practice meditation in the first place, is found to produce so many wobbly waves of thought as to make it impossible to attain the calm vision of contemplation.

Purity and stillness, then, are as critical in astronomical observation as in mystical meditation. This analogy reinforces a point that has been made several times before: spiritual science is no different from the material sciences in regards to the precision of the observing instrument(s) that are utilized in research, and the scientific method used to distinguish between valid and invalid findings. A competent teacher is required to properly learn either mysticism or physics, which entails choosing a center of higher education which promises to provide the student with the knowledge he seeks. As the reader probably has surmised, I am "enrolled" in one such university of spiritual science which—in my opinion—fulfills each of the suggested criteria which such an institution should meet.

Many schools, one teaching

By no means am I suggesting that this particular mystical center is the only source for learning the means of God-realization and knowing the genuine Theory of Everything. Truth is truth, no matter how it is obtained. No physicist would claim that his material science can be studied only at Stanford, or Cambridge, or Tokyo University. The most important requirement for learning any science is a competent teacher. This is especially true in mysticism, for the student does not need any classroom, laboratory equipment, or even textbooks to study this discipline. All that is required is instruction from a perfect mystic in how to carry out the experiment of contemplative meditation in one's own consciousness. Actually, the student of mysticism daily engages himself in the task of becoming less and less, so not only is nothing made of mind or matter needed to pursue this science—what we already possess must be discarded.

> *The Absolute Being works in nonexistence—*
> *what but nonexistence is the workshop of the Maker of existence?*
> *Does anyone write upon a written page?*
> *Does anyone plant a sapling in a place already planted?*
> *No, he searches for a paper free of writing,*
> *he sows a seed in a place unsown.*
> *Be, oh brother, a place unsown,*
> *a white paper untouched by the pen!*
> —Rumi[27]

Proselytizing has no place in any form of science. It is almost unthinkable, for that matter. It is difficult for me to imagine a fervent physics professor shoving a pamphlet about quantum cosmology in my hand, passionately urging me to read it and learn the truth about how the world was created so that my mind can be saved from ignorance. By the same token, neither will you find a perfect mystic enticing people to become students of spiritual science. The teachings of science are available for those who wish to learn about them. Truth is not affected in any manner by the number of people knowing it. Each person has a perfect right to pursue final knowledge about either material or spiritual domains of existence, or remain content with what they already know.

These pages contain many quotations pertaining to mysticism, drawn from Rumi and other sources. Other mystics could have been selected to illustrate the same points, just as the same factual content is found in physics texts around the world. In both spiritual and material science, which references to cite is a practical matter-of-choice. Any genuine science is universal, and transcends the manner by which it is described or the setting of that description. No more distinction is made between "Eastern philosophy" and "Western philosophy" in pure mysticism, as there are geographic differences in the laws of physics: none, in either case.

In previous pages this book has included quotations from mystics, and mystically-inclined students of spirituality, who lived in many parts of the world—including the United States (Huxley), Germany (Eckhart), Italy (Aquinas), Palestine (Jesus), Arabia (Muhammad), Turkey (Rumi), India (Kabir), and China (Lao Tzu). Even though the spiritual level of attainment, native language, and means of expression of each may differ from the others, the essential message does not. Thus, as I have noted before, we should not be led astray by the place of birth of a mystic or the geographic location of a center of mystical teaching. Existing as they do in this physical domain of time and space, mystics have to live somewhere. And their symbolic teachings—spoken words and written prose—will be adapted to fit those surroundings. However, the non-symbolic truths of mysticism which are perceived directly on higher planes of consciousness are changeless and eternal.

> *Although the words of the great saints appear in a hundred different forms, yet, since God is One and the Way is one, how should their words be two? They appear different in form, but they are one in meaning. In form there is diversity, in meaning all is concord.*
>
> —Rumi[28]

Similarly, Lekh Raj Puri writes: "There may be and is much difference among various religions, for they rely on Intellect; but there is never any difference among true mystics for they know with the transcendent power of transport. What they experience is objective Reality, which is the same for all. . . . In true mystic knowledge, there is no scope for difference or doubt; for it is not a matter of opinion and guess work; it is a transcendent perception of objective Reality on a plane of consciousness higher than the intellect and subtler than the senses."[29] Thus in mysticism, direct perception by a refined consciousness results in knowledge of pure truth. This knowledge necessarily is diluted and diminished significantly when observed laws of spiritual planes of existence are described by conceptual symbols of the physical world.

Ultimate reality is not a theory

This direct perception is essentially the opposite of what occurs in the material science of physics. In physics the preferred approach is to: (1)

collect observations of physical phenomena, or create mathematical representations of unobservable phenomena; (2) examine those observations for underlying regularities; and (3) deduce theories that explain the laws of nature which are thought to be responsible for producing the regular phenomena. Physicist Stephen Hawking says, "a theory is just a model of the universe, or a restricted part of it, and a set of rules that relate quantities in the model to observations that we make. It exists only in our minds and does not have any other reality (whatever that might mean)."[30]

As has been noted before, the investigative approach of physics works quite well when applied to discrete parts of the physical universe, but is nonsensical if one's goal is knowledge of *Ultimate Reality*. For Hawking points out that material science starts with observations which appear to be part of objective reality, and ends up with a subjective theory which "exists only in our minds." This does not appear to be progress toward final truth, unless it is assumed from the outset that there is no reality apart from the content of our minds. This apparently is why Hawking says that theories "do not have any other reality." His aside, "(whatever that might mean)", indicates that he rules out *a priori* the possibility that the substance of final truth might consist of something other than matter and mental abstractions—both of which presumably are produced by physical phenomena.

In spiritual science direct—or non-symbolic—perception in contemplative meditation results in knowledge of highest truth, whereas in material science sensory observations are considered to be largely meaningless until embedded in a symbolic theory. This is important to keep in mind when we begin examining specific principles of mysticism, and how these relate to findings of the new physics. Those principles (as interesting and thought-provoking as they may be) merely are symbolic reflections of the higher laws of existence that are perceived directly by perfect mystics. The principles definitely should not be considered *theories* as physics uses that term.

Stephen Hawking writes, "Any physical theory is always provisional, in the sense that it is only a hypothesis: you can never prove it. No matter how many times the results of experiments agree with some theory, you can never be sure that the next time the result will not contradict the theory."[31] This line of reasoning follows naturally from the fact that physicists are able to perceive lower-level phenomena directly with their five senses, or indirectly through mathematical models, but are forced to infer—through conceptual theories—the underlying laws of nature which produce physical phenomena. Thus those higher-order laws always are known tentatively, indirectly, provisionally. The teachings of perfect mystics, on the other hand, can be relied upon conclusively. William Chittick, a translator of Rumi, writes:

Since they are inwardly identified with the Universal Intellect, the prophets and saints have passed beyond the limitations of discursive knowledge and "rational thought," for these are the workings of the partial intellect. The prophets and saints do not seek for knowledge, since they themselves are its source; they know nothing of supposition and opinion, since they dwell in certainty and immediate vision.[32]

Can a snail ever comprehend a garden?

A simple metaphor will illustrate why the relative importance of theory and observation differs in mysticism and physics. Picture a snail crawling across the ground. This creature's physical form constrains it to move at all easily—if slowly—in only a two-dimensional world. When it comes to an obstacle of even modest height—such as a garden hose—the snail must pause and laboriously crawl over the top of it. Further, the snail's confined perspective would make it difficult for it to learn all about that hose. If so inclined, it would have to crawl inch by inch along the length of the hose all the way to the faucet, explore the nature of that connection, and then creep to the other end of the hose—where the homeowner has attached a sprinkler and is watering his vegetable garden.

That snail, as intelligent as it might be compared to others of his species, would find it difficult to fit together all of the facts that he has collected about the hose. Hopefully he would have the equivalent of a snail notebook in which to jot them down. It is almost impossible for human beings to visualize the snail's point of view. If *we* come upon a hose on the ground, it barely is noticed as we stride over it. Being five feet or more above the surface of the earth, our eyes easily take in the sinuous path of the hose as it snakes from the faucet across the lawn to the garden, where the sprinkler provides water to keep carrots and lettuce alive.

The snail, however, does not have the benefit of our all-encompassing vision. So it is forced to come up with a theory that ties together each of the observations that it made in the course of its travel up and down the length of the hose. This certainly would stretch the capabilities of a snail mind to the limit. Similarly, Stephen Hawking says that in physics:

> It turns out to be very difficult to devise a theory to describe the universe all in one go. Instead, we break the problem into bits and invent a number of partial theories. Each of these partial theories describes and predicts a certain limited class of observations, neglecting the effects of other quantities, or representing them by simple sets of numbers. It may be that this approach is completely wrong. If everything in the universe depends on everything else in a fundamental way, it might be impossible to get close to a full solution by investigating parts of the problem in isolation.[33]

After all, even an exact description of the physical characteristics of the faucet-hose-sprinkler system will not lead that snail very far down the path of truly understanding these entities. Recall that physicist Joe Rosen said that science is our attempt to know the reasons for phenomena, to understand them, and not merely to describe what is observed. Without grasping the concept of a "garden," which entails knowing about human "food," the purpose of that watering system never will be understood by our scientifically-inclined snail. And this, unfortunately, requires a more highly evolved consciousness than the snail possesses.

As a final commentary on this metaphor, a person hovering in a helicopter a hundred feet above the garden hose would have an even broader perspective than someone standing on the ground. At a glance, the pilot could see the water tower that serves as the supply source for the garden hose, as well as the vegetable stand on the road where the gardener sells produce to other people, who transport the carrots and lettuce to their homes in the city (and make salads which their children are forced to eat before they can watch television). The broader our vision, the deeper our understanding of the place that garden hose holds in the larger fabric of existence. Each level of knowledge—that of the snail, the person on the ground, the pilot in the air—is true enough in its own right. But anyone who wants to know the most complete truth about the hose and its role in the universe would be advised to seek the highest perspective.

This brings to mind an often-heard criticism of mysticism: that the descriptions mystics give of spiritual realms of existence are not in agreement. The obvious answer to this objection is to picture yourself talking by radio to the above-mentioned helicopter pilot. "What do you see up there?," you ask from your seat on the ground. He answers, "Lots of trees. Cars on the road. A water tower." You respond, "Is that all?" "Of course not," the pilot says, "there are cows in a field, a man working his garden, children playing hide and seek." "But is *that* all you can see?" "My friend," comes the pilot's reply, "it would take too long to tell you everything I'm looking at, and even then it wouldn't be the same as being up here. The sky is clear and my vision extends all the way to the horizon. Next time, come along and fly with me instead of asking so many questions from the ground."

The principles of spiritual science which will be described in a following section are by no means all that a perfect mystic "sees" with his God-consciousness. These representations of higher truths are symbolic reflections of Ultimate Reality, and nothing more. Yet they point toward a final truth which is far beyond the physical laws of nature known to material science. This should not diminish our appreciation for the accomplishments of the new physics. Rather, we should be encouraged to set our sights more firmly toward that God who created both the laws of nature which reflect His divinity indirectly, and man's capacity to see His countenance face-to-face.

> *You name His name; go, seek the reality named by it!*
> *Look for the moon in heaven, not in the water!*
> —*Rumi*[34]

Meaning and Form

The distinction between meaning and observation was discussed in a previous section, so we need not spend much time on it here. Walking into your home and seeing a cake with candles burning on the table simply is an *observation* of some material forms. When your friends jump up from hiding places and yell, "Surprise!," you understand that the *meaning* of the cake is an unexpected birthday party. Meanings are what supply the richness and purpose of life. When they are lacking, people become depressed, listless, even suicidal. Bare facts are not enough for us. We want to know what they mean.

To some, this is an unscientific attitude. They believe that science means facing squarely the stark reality of the universe, stripping away all personal prejudice, subjective philosophizing, and wishful thinking from the objective bare bones of existence. Yet actually this is not the way of material science, which searches for understanding of the "whys" that lie behind the "whats" of the physical universe. The hoped-for Theory of Everything is intended, as astronomer John Barrow said, to explain not only how the universe came to be as it is, but why it could not be anything other. Meaning thus is as much a part of science as of religion or ethics.

Two meanings of the universe

Still, it cannot be denied that many material scientists have come to the conclusion that our world is nothing other than an accidental result of essentially mechanical forces of nature. At the end of physicist Steven Weinberg's book, *The First Three Minutes*, he relates how he is writing those final pages while flying in an airplane across the United States. Observing how "soft and comfortable" the landscape looks with its clouds, snow, and signs of human habitation, he nevertheless comes up with these conclusions:

> It is very hard to realize that this all is just a tiny part of an overwhelmingly hostile universe. It is even harder to realize that this present universe has evolved from an unspeakably unfamiliar early condition, and faces a future extinction of endless cold or intolerable heat. The more the universe seems comprehensible, the more it also seems pointless. But if there is no solace in the fruits of our research, there is at least some consolation in the research itself. . . . The effort to understand the universe is one of the very

few things that lifts human life a little above the level of farce, and gives it
some of the grace of tragedy.[1]

Thankfully, Weinberg's views are by no means shared by all in his
profession. Gerald Schroeder, an applied physicist and theologian, provides
an alternative interpretation of the meaning of creation in his book, *Genesis
and the Big Bang*. Here the author describes his own philosophy:

> We live in a world in which existence in and of itself is good. The less
> than perfect condition of our existence is not an inherent aspect of the
> world. The entire flow of matter over the eons has been, in our corner of
> the universe at least, a journey from disorder toward harmony. So much is
> already good that the bad stands out by contrast. This gives cause for
> optimism . . . we members of humanity are endowed with the sensitivity
> and the special knowledge that enables us to distinguish the true from the
> false and, at times, the good from the evil.[2]

We are left, then, with a central question: Why is it that physics, which
has had such success in laying bare the facts about objective physical reality,
is incapable of understanding the meaning of this world? In a previous
quotation Steven Weinberg said that "the lessons of religious revelation"—
in contrast to those of physics—"point in radically different directions." Yet
the conclusions of Weinberg and Schroeder about the meaning that
underlies physical phenomena reflect equally divergent points of view.
How is it that two highly-trained professional physicists, each of whom is
privy to essentially the same factual knowledge about the material laws of
nature, vary so greatly in their interpretation of the message those laws hold
for mankind?

"Whats" are meaningless without "whys"

The reason why physicists have a common understanding of the *form* of
our universe, but do not agree about its *meaning*, is that the research tools of
physics are unable to reveal the purpose of creation. Without knowing the
"why" of material reality, the "whats" and "hows" of the laws of nature
remain a skeleton with no flesh, answers without a question, aimless tracks
in a featureless desert. Physicists who focus solely on learning more and
more details about the workings of this universe are in the same position as
the aforementioned scientific snail. Making their way up and down the
winding garden hose of material existence, from the tiniest recesses of the
subatomic world to the furthest reaches of intergalactic space, the myriad
facts that are collected are incapable of answering the most important
question: why do these phenomena exist at all?

> *Forms are the oil, meaning the light—*
> *otherwise you would not keep asking why.*
> *If form is for the sake of the form itself,*
> *they why ask "Why?". . .*
>
> *So wisdom cannot allow that the outward form*
> *of the heavens and the inhabitants of the earth*
> *should exist for this only.*
>   —Rumi
>
> *Pass beyond form, escape from names!*
> *Flee titles and names toward meaning!*
>   —Rumi[3]

Let us suppose our snail learned every detail about the size and construction of the garden hose, the mechanical workings of the faucet, the number of holes and spray pattern of the sprinkler head. What then does it know, and what has escaped its understanding? It knows that water comes out of the faucet when the handle is turned, flows through the length of the hose, and is emitted from the sprinkler in a particular pattern. That is all. The purpose of these interconnected entities escapes our snail, assuming it even cares—or is able—to ponder the question of "why all this?" with its limited consciousness. Only by knowing the mind of the gardener who is growing the vegetables can the meaning of the watering system be understood.

> *Whoever looks upon secondary causes is for certain a form-*
> *worshipper.*
> *Whoever looks upon the First Cause has become a light which discerns*
> *Meaning.*
>   —Rumi[4]

Breaking out of circularity

The new physics has begun to take a few faltering steps toward resolving this problem of meaning in material science, but physicists' suppositions are still in their infancy compared to the mature understanding of perfect mystics. Sir Arthur Eddington, a pioneering twentieth-century physicist, aptly pointed out how material science has become trapped in a factual maze which has no exit to the freedom of meaning. Physicists incessantly wander along the same corridors, forever bump into the same blind alleys, continually learn more details about the same things.

Eddington noted that quantum physics has left far behind the conception of atoms as tiny billiard balls, and "now we realise that science has nothing to say as to the intrinsic nature of the atom. The physical atom is, like everything else in physics, a schedule of pointer readings. In science we study the linkage of pointer readings with pointer readings. The terms link together in endless cycle with the same inscrutable nature running through the whole."[5]

As a result, physics is forced to define one property in terms of another property, with no connection to anything outside of the arbitrary confines of this material science. This is the effect of Gödel's Theorem: physics has chosen internal consistency over complete understanding. It survives as an independent discipline, but at the cost of access to final truth, the Theory of Everything sought by physicists. Attachment to a means precludes attaining the goal. Eddington says:

> And you can see how by the ingenious device of the cycle physics secures for itself a self-contained domain for study with no loose ends projecting into the unknown. . . . Electric force is defined as something which causes motion of an electric charge; an electric charge is something which exerts electric force. So that an electric charge is something that exerts something that produces motion of something that exerts something that produces . . . *ad infinitum.*[6]

Mysticism, the spiritual science of consciousness, is the only way out of this endless loop. That inscrutable nature of physical reality referred to by Eddington is the mystery of consciousness—God's Spirit dimly reflected in materiality; His Voice echoing down the corridors of time and space, calling us to return to Him. A true scientist should not be content with the knowledge of forms, but should seek the inner meaning of creation. The new physics has brought us to the very edge of material existence, the farthest reaches of physical reality. This is a great accomplishment. For one more step takes us into the spiritual realms of existence, and the beginning of the path to knowledge of God. That step cannot be taken, however, with the legs of logic and reason which have enabled us to travel to this point. "You have served me well," says the mystic to his intellect, "but to cross the ocean of Spirit I need a more worthy craft." Charan Singh says:

> The mind can neither explain nor understand why this whole universe has been created. He knows best why he has created it, because in the beginning nothing existed besides Him. . . . In order to know why He has created the Universe, we should make our way up to reach Him. . . . With our limited knowledge we may go on trying to analyze, but by mathematics or by scientific means, we can never reach any decision at all. That is why we are working day and night in meditation—to know why He has created this universe.[7]

Is meaning made, or discovered?

To do so, of course, requires reaching the level of consciousness of the Creator. Otherwise, we will remain like our snail who knows all the whys and wherefores of snail reality, but little or nothing of human reality. Similarly, as human beings we are acquainted directly with the world of everyday existence, while only the merest hints of Ultimate Reality filter through the dense veils of mind and matter. Thus most people mistakenly believe that they bear a personal responsibility to ascribe meaning to existence. We search for a philosophy of life, a religion, a belief system, a profession, a relationship, a social or political cause—something that will provide the reason for being. Our attitude, conscious or unconscious, is that physical existence is an empty stage upon which we make up our own lines, and act out our chosen role in the completely individualistic play, *My Life's Meaning*.

Yet mysticism asks: "Is it possible that meaning already is present in every atom of matter and each thought of mind? Could it be that this material universe is the product not of blind mechanical forces of nature, but the conscious will of God?" If so, the meaning of creation does not need to be brought into being—only discovered. The meaning of both physical existence and human life lies in their creator, just as the meaning of a garden is not in the fruits and vegetables themselves, but proceeds from the intention of the gardener. Far from being a denial of the primacy of human consciousness—as some might argue—this perspective equates the ultimate purpose of man with the divine will of God. Yes, we give up some of our purported "humanness," such as the supposed capacity to freely choose the meaning of our life, but this is exchanged for the reality of Godliness. In the words of the perfect mystic Charan Singh:

> Human life is a great boon bestowed upon us by the Lord, but its sole purpose is to regain our lost heritage and once again merge into the Flame of which we are a spark. . . . The uniqueness of the human form consists in its ability to realize God as long as it is activated by life. Towards this end we must bend all our energies. This is our real work. The rest is all to no purpose. . . . Our stay in this world is not intended to be for attachment to its ephemeral joys and possessions, but is for the sole purpose of realizing God . . . the real purpose of human life is to realize and merge in the Ultimate Reality.[8]

Similarly, Rumi says:

> *The universe displays the beauty of Thy Comeliness!*
> *The goal is Thy Beauty—all else is pretext.*[9]

Science of shadows, science of light

Mysticism teaches that the physical universe is a reflection of higher planes of consciousness. The sun shines brightly in the heavens; its rays fall upon a pool of water on this earth; that reflection casts an image on an adjoining wall. The "sun" on the wall bears little resemblance to the shining orb in the sky; it has little of the heat and brightness of the original sun, and appears as a blurry patch of light, not a perfect sphere. Just as in Plato's famous Allegory of the Cave, the images cast upon the walls of physical reality should cause us to turn around and seek out the source of those reflections. Physics is the material science of shadows, mysticism the spiritual science of light.

From the perspective of learning the truth about Ultimate Reality, shadows are valuable only insofar as they direct our attention to the Sun of Creation. On an overcast day everything appears gray and muted. You find yourself walking along with your head down, paying little attention to the world around you. Suddenly those surroundings are thrown into sharp relief. Colors become brighter, outlines clearer. Each tree and fence post throws its shadow upon the ground. This causes you to turn your eyes upward. *The sun has come out.* Those shadows were messengers of light.

> *Sunlight fell upon the wall;*
> *the wall received a borrowed splendor.*
> *Why set your heart on a piece of earth,*
> *O simple one? Seek out the source*
> *which shines forever.*
> —Rumi[10]

Fact one: in 1994 it was reported that astronomers have detected bursts of energy that "might be, for the few seconds they last, the brightest objects in the universe."[11] These emissions of gamma rays, apparently from sources halfway across the known universe, produce more energy in several seconds than our sun puts out in a thousand years.

Fact two: in 1986 a group of astronomers known as the "Seven Samurai" discovered that our Milky Way galaxy and a large chunk of the nearby universe—a hundred thousand galaxies, a thousand trillion suns—is moving together as a unit at two million miles an hour toward some unknown attractive force. Dennis Overbye writes that "It was as if somebody had taken most of the known universe and set it on an ice floe adrift in some mysterious current."[12] Since, other researchers have found evidence for a much larger flow of galaxies that extends two to three times beyond that Great Attractor (as it came to be called). So the Great Attractor apparently is being pulled toward a Greater Attractor. And earth is going along for the ride.

Astonishing findings such as these abound in physics, astronomy, and other material sciences. We will encounter many more in the next portion of this book. The frontiers of science are being pushed outward by the day, by the hour, by the minute. Yet for most people life goes on as always. They are untouched by the discoveries of the new physics, uninterested in findings of modern cosmology. I know this to be the case, for I am a connoisseur of popular science. It interests me deeply, and I often mistakenly believe that others share my passion. I wish that I had a dollar for every dinner table conversation that stopped dead in its tracks after I passionately shared some tidbit of material science.

I may say excitedly, "Did you know that, according to Neil McAleer, our huge Milky Way galaxy still is only about one trillionth of the universe, which is like comparing a very small metal screw with the mass of a hundred-thousand ton ship of which it is a part?"[13] "Oh, really," my companion replies, "could you please pass the salt?" Clearly facts such as these have different meanings for different people. Many scientists share the perspective of most lay persons: facts are facts, and there is not much else that can be gleaned from them. The atomic structure of helium stands on its own. It is useful information for those who make lighter-than-air balloons, but does not possess any hidden meaning.

Veils have a purpose

Others, however, consider many of the findings of modern science to be akin to religious revelation. In their view the mysteries of life are being revealed through quantum theory and other aspects of the new physics. This attitude was reflected in John Updike's novel, *Roger's Version*. The main characters are Roger Lambert, a divinity school professor, and Dale Kohler, a young computer hacker. Kohler comes to Lambert seeking a grant to support his work on a computer program that will prove the existence of God by tying together all kinds of scientific facts. In the course of trying to convince Lambert of the soundness of his proposal, Kohler speaks about what is happening in physics:

> The most miraculous thing is happening. . . . The physicists are getting down to the nitty-gritty, they've really just about pared things down to the ultimate details, and the last thing they ever expected to happen is happening. God is showing through. They hate it, but they can't do anything about it. Facts are facts. And I don't think people in the religion business, so to speak, are really aware of this—aware, that is, that their case, far-out as it's always seemed, at last is being *proven*. . . . God is *breaking through*. They've been scraping away at physical reality all these centuries, and now the layer of the little left we don't understand is so fine God's face is staring right out at us.[14]

Kohler has much more to say on this subject, including his view that most people—including Lambert—don't *want* God to break through. They would prefer to remain comfortable in their own lives, "being human, and dirty, and sly, and amusing, and having their weekends with Michelob [beer]."[15] The teachings of mysticism are in accord with Kohler on that latter point, but do not agree with him that God is about to make an imminent and unmistakable appearance in the discoveries of the new physics. Perfect mystics say that this world is more an apprenticeship than a permanent profession, an overnight inn and not a homestead. In order for this physical universe to carry out God's intended purpose, spiritual blinders and earmuffs must prevent most people from clearly seeing His countenance and hearing His voice.

> *The world is kept standing through heedlessness. If there were no heedlessness, this world would not remain. Yearning for God, recollection of the next world, spiritual intoxication, and ecstacy are the architects of that world. If all these displayed themselves, every one of us would go to that world and not remain here. But God wants us here so that the two worlds may exist. So He has appointed two magistrates, heedlessness and heedfulness, so that both houses may flourish.*
> —Rumi[16]

Until God desires otherwise, the vast majority of people—including physicists—will neither be interested in returning to Him, nor in understanding how His attributes are reflected in this physical world. In one of his discourses Rumi said that there are seven hundred veils of darkness and seven hundred veils of light,[17] meaning that devotees of both material science and the lower forms of spiritual science never will find the final truth about Ultimate Reality through these less than perfect means of investigation. Physics, perhaps, has lifted some of the veils of darkness, but is not even close to the veils of light, much less being able to remove them. The following, therefore, is the proper way of relating mysticism and the new physics:

The findings of material science should be viewed as echoes, or reflections, of the truths of spiritual science.

Seek the source, not an echo

As was noted previously, perfect mystics tell us that the physical universe is a reflection of higher domains of existence. Thus the precept, "as above, so below." But when the sun is reflected in a pool of water, and thence onto a wall, the resulting image is a pale copy of the original. The same is true with creation as a whole. Seth Shiv Dayal Singh says that "all the different regions were created by the Supreme Being as reflections of the Real Region, so that the lower regions also share to some extent the features and conditions of the highest, but there is a lot of difference in regard to permanence and other conditions. Each region has its own distinct creation, marked by different degrees of subtleness and purity. Only he who has seen all the regions can appreciate the difference."[18]

The principles of mysticism stand above, and contain, the findings of physics just as a portion of the rays of the sun is in each reflection. However, a reflection cannot be equated with the original. Neither is an echo coequal with the source of sound. Take away a reflection or an echo, and whatever produced those effects is not affected. The causative chain works only in one direction: from higher domains of consciousness to the lower. God creates; existence is created. This makes it impossible to prove the existence of God, or know final truth, through intellectual analysis of His creation. When an echo fades away into nothingness, so goes that proof of the source of the sound. Yet the source remains as it was.

> Though in the world you are the most learned scholar of the time,
> behold the vanishing of this world and this time!
> —Rumi[19]

If we seek knowledge of Ultimate Reality, our attention must be focused on permanent truth and not on ephemeral knowledge of physical existence. Astronomers tell us that this universe began with the Big Bang, and it likely will end with a Big Crunch[20] (billions of years from now, so no need to worry about it now). The galaxies which are still rushing apart in the expansion of the universe one day will begin moving together under the inexorable force of gravity. At some point what began as a minuscule point of unimaginably potent energy, and evolved after the Big Bang into the vast physical universe, will return to nothingness. The lights of every star will blink out. A curtain will come down on this domain of creation.

God will remain unchanged, of course. So the foundation of eternal truth cannot be found in a universe which will not last. Nevertheless, the creator is not separate from his creation. Every particle of this universe, which includes each cell of our bodies and every thought of our minds, is part of God. Yet the density of mind and matter permit His attributes to be only dimly reflected in physical creation. The teachings of spiritual science are the sole means by which a reflection of ultimate truth can be seen in its true light, and the source of final knowledge recognized in its confused echo.

> That voice which is the origin of every cry and sound:
> that indeed is the only voice, and the rest are only echoes.
> —Rumi[21]

Mastering the universe at a snail's pace . . . or a leap

Here is another way of considering the matter of meaning and form. You and I live on this planet with six billion other people. Almost everyone's daily concerns revolve around mastering this life we are living: keeping a roof over our heads, maintaining cordial relationships, learning the arts and sciences, getting ahead in our profession. Let us assume that you and I are successful beyond our wildest dreams, and are able to master the entire world. (I leave it to the reader to decide what master means to you: "know," "control," "own," "love," whatever you like. I will use this term in the sense of "understanding," or "knowing.")

No, even more. We master the entire solar system. This is not an impossibility, since space probes already have passed beyond the confines of the nine planets, and it appears certain that human beings are the most advanced life form in our immediate neighborhood. This means that out of a hundred billion stars in the Milky Way galaxy of which our sun is a part, we fully understand *one* of those star systems. Thus we possess a one-hundred billionth part of the galactic knowledge. Now, I realize that it could be argued that by knowing the laws of nature that govern our own star system, we know about all of them. But this is a sol-centric prejudice. Not having taken one step onto even another planet, how can mankind believe that we know all the secrets of the galaxy from our cramped perspective on the outer edge of one of the Milky Way's spiral arms?

Still, to illustrate my point, I will accede to this argument. You and I now possess complete understanding of the entire Milky Way galaxy. A wonderful accomplishment, and perhaps it is time to rest on our laurels. But wait. There are at least a hundred-billion other galaxies in the physical universe, so we still are only one-hundred billionth of the way to our goal. We must master each of those galaxies as well if we are to be sure that no aspect of complete knowledge has escaped us. At this point it becomes almost impossible to visualize the magnitude of the task we are facing. In spatial terms, one-hundred billionth of the earth's circumference is a little more than a hundredth of an inch. So if ultimate understanding of the universe is viewed as equivalent to walking around the globe, mastering our entire galaxy means that we have moved forward only a tiny fraction of an inch.

This is the problem of believing that knowledge of material forms can lead to the truth about Ultimate Reality. Even if understanding of the *meaning* of the universe proceeds in concert with knowledge of its *form*, our progress will be at a snail's pace. Worse, we could continue that exceedingly slow movement of incremental knowledge for countless eons and never, ever, reach our goal. Andrew Harvey says that "The entire cosmos cannot contain God because God goes beyond the cosmos."[22] So if one wants nothing less than complete knowledge, little is achieved even by learning every detail about the physical universe.

There must be a better way to attain final truth. And there is. Mysticism teaches that since consciousness is the essence of Ultimate Reality, each human being possesses the means to know God directly—and to understand the meaning of every domain of existence. Perfect mystics explain that this knowledge is "within us." Through the research tool of contemplative meditation, every nook and cranny of creation can be explored without leaving one's chair. It is not necessary to laboriously piece together the fabric of final truth. Ultimate reality is a whole cloth. Thus mystic understanding, it is said, is not taught but caught. It is acquired in large leaps, not small steps.

Admittedly, this ability is not easy to acquire. There is no guarantee that a student of spiritual science will perfect his consciousness to such a degree even after a lifetime of meditative practice. But not every physics undergraduate earns a Ph.D. in that field either. The truths of any science have nothing to do with how many people are capable of fully understanding them. Complete knowledge of Ultimate Reality—God—is reached by only a few people in one lifetime, but anyone can make significant progress toward this goal by following the teachings of mysticism. A perfect mystic has reached the endpoint of this course of study, and is able to enter higher domains of consciousness as easily as you or I walk through our front door. Julian Johnson, an American physician and student of mysticism, wrote about the potential of man to know the creation:

> . . . the Creator has so constructed man that he is able, when properly informed and trained, to place himself in conscious communication with the entire universe. At first thought, this would appear but a flight of fancy, but it is literally true. It can be done. This fact is due to the way man has been constructed. . . . He is in fact a *replica* of the entire universe of universes on a very small scale, and for that very reason he is able to reach consciously the entire universe lying outside of himself. . . . We have the ability, when our faculties are awakened, to actually hold conscious communication with the most distant heavens, to explore the utmost regions of space. This applies not only to all physical worlds of starry galaxies, but to all the astral and the higher spiritual regions. There is no limit.[23]

So when the student of spiritual science is able to properly focus his consciousness on the dimensionless point of the eye-center, new vistas of both meaning and form open up.

Though the worlds are eighteen thousand and more,
not every eye can see them.
Every atom is indeed a place of the vision of God,
but so long as it is unopened,
who says, "There is a door."
                    —Rumi[24]

Principles of spiritual science, and echoes in the new physics

Perfect mystics are the teachers of spiritual science who open the door of God's knowledge and love to seekers of Ultimate Reality. Each of the following seven principles of mysticism will be introduced with a quotation from Rumi—a perfect mystic—followed by an elucidation of that principle from the standpoint of spiritual science. I cannot emphasize strongly enough that *these principles do not encompass all of mysticism.* They have been selected from the much wider scope of spiritual science to match with the findings of the new physics. Much of mysticism is completely beyond the ken of material science: there is no reflection at all of these higher truths in physics, so it is not possible to discuss the meaning of an echo if none exists.

> His rules are manifest in all creation, because all things are the shadow of God, and the shadow is like the person. If the five fingers are spread out, the shadow too is spread out; if the body bows, the shadow also bows; if it stretches out, the shadow also stretches out. . . . But this awareness of ours in relation to God's knowledge is in the predicament of unawareness. Not everything that is in the person shows in the shadow, only certain things. So not all the attributes of God show in this shadow, only some of them show, for *"You have been given of knowledge nothing except a little."*
>
> —Rumi[25]

Those echoes of Ultimate Reality which *can* be discerned in the new physics necessarily will be discussed in a broad fashion. Since the theme of this book is that material science is unable to come to grips with final truth, and mysticism *is*, it would not make much sense to describe the findings of a limited science in unlimited detail. (Many authors have devoted entire volumes to discussing subjects in the new physics which will be described in only a few paragraphs here. Those books are recommended to those who desire more information about a particular subject. I do not want to run the risk of obscuring the shape of the simple principles of mysticism in a complex fog of details about the new physics.)

Another reason for the organization of the following sections was expressed by John Polkinghorne, a physicist and theologian. He notes that in regards to understanding how the phenomena of nature came to be as they are, "Our thought is constrained to a one-way reading of the story, in which the higher emerges from the lower. In consequence the latter retains its hold upon our mind as controlling the metaphysical picture. It is by no means clear that this is more than a trick of intellectual perspective."[26]

In other words, material sciences generally assume that higher-order phenomena—such as consciousness and life itself—are the result of lower-level causes. In this manner brain researchers are able to come to the conclusion that thinking develops when nerve cells reach a certain degree of complexity, and geneticists can believe that much of human behavior is caused by tiny molecules of RNA and DNA. This is a simplistic philosophy of science, and the flaws of reductionism are becoming increasingly evident to even adamant adherents of materialism. Mysticism, of course, teaches that every perceptible cause in the physical universe ultimately can be traced backward to a Final Cause, God. So from either a spiritual or materialistic point of view, there are sound reasons for eschewing a "bottom-up" approach to understanding reality.

Finally, we should consider for a moment the concept of an "echo." This term has several connotations, each of which applies to the findings of the new physics as they pertain to their source in spiritual science. The dictionary tells us that one definition of an echo is the "repetition of a sound by reflection of sound waves from the surface." In this sense, then, an echo is akin to the source—just to a lesser degree. A sound becomes diminished in clarity as it echoes from one surface to another. As it does so, the sound often appears to be coming from a different direction than the actual source. Thus we can be fooled by an echo. After bouncing off several surfaces, it may enter our ears from the north, whereas the source actually is south of us. While an echo may be an accurate but blurred copy of the source, like a photograph that has been photocopied and then photocopied again many times until its sharpness is lost, some attributes of an echo—such as direction—may be opposite from those of the source.

However, from the mystical point of view the most interesting definition comes from the Greek myth of Echo—where this term originated. Echo was a girl who loved Narcissus, a handsome young man who was excessively proud of his good looks. Totally self-absorbed, he spurned Echo. Terribly hurt by his rejection, she withered away until only her voice remained. Narcissus then was punished by angry gods: gazing into a pool of water, he was made to fall in love with his own reflection and could not bring himself to leave the banks of the pool.

Though a myth, this tale is a warning to those who love reflections. You will become trapped in an illusion of your own making. Echo at least was devoted to someone other than herself, though tragically her love was not returned. Fortunately, genuine devotion to God never is spurned. Yes, like Echo, it is necessary to cease attending to anything other than the voice of Spirit which speaks in the depths of contemplative meditation. But in so doing the student of spiritual science is saved from spending his life staring into the pool of his own thoughts, waiting for release from the torment of narcissism.

> *What sort of Beloved is He?*
> *As long as a single hair of love for yourself remains,*
> *He will not show His face;*
> *You will be unworthy of union with Him,*
> *And He will give you no access.*
>   —Rumi[27]

Principle I.
God is one, and present everywhere.

I have dispensed with duality, and seen the two worlds as One;
One I seek, One I know, One I see, One I call.
He is the first, last, the outward and the inward.
I know none other than He, and He Who Is.
— Rumi[1]

The most important principle of mysticism—in a real sense the *only* principle—is that God is all that there is. Philosophically, spiritual science is absolutely monistic. All of existence is composed of a single substance: God. Little more can be said. For all perfect mystics teach that this Ultimate Reality is beyond all words, all symbols, all thoughts. He is formless, nameless, without attributes. Any description of the absolute serves only to divide what cannot be divided, and so contributes to the illusion of separateness in which we find ourselves.

Without composition and one in substance, we were all on yonder side, headless and footless.
Like the sun we were one in substance, like water we were pure and without ripples.
When that pure light entered into form, multiplicity appeared like the shadow of a battlement.
— Rumi[2]

In this physical universe apparent division is the order of the day. I exist in one body, you in another. The worlds of nature and human technology both are composed of a myriad variety of separate entities. We speak of love and living together as one family of man, but murder and mayhem stalk the streets, prejudice and pride suffuse our minds. Given such conditions, it becomes difficult to believe that everything outside of us and everything inside of us is one and the same—God. Our discussion of the following principles will help us understand how this can be true, for it is.

The perfect mystic Sawan Singh says, "All this creation is the unfolding of the One only. There is no one except Him. He is himself the warp and the woof."[3] Thus the qualities of God perceived by our limited understanding—omnipotence, omniscience, omnipresence—flow from this

Oneness. When God is everything and everywhere, how could He *not* know all, be all, control all? There is nothing else in existence which could possess these properties, or indeed any qualities separate from God. Whatever we see in this world, comes from Him. Whatever we ourselves are, comes from Him. Not only does it come from Him, it is Him. Our inability to realize this fact stems from the dense lenses of mind and matter through which human consciousness normally is forced to view reality. For while composed of God's essence, all of physical existence is a "frozen" semblance of that divine unity.

> *When I filled the ice-vat with water, it melted—no trace of it remained.*
> *The whole world is indivisible, the world's harp has but a single string.*
> —Rumi[4]

Mysticism teaches that since that One cannot be conceived by thought, it is useless to try to analyze one's way to God-realization. As was noted previously, it is not difficult to say, "God is one. I am God." Try it now. Have you changed into the Supreme Being by reciting these words? While thinking about this concept may be reassuring, or make you feel better about your situation in life, any abstract notion of God is worlds apart from the direct realization of perfect mystics. Seekers after final truth should direct their efforts to becoming that Ultimate Reality through contemplative meditation. Since no separate words, or thoughts, or images exist in that unity, they must be discarded before the drop of our soul returns to the ocean of God.

> *Whoever has been parted from his source*
> *longs to return to that state of union.*
> —Rumi[5]

Thus it is unproductive to be concerned about such questions as, "Why do mystics usually refer to God as *He*, if the supreme being is beyond the attribute of sex?" Some people object to this practice of speaking as if God had a gender, a practice which I have followed in this book. They have a reasonable point, for the vagaries of symbolic expression admittedly bear no resemblance to the ineffable nature of the Lord. Yet if a particular form of expression is limited, similarly limited is any objection to that form. Perfect mystics advise us to go beyond the divisive domain of names and focus solely on becoming one with the Named. Charan Singh says:

The Lord, the Supreme Father, God, whatever name you give that Power is One here, there and everywhere. . . . It transcends all comparison and relativity; it goes beyond the difference of unity and manyness. It transcends all description and all qualities. . . . It is beyond everything we can think of, and yet it is in everything and in all things, and with everything. It is everything in itself. Nothing exists except that "It."[6]

Echoes in the new physics.

As was noted above, some of the ways God's oneness appears to human understanding are as *omnipotence, omniscience,* and *omnipresence.* How the first two qualities are reflected in findings of the new physics will be discussed under succeeding principles. Here I will focus on the evidence for omnipresence. Remember that material science is able to observe only echoes and reflections of Ultimate Reality, so this evidence will be subtle in comparison to the perfect mystic's direct perception of divine unity. Still, one noted physicist—Henry Stapp—found the conclusion of "Bell's Theorem" to be so important that he called it "the most profound discovery in science."[7]

Bell's Theorem

Bell's Theorem is part of quantum theory, and grew out of a longstanding disagreement between Albert Einstein and Neils Bohr regarding the nature of the quantum world. Einstein could not believe that God plays dice with the universe, so the apparent randomness of individual quantum events—such as the emission of a single radioactive particle— never was accepted by him. Throughout his life Einstein believed that an underlying determinism was responsible for this seeming causelessness. Even though physicists could not predict when an individual quantum event would occur, this was a result of scientists' ignorance, not a fundamental property of physical reality.

Further, Einstein believed that whether or not the true nature of the quantum world ever could be known, it did possess objective qualities of its own. Physicist F. David Peat says:

> In Einstein's universe, it is as if everything can be characterized by cards of identity. Like international travelers, quantum systems carry identity cards that specify all their properties. Nature may conspire to prevent our reading all the information on these cards, but it nevertheless makes sense to speak of the independent reality of a quantum system. It was this notion of an independent reality composed of a variety of objective elements— cards of identity—that Bohr denied. For him the quantum world had no objective reality apart from our acts of measurement, and different experiments forced it to give different, complementary answers.[8]

In an effort to prove his point, Einstein and several associates devised a thought experiment. The details as they apply to actual subatomic particles are daunting for the layman, but the gist of Einstein's argument is captured in this version of an analogy described by Peat. Imagine that two twins, Bruce and Bob, live in the same house but work at opposite ends of town. After getting up in the morning they always dress alike, with one exception. If Bruce puts on a black hat, Bob chooses to wear a white one—and vice versa. When one of the brothers chooses a hat color, he does so completely at random. If the co-workers of either Bruce or Bob set up an office pool where the object is to predict the color of the hat he will wear that day, flipping a coin would be as good a way as any to decide whether to bet on "black" or "white."

However, even though this may be true in regards to the office pool, someone with more information can see through this indeterminism. For the twins always choose a hat color opposite to that of their sibling. So if a person waited outside the door of Bruce's workplace and saw him coming down the sidewalk wearing a *white* hat, the fact would be revealed that Bob was wearing a *black* hat that day. That person could rush to a telephone and place an office pool bet that Bob's hat color will be black. His chance of being correct will be one hundred percent, not random but certain, given his knowledge of the opposite twin's choice of headgear.

This way of looking at the quantum world made perfect sense to Einstein, but most physicists—including Bohr—remained unconvinced by his argument. After all, this was just a thought experiment. It was not until the 1970s that technological advances permitted an actual experiment to be conducted with polarized photons of light. A physicist named John Bell devised a relatively simple (but still too complex to be described here) approach to reveal more about the nature of the quantum world. When the predictions of Bell's Theorem were confirmed by a series of carefully conducted experiments, quantum reality turned out to be more intriguing than either Einstein or Bohr had imagined.

If we use our imagination to picture the twins obeying the rules of quantum mechanics rather than the strictures of our everyday world, Bell's Theorem revealed that contrary to all common-sense expectations, what actually occurs is this: if Bruce shows up at the office wearing a white hat, then indeed it can be predicted that Bob has a black hat on. But if someone takes Bruce's white hat off his head and replaces it with a black hat, then Bob's hat instantly changes to white—even though Bob is nowhere in the vicinity. This finding, had Einstein been alive to know about it, would have both pleased and disturbed him. For even though evidence of causality had been demonstrated in the quantum realm—altering the polarization of one paired photon changes the polarization of its twin—the price to be paid was "locality." This would not have pleased Einstein at all, because a central foundation of his theory of relativity was that no signal or material entity can travel faster than the speed of light.

Thus a *local* reality means, according to F. David Peat, that "each system or object can be defined and understood in its own particular region of space."[9] Such entities can be influenced only by objective forces that obey the laws of nature, which includes light's unbreakable speed limit of 186,281 miles per second. No signal can travel faster than this. Experiments testing Bell's Theorem, however, have conclusively demonstrated that apparent faster-than-light (or "superluminal") connections exist in quantum reality. That reality actually is non-local. In the words of physicist Heinz Pagels:

> Bell was motivated to find a way of testing if there were hidden variables that exist out there in the world of rocks, tables, and chairs. He showed that the violation of the inequality [of Bell's Theorem] by quantum theory did not necessarily rule out an objective world described by hidden variables but the reality they represented had to be nonlocal. Behind quantum reality there could be another reality described by these hidden variables and in this reality there would be influences that move instantaneously an arbitrary distance without evident mediation.[10]

Photons can change the color of their hats, so to speak, in an "unmediated, unmitigated, and immediate" manner—as Nick Herbert puts it.[11] Unmediated, because nothing material links the two photons. Unmitigated, because the connection between them cannot be shielded by any type of matter or energy. Immediate, because the strength of their linkage does not diminish with distance. Paired photons can alter their polarization simultaneously no matter whether they are two inches or two billion light years apart. This does *not* mean, however, that some sort of faster-than-light signal is traveling between them. What kind of signal takes no time to travel and cannot be detected by any material means?

Unity underlies separateness

Physics agrees with mysticism that this quantum interconnectedness reflects a unity that underlies the separate phenomena of the universe. Whether termed a "single quantum system," or "God," it exists as a fact of both material and spiritual science. Physicists Davies and Gribbin write: "Assuming that one rules out faster-than-light signaling, it [Bell's Theorem] implies that once two particles have interacted with one another they remain linked in some way, effectively parts of the same indivisible system. This property of 'nonlocality' has sweeping implications. We can think of the Universe as a vast network of interacting particles, and each linkage binds the participating particles into a single quantum system. In some sense the entire Universe can be regarded as a single quantum system."[12]

Why, then, do we feel so separate and alone? And if nonlocality has been proven, why do I have to rely on the vagaries of the post office to send messages to other people? Where is this unmediated, unmitigated, and immediate communication that has been proven to exist by experimental physicists? Unfortunately, it cannot be perceived in the everyday world. Nick Herbert says that "It's important to remember that Bell's theorem requires *reality*, not *phenomena* to be superluminally linked. . . . These patterns [of quantum phenomena] are bound to be local, by Eberhard's proof, and have never been observed to be otherwise. On the other hand, the quanta themselves—the unpredictable alphabet which spells out the words and paragraphs of the world's phenomena—must be non-locally connected, according to the theorem of John Bell."[13]

If this seems confusing to you, do not be concerned. Physicists also do not understand how the "unpredictable alphabet" of quantum reality turns into the well-written phenomena of physical existence. The following principles of mysticism, however, will be found to provide the answer to this perplexing problem. It will be shown that the aforementioned distinction between *meaning* and *form* is key to understanding both spiritual and material domains. The meaning of the physical universe cannot be found in its form. The essence of God, or Ultimate Reality, is infinitely more subtle than phenomena, even though He is the substance of everything. Being everywhere, and controlling all, God does not show himself through the five senses, or mental cognition. Nick Herbert says:

> . . . nature undoubtedly uses superluminal links to accomplish her inscrutable ends but these deep quantum connections are private lines currently inaccessible—and perhaps permanently inaccessible—to humans for communication purposes. Bell's theorem shows that the world is built in a most curious fashion: To achieve merely subluminal effects, things are hooked together by an invisible underlying network of superluminal connections.[14]

This fact is "curious" only if one believes that nothing exists apart from physical reality, and that the sole purpose of the laws of nature is to produce material phenomena. If no attempt is made to discover the meaning that lies behind forms, then physicists will remain in the condition of our snail who tries to learn about a garden hose and sprinkler head without being able to fathom the notion of a "garden." Mysticism teaches that God has not made the creation without a purpose. That purpose is embedded in every atom of the material world and each law of nature. Simply put, it is to realize that God is One, and that Love is the essence of Ultimate Reality.

Bell's Theorem points toward the truth of this central principle of spiritual science.

Apples, rainbows, and . . . reality

In addition, Bell refuted one of the tenets of the traditional Copenhagen interpretation of quantum physics which was at odds with mysticism. This tenet held that there is no reality in the quantum world apart from observation. In other words, subatomic particles "do not possess definite attributes except under definite measurement conditions."[15] This might be termed the "what you look for is what there is" view of Ultimate Reality, which obviously does not allow much room for God. Rather, it tends to make mini-gods out of each observing entity. An unanswered question, of course, is "If I create external reality by observing it, then who is creating *me*?" This is a central problem facing those, such as Stephen Hawking, who hope to explain the creation of the universe through quantum physics. If measurement brings attributes into existence, then who was around at the moment of the Big Bang to observe that event and bring physical existence into a definite form?

Bell's Theorem revealed that the so-called "rainbow world" of Neils Bohr did not match with scientific fact. Nick Herbert says that while objective, a rainbow appears in a different place for each observer: "For Bohr the search for deep reality—for 'real' quantum attributes which an electron possesses independently of observation—is as misguided as looking for the rainbow's end."[16] However, John Bell himself rejected this solipsistic view that each person in some fashion creates his own reality. When asked by an interviewer whether he would prefer to retain the notion of objective reality and throw away the tenet of relativity that signals cannot travel faster than the speed of light, Bell replied:

> Yes. One wants to be able to take a realistic view of the world, to talk about the world as if it is really there, even when it is not being observed. I certainly believe in a world that was here before me, and will be here after me, and I believe that you are part of it![17]

If one accepts that God is the sole substance of creation, it indeed would be strange if He exists only when a created being is looking at Him. This would be like a pregnant woman existing only when her unborn child is awake. If the mother exists by virtue of the child, and the child by the mother, we have circularity—not a scientific explanation. Bell's Theorem proved that whatever the nature of quantum reality is, it cannot be the individualistic "rainbow world" favored by Neils Bohr. Returning to the twin analogy, it is not Bob—or even Bob's nearby coworkers—who are responsible for turning the color of his hat from black to white. Rather it is the act of exchanging *Bruce's* hat for a black one. This event could take place halfway across the universe, and quantum theory holds that Bob's hat will change color just as surely as if Bruce were standing right next to Bob. So it makes no sense to talk about an observer-created reality if that observer can be affected by unseen influences from anywhere in the universe.

On the other hand, Nick Herbert observes that an "apple world," in which ordinary objects possess attributes all their own even when not being measured, is equally inconsistent with Bell's Theorem. The condition of apples, and everything else, actually is influenced by the ubiquitous non-local connections which have been proven to exist beneath the surface of phenomena. Herbert concludes that if quantum theory forbids the existence of apple and rainbow worlds, a possible alternative is what he terms a "non-local contextual world." This means that the physical universe possesses an objective reality, but how it *appears* will depend upon the context—or conditions—in which it is being observed. And that context is non-local. It is possible for anything in existence to affect one's perception of reality.

This vision is compatible with spiritual science. For mysticism holds that the material world indeed appears differently to different people. The state of our consciousness largely determines the content of our observations. A loving person sees love wherever he looks. Angry people view the world as filled with infuriating influences. The same applies to scientific forms of observation. However, while the location of our "seat" of consciousness influences how existence appears to us, reality is maintained by the objective presence of God—not our subjective mind, or sense perception.

Further, generally we are not aware of how God's Spirit brings about the changes in our surroundings. Life often seems to be a randomly fluctuating pattern without much rhyme or reason to it. Things just seem to happen. Even when we put forth our best efforts in a certain direction, we frequently end up facing the opposite way. Mysticism teaches that this seeming randomness is actually God's "private line," to use physicist Nick Herbert's term. There is a divine substrate to the material world which can be perceived only through the observational tool of contemplative meditation. This fact is reflected in findings of the new physics. Nick Herbert says:

> A universe that displays *local phenomena* built upon a *non-local reality* is the only sort of world consistent with known facts and Bell's proof. . . . Bell proves conclusively that the world behind phenomena must be non-local. . . [yet] the world's superluminal underpinning is almost completely concealed—non-locality would have been discovered long ago if it were

more evident; it leaves its mark only indirectly through the impossibly strong correlations of certain obscure quantum systems.[18]

Non-locality simply is another way of speaking about Unity. Both mysticism and physics know that when the surface of phenomena is penetrated, the universe is seen to be one. Yet, as was noted in our discussion of monism, these disciplines disagree about what that single substance is. This brings us to the next principle of spiritual science.

Principle 2.
Spirit is God-in-action,
and of His same essence.

When God sings and recites over parts
of non-existence that have no ears and no eyes,
that second they dance into existence.
When God sings over beings in existence,
they run quickly back where they came from.
God talks to a rose. It blooms.
To a rock. It changes to transparent crystal.
                                        —Rumi[1]

God *is* the Creation. But the Creation is not all of God. This is a central tenet of mysticism which cannot be understood through logic alone. Logic, mind, thoughts, concepts—all these and so much more have been created by the Lord. Since He was prior to the creation, mind and matter cannot be aware of what preceded their existence. Only secondary causes are perceptible to us in this physical universe that is so far removed from the essence of God. To understand even a hint of the purpose and means of creation, we must rely on the symbolic explanations of perfect mystics, or—better—reach the level of consciousness where the non-symbolic truth of Ultimate Reality can be perceived directly.

Insofar as words are capable of describing what is beyond words, mysticism teaches that the absolute state of God's being is beyond time and space, and is nameless and formless. Only when God desires to bring creation into existence does a ray separate from His light, a voice from His silence. Charan Singh says: "Before the creation, God was the only light and life. Nothing else existed. So, in whatever He has created, He has put His light and life; and that life and light is in His creation now. . . . There is nothing in existence which was not created by this Word of God and which is not sustained by It."[2]

Spirit, God's command

This "ray" or "voice" goes by many names in different languages: Word, Holy Ghost, Tao, Spirit, Shabd, Nad, Kalma, and others. In this book I have chosen to refer to this manifestation of God-in-Action as **Spirit**, largely because I am writing in English. Absolutely no significance is attached to the particular symbol used to describe the indescribable essence of God. Close reading of the scriptures of every major religion, and the writings of each perfect mystic, reveals that the divine creative energy referred to by different names is the same thing in reality.

In some of the quotations which follow I have substituted "Spirit" for an equivalent term in another language. Synonymous terms in English remain unaltered. This will help to keep us focused on the meaning of this all-pervading conscious energy, and not become entangled in names and forms. As was illustrated earlier by Rumi's story regarding the men who were arguing about what they should buy to eat—failing to realize that each wanted some grapes because they were calling their desire by different names—it is a waste of time to debate the best way of addressing, or describing, God. Such theological disputations are a sign that we have failed to reach the kernel of truth and are content with its shell.

> *It is like children playing with walnuts; if offered walnut oil or walnut kernels, they will reject them because for them a walnut is something that rolls and makes a noise, and those other things do not roll or make noises.*
>
> —Rumi[3]

Sawan Singh explains the difference between God-as-God, and God-in-Action: "Before the creation, Spirit was unmanifested and nameless. It then existed in itself. In that state it was called indescribable, nameless, invisible, unfathomable, unutterable, and inexpressible. When it became manifest it became known as Spirit. . . . Nothing can manifest without its help. Spirit is the life, the essence, the root and the quintessence of every created thing."[4] While the ultimate truth is that God is One, and there is nothing other, from the standpoint of relative knowledge (where we exist now) three aspects of the divine Unity can be discerned: God-as-God, God-in-Action, and God-in-Creation. I have generally referred to this trinity as God, Spirit, and Creation. In this translation of Rumi, the terms "Commander," "Command," and "Creation" are used:

> *The world of creation possesses quarters and directions, but the world of the command and the Attributes are without directions. Know that the world of command is without directions, oh friend! Therefore the Commander is even more directionless.*
>
> —Rumi[5]

The command of God is Spirit. This is His projection, which might also be called All-Pervading Conscious Energy. I am not seriously suggesting this as a term to be added to the nomenclature of spiritual science, but it has the merit of encapsulating three of the divine attributes in a single description. In the preceding principle we discussed the attribute of all-pervasiveness, or unity. In this and succeeding principles the qualities of consciousness and energy will be examined, even though these three attributes cannot be separated in actuality. As we shall see, many perplexing enigmas in the new physics are resolved when it is realized that the energy which pervades the universe is *conscious*, not mechanical.

Purpose of creation

But before the nature of Spirit is examined in more detail, this question must be addressed: why did God-as-God choose to become God-in-Action and project as God-in-Creation? The mind of man is as capable of understanding the answer as is an infant able to fathom how it came to be born. Nevertheless, to satisfy our intellectual curiosity perfect mystics have attempted to explain the reason for creation—with the proviso that only a consciousness capable of creating a universe can fathom why *ours* is here. In other words, attaining God-consciousness not surprisingly is the only means of knowing fully the mind of God. We can, however, begin to grasp that one of the purposes of creation is Union.

> God says, "I was a Hidden Treasure, so I wanted to be known." In other words, "I created the whole of the universe, and the goal in all of it is to make Myself manifest, sometimes through Gentleness and sometimes through Severity." God is not the kind of king for whom a single herald would be sufficient. If all the atoms of the universe were His heralds, they would be incapable of making Him known adequately.
>
> —Rumi[6]

In some sense which is far beyond our comprehension, God wished to reveal Himself. Being formless and nameless in the state of God-as-God, this required Him to take on forms and names and manifest as God-in-Creation. Spirit is the means by which God projects Himself as the creation, and also serves as the avenue for returning to Him. If you take an elevator *down* from the top of a building to ground level, the same mechanism is available to transport you *up* again. There are, however, certain conditions that must be met before one's consciousness is able to ascend to the domain of Ultimate Reality via the power of Spirit. Basically, attachment either to the external creation or to our internal sense of "I-ness," or ego, prevents the soul from catching the wave of Spirit and returning to the ocean of God.

This is for our own good. If you reach through a window and hold onto a street lamp post after boarding a bus, one of three things is going to happen (all bad, if you want to get to your destination): either (1) the bus driver will refuse to start moving until you let go of the post, (2) the bus starts off and you are pulled out of the window back onto the street, or (3) the bus begins to move and your arm stays attached to the post while the rest of your body travels on. Since this last option would be painful, perfect mystics advise that it would be better to cast off your attachments to this world before trying to leave it.

Each of us will depart one day on the bus of death, and then we will learn the extent of those attachments—for after leaving our body we will end up where our desires take us. Even though God is the "driver" for everyone, each soul takes an individual route after leaving the physical body. Thus contemplative meditation is a rehearsal for death: by gradually loosening our grip on external phenomena and internal imaginings, we learn how to merge our consciousness with God's Spirit at the moment of death instead of clinging to creation.

Spirit—the soul of the universe

Spirit sustains the physical universe in the same manner as soul enlivens living beings. When either Spirit or soul is withdrawn, "death" results. Charan Singh says:

> So long as the soul is within the body, we walk and move and do all the business of the world. At the time of death, when the spirit current is withdrawn from it, the body is still there but becomes useless and inactive. In the same way, this world—the sun, the moon, the earth and the stars—are all sustained by Spirit. When the Creator withdraws that current of the Word out of it, the whole universe crumbles down like dust.[7]

This Word is not something spoken or written in human language. It is the command of God, and often is referred to as the Word by mystics. As was shown in a previous chapter, the scriptures of every major religion refer to creation as having been brought about by a word, or a saying, or a name. Here are those quotations again:

Sikh *Adi Granth:*	One Word, and the whole Universe throbbed into being.
Christian *Bible:*	In the beginning was the Word, and the Word was with God, and the Word was God.
Islamic *Koran:*	Creator of the heavens and the earth from nothingness, He has only to say when He wills a thing: 'Be', and it is.
Jewish *Old Testament:*	And God said, Let there be light: and there was light.

Taoist *Tao Te Ching:*	The nameless is the beginning of heaven and earth.
	The named is the mother of ten thousand things.
Hindu *Upanishads:*	Accordingly, with that Word, with that Self, he brought forth this whole universe, everything that exists.

Is it simply a coincidence that in each scripture the mechanism of creation involves a *sound* rather than some other means such as a gesture or touch? If—as some believe—the core of religion is subjective imagination, why are all of these "fantasies" so alike? The reason is that they reflect an objective reality that can be known directly through spiritual science. Spirit is an all-pervading conscious energy that can be heard and seen on higher domains of consciousness. Spirit created this physical universe and continues to sustain it at every moment, just as soul—a drop of Spirit—enlivens the bodies of living things. Far from being fantasy, this is scientific fact. From God came a wave of Spirit, and that wave created every plane of existence. As a divine essence, Spirit has the same attributes as God, including omnipotence, omnipresence, and omniscience. Every atom of material reality echoes God's grandeur and speaks of His presence.

> *Thou art the source that causes our rivers to flow.*
> *Thou art hidden in Thy essence, but seen by Thy bounties.*
> *Thou art like the water, and we like the millstone. . . .*
> *Our every motion every moment testifies,*
> *For it proves the presence of the Everlasting God.*
> *So the revolution of the millstone, so violent,*
> *Testifies to the existence of a stream of water.*
> —Rumi[8]

Echoes in the new physics

Material scientists agree about the event which created our universe: the Big Bang. However, there is no consensus about its cause. So I will begin with what physics and cosmology believe to be facts about creation, and then turn to speculative theories regarding what initiated that process. As was noted previously, it was not until the 1920s that astronomers realized that the Milky Way galaxy, made up of our sun and at least one hundred billion other stars, was not the entire universe. Even now, many people believe that when they look up at the night sky they are viewing all of the cosmos. Actually, every point of light which can be seen by the naked eye is from a source within our Milky Way galaxy, with the exception of Andromeda and the two Magellanic Clouds.

A telescope is needed to perceive other galaxies, which are estimated to total one hundred billion or more. These have been found to be rushing apart from each other at many miles per second. This alone is fairly conclusive evidence for the Big Bang, since if the galaxies are moving apart now, they must have been closer together in the past. Extrapolating backward from the known expansion rate has led to an estimate that some ten or fifteen billion years ago everything in the universe was condensed into an unimaginably small area—less than the size of an atom, in fact. The Big Bang was what formed the raw material of the universe and set it into motion, a process of creation which continues at this very moment.

The Big Bang is still "banging"

Every four or five seconds the universe adds on to itself a volume equivalent to the Milky Way galaxy.[9] This leads to the question: what is being added, and where is it coming from? Astronomers tell us that the Big Bang was (and *is*) not like an explosion in some existing place. Rather, it is the actual creation of time and space. What is being "added on" at the edges of the expanding universe is not some form of energy or matter, but raw physical existence. This is, of course, most difficult to visualize. When we try to picture the universe expanding, it is almost impossible not to see it blowing up like a balloon into some existing space. What could be on the outside of the physical universe if time and space as we know it are on the inside, and rushing onward, forming material existence out of . . . what?

Cosmology has no final answer to this question. Mysticism does. Spiritual science teaches that the physical world is but one of several stages of creation or levels of consciousness. These exist not as separate material spheres, or different parts of the same physical stage—but as varying dimensions of reality, much as dreaming coexists in our mind with the waking state.

Similarly, physicist Anthony Zee writes that "we might imagine our universe expanding into a surrounding higher-dimensional space . . . no physical principle forbids the possibility that our universe may be contained in a larger space. As an analogy, picture a village in a long narrow valley surrounded by steep mountains. To get out, one has to be energetic enough to climb over the mountains. You can easily speculate, as some physicists have, that we are also kept from getting outside the universe by an energy barrier."[10] And F. David Peat says that superstrings—one of physics' leading candidates for a Theory of Everything—"suggests that our universe may have evolved out of a higher dimensional space during the first instants of the big bang."[11]

It almost seems as if physics and cosmology, which are virtually one and the same in modern science, are coming to accept the existence of higher domains of reality—a central tenet of mysticism. However, there are barriers, some technical and some philosophical, which prevent physicists from becoming mystics. On the technical side, theorists find that mathematical models which attempt to trace the process of creation back to the beginning of the Big Bang come to an abrupt halt at a mere 10^{-43} seconds after that event, or one millionth of a trillionth of a trillionth of a trillionth of a second prior to the moment of creation. It is difficult to conceive how short this "Planck time" is. If a second were divided into 10^{43} parts, this would just about equal how many electron radiuses it would take to reach to the edge of the known universe, fifteen billion light years away. That is, half the width of an electron bears the same proportion to the size of the entire universe, as Planck time bears to a single second.

So close, yet so far. For this minuscule time, or equivalent distance (the universe at this time was a mere 10^{-33} centimeters across), stands between physicists and their ability to understand the Big Bang with current theories of quantum physics and relativity. At this instant of creation the energies involved were so intense that those theories become meaningless. Paul Davies says, "at the Planck scale the separate identities of space and time can be smeared out . . . time begins to 'turn into' space. Rather than having to deal with the origin of space-time, therefore, we now have to contend with four-dimensional space. . . . "[12] To put it simply, something strange occurs at the Planck time which physicists cannot deal with. So they tend to ignore it. "Sweep it under the table," goes the thinking, "if logic and mathematics cannot handle the problem."

Could no cause be the cause of creation?

Hence, many practitioners of the new physics have a fascination with the notion of causelessness being a fundamental property of the universe. Not only that, causelessness is held by many material scientists to be the cause of creation. "But wait," you might say, "isn't *causation* one of the properties of a valid scientific theory? How can not knowing a cause of something come to be viewed as a triumph of the new physics?" Well, it can't. This indicates that some areas of the new physics have become so esoteric as to be more akin to theology than science. After discussing the concept of quantum cosmology—in which the Big Bang is held to be a random quantum fluctuation—Paul Davies notes:

> Whatever the truth about these deep conceptual issues, the universe must have come into existence somehow, and quantum physics offers the only branch of science in which the concept of an event without a cause makes sense. When the subject at issue concerns spacetime it is in any case

meaningless to talk about a cause in the usual sense. Causation is rooted in the notion of time, and so any ideas about an agency creating time, or causing time to come into existence, must appeal to a wider concept of causality than is at present familiar in science.[13]

Let me state the issue in straightforward terms: Physics has determined that the Big Bang created time and space as we know them. Not surprisingly, time and space "smear together" into some other entity when mathematical models approach the instant of the Big Bang. Commonsense notions of causation require time and space to work, so they cannot be used to explain how time and space came to *be*. Up to this point, mysticism and physics are in close agreement. But many physicists attempt to resolve this dilemma by denying that there was any cause at all of the Big Bang. It just happened on its own. Mysticism takes a more scientific approach. "Yes," say perfect mystics, "this universe was created by a cause. That cause was the all-pervading conscious energy of God's Spirit, which emanated from a higher dimension of existence."

Call me biased if you like, but even if I did not believe so strongly in mysticism, this explanation would make more sense to me than the one offered by the new physics: the cause of physical existence is causelessness. Referring back to the properties of an acceptable scientific theory, this tenet of quantum cosmology strikes me—as well as many others—as being the opposite of beautiful: *ugly*. It grates on the nerves. Being unable to escape from the self-referential confines of Gödel's Theorem, some theorists now view circular explanations as the highest truth about Ultimate Reality, rather than as a barrier to final knowledge. Thus physicist Joe Rosen writes about the self-generating universe scheme:

> It possesses a beautiful completeness: The universe is the cause and effect of its own existence. . . . The picture provides an answer to the age-old question of what "was before" the coming into being of the universe (also called "the Creation"). The universe was preceded by and evolved from a quantum fluctuation of space-time, whereby a Planck-sized region of space-time closed up upon itself, became disconnected from the universe, and formed a distinct and separate universe in itself, the baby universe that is the universe. . . .

> When the universe is identified with a baby universe born of itself, it can then be said to have been preceded by a space-time fluctuation and its birth to have occurred immediately following the occurrence of that fluctuation. The preceding, following, and occurring are with respect to the time of the universe considered as a metatime in which (baby) universes, including the universe itself, form.

That reasoning must seem quite tortuous. In fact it is, due to the circularity inherent in any self-referential situation. And the self-generating universe is self-referential. It takes getting used to.[14]

This is an understatement, but I do not wish to belittle adherents of quantum cosmology. I only want to point out that *simplicity* and *logical implication* were two other properties of a valid scientific theory which Rosen himself proposed. After reading the above description of how "baby universes" such as ours might spring randomly from a "parent universe," which somehow also is the parent of itself, does it seem that those speculations are simpler and more logical than the statement, "God created the universe through Spirit"? This one-line "theory" has the merit of retaining a causative link between the creation and its Creator, and explains in a more direct manner how consciousness and the precisely formed laws of nature came to be as they are.

Stephen Hawking relates a story about a well-known scientist who gave a public lecture on astronomy. He described how the earth revolves around the sun, and other facts about the universe. At the end of his talk, "a little old lady at the back of the room got up and said: 'What you have told us is rubbish. The world is really a flat plate supported on the back of a giant tortoise.' The scientist gave a superior smile before replying, 'What is the tortoise standing on?' 'You're very clever, young man, very clever,' said the old lady, 'But it's turtles all the way down!'"[15] We laugh at this superstition, but the new physics has come up with only a slightly more believable replacement for this turtle-tower: the vacuum state.

Fullness of the energetic vacuum

As will be discussed in more detail under Principle 6, the vacuum plays an important role in modern physics. In fact, the energetic vacuum is believed by some to be responsible for the creation of the universe. Under the rubric of quantum cosmology, many theories are vying for acceptance. In all that I am aware of, the vacuum provides the energy needed to power the Big Bang. This will be a surprise to most people, since a vacuum commonly is seen as something empty and lifeless.

Such is not the case in quantum physics. Physicist Heinz Pagels has noted that Aristotle's view was that nature abhors a vacuum, and matter rushes in to fill any void which comes to exist. Actually, Pagels says, the modern conception is just the opposite:

Matter is the exception in the universe. Most of the space between the stars is empty or nearly so, and even solid matter is mostly empty space with all the mass concentrated in the tiny atomic nuclei. Almost everything is vacuum . . . it is not empty; it is a plenum [fullness]. The vacuum, empty space, actually consists of particles and antiparticles being spontaneously created and annihilated. All the quanta that physicists have discovered or ever will discover are being created and destroyed in the Armageddon that is the vacuum. . . . the view of the new physics suggests, "The vacuum is all of physics." Everything that ever existed or can exist is already potentially there in the nothingness of space . . . that nothingness contains all of being.[16]

Over seven hundred years ago, long before particle accelerators or the mathematics of quantum physics, a perfect mystic from Persia said much the same thing:

> *Nonexistence is eagerly bubbling in the expectation of being given existence. . . . For the mine and treasure-house of God's making is naught but nonexistence coming into manifestation.*
> —Rumi[17]

There is, however, a vast difference between the material vacuum of the new physics, and the "nameless and formless" state of God known to perfect mystics. The former lies near the bottom of the levels of creation, and the latter is the highest state of divinity. The physical vacuum is a manifestation of conscious Spirit, and material science has correctly recognized that it is all-pervading and full of infinite energy. However, physicists are at a loss to explain how the vacuum came to exist in the first place, so it is viewed—like God—as eternal. The new physics is on the right track, but unfortunately it is considering the physical vacuum to be the final cause of existence rather than what it actually is: the last link in a causative chain that extends from God, through the action of Spirit, to the material creation.

Physicist Willem Drees observes that even though the origination of the universe can be theorized as a quantum event, this still requires the existence of a physical vacuum with certain properties. So from the standpoint of the new physics, the question of "What created the physical universe?" remains in only a slightly modified form: "What created the physical vacuum which created the physical universe?" Drees says, "Even the most extreme 'nothing' of the physicists is not an absolute Nothing devoid of any properties and measures. Even if theories are perfect and complete, they do not answer the question of why there is anything which behaves according to those theories. The mystery of existence is unassailable. It remains possible, therefore, to understand the Universe as a gift, as grace."[18]

Physics gives us the God who burps . . . infinitely

To restate the points of agreement and disagreement between mysticism and the new physics: both spiritual and material science teach that our universe was created through the action of an all-pervading energy. Mysticism calls this energy Spirit, and physics terms it the vacuum. However, Spirit is conscious, and according to the physicists, the vacuum is not. Thus physics must take pains to explain how consciousness came to be produced by unconscious means, and how a universe with such precise and well-formed laws of nature resulted from a random quantum fluctuation. Physicists' speculations concerning these questions take on various forms, but most are based on what might be called the "infinite number of typing monkeys" principle.

If you were able to secure an infinite number of monkeys, and set them in front of an infinite number of typewriters, with an infinite supply of paper and ribbon, and trained them to pound away at the typewriter keys for an infinite length of time, then eventually one of the monkeys would type out the entire works of Shakespeare word for word—with no spelling errors. Not only that, mathematician Ivar Ekeland says that eventually those entire works would be typed out by some monkey not just once, but twice in an row.[19] Given infinity, anything can. happen, and everything will happen. Thus even though we do not observe universes popping in and out of existence now, physicist Pagels says:

> It seems that a vacuum is stable. Likewise it once seemed that atoms were stable but now we know that they are not. . . . According to the laws of quantum theory there is a probability that otherwise stable nuclei will decay. I believe it is possible that a vacuum is similarly unstable—there is a tiny quantum probability that a vacuum will convert itself into a big bang explosion. No explanation exists for a specific nuclear decay—only a probability can be given. Similarly, no explanation would be needed to account for the specific event of the big bang if this idea is right. Since no one is waiting for the event to happen, even if it has an infinitesimal but finite probability, it is certain to happen sometime. Our universe is a creation by the God that plays dice.[20]

This conception of God is not one which will appeal to most people. It certainly is at odds with the tenets of mysticism, and every major religion. For this God apparently is manifested only in the physical vacuum. Taking on that form—and I do not mean to be sacrilegious here, but only reflect the theology of the new physics—He "burps" quantum fluctuations uncontrollably, and for infinity. This is His only purpose, producing random bursts of vacuum energy, billions upon billions upon billions of them. Quantum cosmology says that once in a great while, entirely by accident, one of those fluctuations "takes hold" in some mysterious way and grows into a stable universe. We are existing, obviously, in one such universe or we would not be here to speculate about why we are here.

Such is the prevailing metaphysics of the new physics. And metaphysics it is. For even if one grants that over an infinite length of time (if one can even speak of time being "infinite") anything can happen in the creative vacuum, there still are limits on that anything. Physicists assume that the vacuum, and existence as a whole, is governed by the principles of quantum physics—and possibly other laws of nature. So the Big Bang cannot happen any old which way, for this would not be science, but imagination. That is, even given just the right kind of quantum fluctuation, physics is faced with explaining how that random event grew from subatomic size to become the vast expanse of our universe.

A big problem with the Big Bang

A core problem in the traditional Big Bang theory is this: it was noted previously that relativity theory does not permit any physical influence to travel faster than light. "Consequently," says Paul Davies, "different regions of the universe can come into causal contact only after a period of time has elapsed. For example, at 1 second after the initial [big bang] explosion, light can have traveled at most one light-second which is 300,000 km. Regions of the universe separated by greater than this distance could not, at 1 second, have exercised any influence on each other. But at that time, the universe we observe today occupied a region of space at least 10^{14} km across. It must therefore have been made up of some 10^{27} causally separate regions, all of them nevertheless expanding at exactly the same rate."[21]

Physicists know that every part of the universe expanded at the same rate after the Big Bang because traces of that event, in the form of microwave radiation, still are pervading the cosmos in a remarkably even manner. Edward Harrison notes that when you hold up your hand to any part of the sky, day or night, a thousand trillion particles of that cosmic radiation will fall on it in one second.[22] The moment of creation still enfolds us, albeit in a vastly reduced concentration of energy due to the expansion of the universe. At the instant of the Big Bang, the energy unleashed may have equaled the total energy of ten million billion quasars—or young galaxies—*each* of which equals three hundred billion suns in light energy alone. If a pinhead-sized amount of the one-second old universe could be brought into our solar system, it would equal over eighteen times the entire energy output of the sun—which is 865,000 miles in diameter.[23]

As this unimaginably potent surge of energy began to form time and space out of nonexistence, it did so at a rate far in excess of the speed of light. Only material entities are subject to light's speed limit, not the formation of materiality itself. Yet now astronomers observe a remarkably uniform and consistent universe, in which both the laws of nature and the texture of galactic structures are the same in every direction. This perplexes theorists, even though Bell's Theorem proved that there is a wholeness to reality which underlies phenomena, and that "quantum connection" is not bound by the speed of light but propagates influences instantly. Nevertheless, this reflection of the omnipresence of Spirit can be perceived only indirectly by the means available to material science, so another answer to what is termed the "horizon problem" in Big Bang cosmology was sought. Why is the universe so similar, when it seemingly was divided into so many separate parts soon after the moment of creation, which then would have been "over the light horizon" and thus out of touch with each other?

Inflation comes to the rescue

A theory which claims to explain away this problem has found much favor in the new physics. Termed inflation, it is complex in detail but simple in essence. Basically, this notion holds that a mere 10^{35} part of a second after the Big Bang there was an extremely rapid expansion of the newborn universe. John Boslough says: "far smaller than the nucleus of an atom when it began, the universe would have inflated to a diameter of about 10 centimeters, about as big as a softball. This was equivalent to something the size of a grain of sand growing bigger than our universe in the same span of time [10^{-32} of a second]."[24] Before inflation began, all the regions of the universe that are observed today would have been in contact with each other, while innumerable other parts of the nascent cosmos would have flown off in their own directions, never to be perceived again. Joe Rosen writes:

> Those universes carried on the cosmic inflation by flying away from each other at such stupendous speeds that no interaction among them was possible, since no influence emanating from any island universe could propagate fast enough to overtake any other. They are now supposed to be so far apart and dispersing so rapidly that still no interaction among them is possible or ever will be. . . . There is no scientific answer even to the question of how far and in what direction our nearest neighbor island universe is supposed to be.[25]

I have spent some time discussing Big Bang cosmology to illustrate both how close—and how far away—are mysticism and the new physics in this area. Each agrees that our universe was created by a force so powerful that it is far beyond our capacity even to imagine, not to speak of actually reproducing with human technology. Spiritual science teaches that this was an emanation of God's Spirit, which consciously brought physical existence into being. Physics, though no consensus exists, is inclined to believe that the universe resulted from a random quantum fluctuation of equally potent and all-pervasive vacuum energy. Because material science finds no consciousness behind the creation, it is forced to explain how order evolved from a chance fluctuation.

Schemes such as the "inflation theory" have been devised for this purpose. However, while answering some questions, they create other mysteries. As the Rosen quotation indicates, if inflation *was* responsible for forming a coherent universe, then there should be innumerable other physical universes that likely have laws of nature vastly different from ours. We can never know anything about these other domains of material existence, because they have receded from our universe at a speed greater than light, and we will never be able to catch up to them. Hence, one of the leading theories of the new physics now posits other domains of existence which are as objectively real as ours—but *cannot* be known by mankind. Mysticism, on the other hand, teaches that higher levels of consciousness exist which *can* be reached through contemplative meditation.

So, in a sense, physics is becoming more "mystical" than mysticism. Theoreticians are busily engaged in developing theories about the process of creation that can never be tested, and imply the existence of other worlds—or dimensions of existence—that never can be perceived directly. Astronomer David Lindley writes:

> Inflation is a nice idea; it would be pleasing if particle physics worked in such a way that it made the universe large and uniform. But there is no substantial evidence that inflation actually occurred. . . . Nevertheless, so enamored are particle physicists of the idea of inflation as a cosmic panacea that they have taken to inventing theories that do nothing except make inflation work. The argument is circular—cosmologists like inflation because particle physicists can provide it, and particle physicists provide it because cosmologists like it—and has proved, so far, immune to test.[26]

Modern creation myths

Cosmology, which attempts to explain how the universe came into existence, *always* will be untestable as long as its investigative methods utilize the traditional tools of material science: reason, logic, and mathematics. For these tools are intended for use in an experimental setting where the predictable and reproducible aspects of the world can be revealed. The Big Bang was a once-in-a-universe event. It can neither be predicted nor reproduced by physicists and astronomers. Thus cosmology is nothing other than a form of theology with numbers attached. Those equations should not mislead us into believing that the assumptions which underlie them are any less metaphysical than the creation myths of our ancestors.

As long ago as 2000 years B.C., the Babylonians had developed their own cosmology in the form of an "Epic of Creation."[27] It begins with two gods who beget four generations. These offspring become extremely rambunctious, noisy, and disturbing to the original parents, and plots are laid to get rid of the progeny. A long process involving battles, schemes, and negotiations between various parties finally leads to the world's creation when one of the parent gods is killed and split in two, with half of her becoming the sky and half becoming the earth.

Such myths are greeted with a smile by present-day cosmologists, for their content clearly is not in accord with scientific fact. But there are considerable *structural* similarities between this ancient epic of creation and the modern Big Bang theories I have just summarized. Each requires a large cast to carry out the task of creation: the Babylonian myth describes by name at least eight separate gods, and the existence of many other entities is implied; quantum cosmology requires at the very least time and space, a physical vacuum, laws of probability, and principles of quantum physics to explain how our universe came to be. In either case, one is left wondering: what—or who—created all of the entities which created the world?

In a myth such as the "Epic of Creation" it is tempting, and probably correct, to say "These gods and their machinations are reflections of human nature. Those who created the myth used their own personal traits to explain the behavior of the gods." Joseph Campbell, a scholar of mythology, notes that in the Babylonian period gods became "omnipotent, freely willing creators of a comparatively arbitrary order—personifications of fatherhood writ large, subject to whims, wrath, love, and all the rest...."[28] Many early Greeks, of course, viewed their gods in a similar fashion. Divinity was in large part a projection of purely human attributes, some good and some bad.

The same tendency to take an anthropomorphic approach toward cosmology is evident in theories of the new physics, albeit in a much more sophisticated guise. A final cause of creation is sought in concepts immediately accessible to the mind of man: randomness, causelessness, uncertainty. To physicists these attributes appear to constitute the bedrock of the subatomic quantum world, so it is assumed that they are the foundation of *all* existence. However, material scientists recognize that what is perceived in the quantum realm is a function of both the means of observation and whatever reality exists in that realm itself. This implies that the principles of quantum cosmology in part are a reflection of the limited human mind, and in part are objectively real.

As the seemingly firm ground of materiality fades away—in the subatomic world or at the moment of the Big Bang—it becomes increasingly difficult to separate objective reality from the subjective human mind. Earlier we found that thinking is part of the subjective, yet symbolic, domain of being. Certainly thoughts—in the form of either words or numbers—can describe objective material phenomena quite well. But in the absence of materiality, thinking quickly becomes imagination. There is nothing for our thoughts to grasp, no foundation to support our concepts. Hence myths and mathematics are equally ineffective in understanding the truth about Ultimate Reality.

Spirit created the universe

That truth is Spirit. While formless and imperceptible to the physical senses, Spirit possesses all of the attributes of God, and indeed *is* God-in-Action. The stupendous force of the Big Bang was produced by this cause. Regarding the domain from which Spirit emanates, the perfect mystic Jagat Singh says that "Millions of suns pale into insignificance before the effulgence of even one atom of this region."[29] A single drop from a wave of this omnipotent and omnipresent divine ocean of Spirit was sufficient to bring our universe into being. Is not this a much simpler and more reasonable explanation for the Big Bang than the tortuously complex theories of quantum cosmology? From a higher dimension of consciousness emanated a lower dimension. Neat, clean, precise—no loose ends to be explained away by chance vacuum fluctuations or baby universes flying out of sight.

To reiterate, spiritual science and material science are in complete agreement that the creation of our universe was brought about by a force which was all-pervading and possessed essentially infinite energy. If the new physics chooses to call this force "vacuum," and mysticism terms it "Spirit," the reality is deeper than the words. The primary point of disagreement between these scientific disciplines lies in whether this creative force was conscious, or unconscious. For if the vacuum is incredibly strong but stupid, only chance will enable it to produce a marvelously well-formed universe such as ours. By the same reasoning—that is, by the "infinite number of typing monkeys principle"—perhaps a gorilla with a sledgehammer eventually could carve a perfect replica of Michelangelo's "Pietá" with a few lucky blows, but only after smashing many, many other slabs of marble into random fragments.

Therefore, it is time to move to an examination of *consciousness*. What is its true nature, and where does it reside: Everywhere? Only in the mind of man and possibly other living things? Or nowhere, being an illusion of the material brain?

Principle 3.
God, Spirit, and creation;
all are consciousness

This world is a single thought of the Universal Intellect;
the Intellect is like the king and forms are its messengers.
—Rumi[1]

This quotation from Rumi bears careful study, for these two lines—when correctly understood—lay bare the mystery of consciousness. He tells us that this world, and our entire universe, is a thought of the Universal Intellect. A thought? How can this be when everything around us seems so substantial? Thoughts come and go, while the physical creation has lasted for some fifteen billion years. Thoughts appear to be wispier than the most delicate cloud, while earth and rocks blunt the most powerful drilling tools. So what do perfect mystics mean when they tell us that the essence of *everything* in existence, including ourselves, is consciousness?

To begin to understand the answer to this question, it is necessary to lay aside a common sense conception regarding the nature of consciousness. This is *not* the same as simply being awake and aware, as when a doctor says after a patient's anesthesia wears off: "He is conscious now. You can talk to him." For this means only that a person is able to perceive *contents* of consciousness. These can be either external, such as sights, sounds, smells, tastes, and tactile sensations, or internal—emotions, thoughts, images. Most people, including professional neuroscientists, confuse what is floating in the stream of consciousness with the "water" itself. This is understandable, since that stream is a wave of formless Spirit and cannot be perceived by the senses or intellect. Emotions, thoughts, perceptions and the other contents of consciousness are carried along by that current much as debris is washed down a swiftly flowing river.

The surface of thought's stream
carries sticks and straws—
some pleasant, some unsightly.
Seed-husks floating in the water
have fallen from fruits of the invisible garden.
Look for the kernels back in the garden,
for the water comes from the garden into the riverbed.

If you don't see the flow of the water of Life,
look at the movement of weeds in thought's stream.
                                                     —Rumi[2]

Thoughts and perceptions are not pure consciousness

Modern material scientists who study the nature of consciousness are acting in accord with Rumi's closing piece of advice: if you are unable to see the water of Spirit, all that can be discerned are the weeds of thoughts and perceptions. Francis Crick, a physicist and biochemist who won a Nobel prize for discovering the structure of DNA, has turned his attention to studying consciousness. In an opening chapter of his book, *The Astonishing Hypothesis*, he says "there are many forms of consciousness, such as those associated with seeing, thinking, emotion, pain, and so on. Self-consciousness—that is, the self-referential aspect of consciousness—is probably a special case of consciousness. In our view, it is better left to one side for the moment."[3] Crick then goes on to explain why he chose visual awareness as his focus of study, and by the end of the book argues that science is on the road to proving that there is no such thing as the "disembodied soul."

It is quite a jump from explaining visual perception to demonstrating the non-existence of the soul. This illustrates how far removed popular ideas about consciousness are from the facts of spiritual science. In mysticism, consciousness has nothing to do with being aware *of* anything. J.R. Puri has described what occurs in contemplative meditation after all external physical sensations have been eliminated from awareness:

> . . . the mystic next excludes from consciousness all sensuous images, and then all abstract thoughts, reasoning, processes, volitions and other particular mental contents. One may ask, what then would be left of consciousness? In the absence of any mental content whatsoever, there would be a complete emptiness, a void, a vacuum. One would suppose *a priori* that consciousness would then entirely lapse and one would fall asleep or become unconscious. But the introvertive mystics unanimously assert—and there are thousands of them all over the world—that they have attained to a complete vacuum of particular mental contents, and what then emerges is a state of pure consciousness. It is pure in the sense that it is not the consciousness of any empirical content. It has no content except itself. And this self-realization is often eventually spoken of as God-realization.[4]

So Crick incorrectly identified the self-referential form of consciousness as being a "special case." Actually, when fully realized this form *is* consciousness, one with Spirit—and hence with God. This helps us to understand how all of creation can be consciousness. Mysticism teaches that pure consciousness, or Spirit, is the substrate of everything in existence, including the mind. This is why God is said to be "within" us. God is within everything, but since we are human beings, human consciousness is the only place He can be realized by us.

In other words, some entities can be conscious of being consciousness, while others remain ignorant of this fact. Though nameless and formless, Spirit takes on various forms in the course of fulfilling God's creative will. Sawan Singh describes the difference between conscious and inert projections of Spirit: "A conscious being feels its consciousness itself, while in the case of an inert being, its existence is felt by others only."[5] What we call matter thus is unconscious consciousness. Along these lines, there is an important difference between the *essence* of existence, which never changes, and the *effects* brought about by Spirit, God-in-Action.

> *God did not increase by bringing the universe into existence; He did not become what He had not been formerly.*
>
> *But effects increased when He brought the creatures into existence: between these two increases there is a difference.*
>
> *The increase of effects is His making Himself manifest, so that His Attributes and Acts may become visible.*
>
> *But the increase of an essence would be proof that it is of temporal origin and subject to causes.*
>
> —Rumi[6]

This is why mysticism places so much importance on eliminating external perceptions and internal thoughts from one's consciousness in contemplative meditation. It is not necessary, or even possible, to develop a higher form of consciousness. This already exists in the form of eternal and omnipresent Spirit. All that prevents us from being perfectly aware of this Ultimate Reality are the veils which obscure pure consciousness: sensations, emotions, thoughts, images, and other contents of the mind. These, of course, also are composed of Spirit, as is everything made of mind or matter. Charan Singh says: "All that we see is just His own projection. Everything is projected from Him. If we admit that there was something besides Him, then the Lord is not one. He is the only One—always was, is and will be. He is everywhere, and everything is His own projection."[7]

How One becomes many

How, then, is the evident manyness that surrounds us produced out of that divine Oneness? In subsequent sections we will discuss how varying levels of vibrating energy are responsible for both spiritual and material phenomena. For now, an analogy in the form of a simple experiment will serve as a partial answer to that question. Take some ice cubes out of your freezer and place them in a pot on the stove. Begin heating the pot, and soon there will be smaller ice cubes floating in some liquid water. Eventually there will be nothing but liquid in the pot, the ice having melted completely. Raise the temperature of the stove's burner and continue watching while the liquid boils. Steam will begin to form. Finally, there will be no water left in the pot, and the air in your kitchen will be full of invisible vapor.

The ice, the liquid, and the vapor all are the same substance—water—but its form differs depending upon how fast the molecules that make up water are moving. The molecules in ice are far apart and almost motionless; in liquid water they are close together and move about freely; in water vapor the molecules move around violently and bump into each other. Applying, or removing, heat energy from water changes its form—but under normal conditions these changes occur gradually, so several forms may be present in varying proportions.

Through the simple experiment I just described, over the span of a few minutes water can be observed consecutively as: (1) ice, (2) ice and liquid, (3) liquid, (4) liquid and vapor, and (5) vapor. Similarly, perfect mystics describe creation as being divided into various domains, each of which is distinguished by the form of Spirit which is predominate on that level: (1) matter, (2) matter and mind, (3) mind, (4) mind and Spirit, (5) Spirit. Our physical universe is level 1. God-as-God, Ultimate Reality, the Theory of Everything, is realized on level 5. Lekh Raj Puri writes:

> Spirit is truly and always one; but its manifestations on different planes of creation are different. Just as ice, water, and vapour are not three things, but only three forms of the same thing, similarly Spirit is one, but its forms are many. In the very highest transcendent realms, it abides as an extremely fine and subtle entity; but as we descend toward less subtle regions, this Spirit also takes less subtle forms.[8]

This analogy between water and Spirit is not perfect, however. For in physics it is moot to ask, "Which is the *true* form of the molecule H_2O: solid, liquid, or gas?" Given the particular conditions on earth, we find that all three manifestations of water are both evident and plentiful here. On either a frozen or extremely hot planet, however, this would not be the case. So the form water takes depends upon the temperature of its surroundings, none of which can be said to be "truer" or "higher" than another. A fish prefers the liquid ocean; a polar bear, frozen ice and snow. Even from the standpoint of the use to which water is put, the situation is relative. To cool a drink, ice cubes are best. To heat a building, hot steam is required.

However, in mysticism the true form of Spirit is clear: pure consciousness is the only form which is eternal, for this is God's essence. The other manifestations—matter and mind—are impermanent, and so untrue from the standpoint of final knowledge. The highest truth is that which never changes. Since all forms of water are transitory, one form cannot be said to be truer than another. But Spirit is forever Spirit. In other words, pure consciousness is not a relative form of truth. Matter, mind, and Spirit should not be thought of as co-equal manifestations of the same substance, in the same sense as ice, liquid and vapor are all water. Rather, matter and mind are lower forms of Spirit and so stand as barriers to highest truth.

Boiling purifies

One of Rumi's teaching stories illustrates how a student of spiritual science removes those veils of God and becomes fit to merge his personal consciousness with Spirit.[9] As a seeker of that Truth, the student is like the chickpea (a favorite Middle Eastern food) in the pot, which "is at the mercy of the fire. Constantly, as the water boils, the chickpea rises to the top of the pot with a hundred outcries." Why, it asks the cook, are you setting the fire on me and beating me down with a ladle? I used to be green and fresh in the garden, where cool water helped me grow. Now, says the chickpea, this very water is boiling me alive. "Boil nicely now," replies the cook. "Do not leap away from the one who made the fire. I am not boiling you because I detest you; on the contrary, it is so that you may get taste and flavour, so that you may become fit to eat and mingle with the spirit."

Mysticism teaches that insofar as we are capable of understanding the divine will, one reason why domains of mind and matter were created by God was to enable souls to return to Him. Suffering causes us to seek union with His perfect bliss and love. Our "boiling" in this crude manifestation of Spirit—the physical universe—is intended to remove the coverings that obscure our divinity. The cook says,

> Boil on, chickpea, in tribulation, that neither being nor self may remain to you. . . . If you have been parted from the garden of water and clay, yet you have become food fit to enter into the living. . . . In the beginning you grew out of God's attributes; return now swiftly and nimbly into His attributes. . . . The frozen grape is thawed by cold water and lays aside its coldness and frozenness. . . . Formerly I too was like you, a part of earth. . . . When I had quaffed the cup of fiery discipline I became suitable and acceptable. For a space I boiled in Time, for another space in the pot of the body. By this double boiling I became strength to the senses; I became spirit; then I became your teacher.

Spirit is what remains when the spiritual scientist is able to remove the dross of matter and mind from his consciousness through contemplative meditation. Sawan Singh writes: "The Spirit is conscious and consciousness. . . . The Lord is the ocean of super-consciousness, and Spirit is its wave. The Supreme Lord, the soul and the Spirit are a holy Trinity. The One Lord exists in all the three forms. . . . The Spirit is the life, the essence, the root and the quintessence of every created thing. . . . Whatever exists in this creation is Spirit. It is the cause of all creation and dissolution."[10]

Mind—surprise!—is unconscious

Mind is completely separate from Spirit, in somewhat the same sense that liquid water is separate from water vapor. However, a more accurate analogy is to visualize a complex network of water pipes where the liquid's flow is both regulated and energized by steam. Like that steam, Spirit is the motive force behind mind and matter. Thus perfect mystics state unequivocally that the mind is unconscious. This is an important point which bears repeating: the very mind which is reading—and writing— these words is *unconscious*. Those scientists who are trying to lay bare the inner workings of the brain and mind may be engaging in valuable research, but this is not the study of consciousness. Only spiritual science— the study of Spirit—understands how mechanical are mind and matter. Even the vaunted human intellect is inert unless enlivened and guided by Spirit.

Sawan Singh says, "We know about other sciences but we know nothing about our soul, which is the motive force from which we know about the sciences and because of which the mind and the intellect work."[11] Similarly, Julian Johnson observes that many people confuse the mind and Spirit, and so fail to make much headway toward understanding the nature of Ultimate Reality:

> The mind is not self-conscious, nor self-acting. It has no power of automation or of initiative. It is simply a machine, though highly sensitive and extremely powerful when motivated by spirit. Mind and spirit have been greatly confused in western psychology. Mind works only when activated by the Soul . . . there is no divine mind. The supreme divinity is far above all mind . . . every activity in the universe is carried on by spirit, and spirit only. But spirit works through many intermediate substances on these planes. Without spirit, mind is as inert as steel. . . . The chief function of mind is to serve as an instrument of spirit for all contacts with the material worlds.[12]

This surprising conclusion—that mind is not conscious—will be discussed further with other principles. We shall find that in mysticism consciousness and free will are intimately linked, and that the meaning of "freedom" goes far beyond the usual understanding of this term. Mind is considered by spiritual science to be unconscious because it possesses just the slightest modicum of free will. Being so accustomed to our confinement in a low-energy universe where "frozen" matter is the prevalent form of Spirit, even the pitifully small amount of freedom our mind possesses seems liberating to us. To an ice-encrusted mountain rooted to the earth, the motion of a slow-moving glacier must appear miraculous.

When perfect mystics say that all of our thinking, perceiving, conceptualizing, theorizing, visualizing, and the like is the work of an unconscious mind, they speak from the standpoint of one who knows what real consciousness is. Compared to a rock or a tree, the human mind is relatively conscious. Compared to Spirit or a state of God-realization, mind is completely unconscious. More precisely, mind is unaware of the conscious essence which forms and energizes it. Similarly, in a sense even a stone can be said to be conscious, since its deepest reality also is Spirit. However, neither a mind nor a stone is able to *know* what it truly is made of, and so cannot fulfill one of God's purposes for creation: *I was a Hidden Treasure, so I wanted to be known.* Only the conscious soul, a drop of Spirit, is able to merge with the ocean of divine super-consciousness and thus realize God completely.

Mind links Spirit and matter

Before moving to an examination of how consciousness is regarded by the new physics, Julian Johnson's statement above that "spirit works through many intermediate substances" needs to be addressed. The focus of this book is on Ultimate Reality, and how this final truth is echoed in findings of material science. Mysticism teaches that God creates by projecting His Spirit, and Spirit continues to sustain this physical universe. However, there are a number of linkages between Spirit and matter. Mind is one of them.

The workings of these intermediate forces and substances can appear intricate and complex, but understanding of them is not essential for the seeker of Ultimate Reality. (I live in the state of Oregon, where electricity passes through various substations and transformers on its way from a hydroelectric generator on the Columbia river to the light bulb in my kitchen. If I am interested only in learning about the *source* of that electricity, I can drive directly to Bonneville Dam without tracing all of the intermediate paths the power takes before it reaches my house. So detailed explanations of what connects Spirit and the material universe would distract from this book's focus on *Ultimate Reality*.)

Those who are interested in learning more about how conscious Spirit is linked with unconscious mind and matter are encouraged to read one or more of a series of books by John Davidson: *Subtle Energy, The Web of Life, The Secret of the Creative Vacuum,* and *Natural Creation and the Formative Mind.* These works describe in detail how higher spiritual domains of existence are connected to the physical universe, and our own minds and bodies. In addition to expanding upon many of the subjects with which this book is concerned, Davidson fills in the "other" category in this schematic of existence—of which only the capitalized entries receive much attention in these pages:

GOD—SPIRIT—[other]—PHYSICAL UNIVERSE

Rumi said that this world is a thought of the Universal Intellect, or the Universal Mind. Thus mind created the physical universe—but this was not a small mind like ours. If it was, material scientists would not be having so much difficulty understanding with their own intellect why and how creation came to be as it is. The Universal Mind bears the same relation to the individual human mind as the soul bears to Spirit: each is a drop of a vast ocean. Until a drop is reunited with the ocean, it will have a separate consciousness that is much more limited than its source. There is but one way to know the complete truth about Ultimate Reality: *become it.*

> *When you pass beyond this human form,*
> *No doubt you will become an angel*
> *And soar through the heavens!*
> *But don't stop there.*
> *Even heavenly bodies grow old.*
> *Pass again from the heavenly realm*
> *and plunge into the vast ocean of Consciousness.*
> *Let the drop of water that is you*
> *become a hundred mighty seas.*
> *But do not think that the drop alone*
> *Becomes the Ocean—*
> *the Ocean, too, becomes the drop!*
> —Rumi[13]

In summary then: God projected Himself as Spirit, which was formed into Universal Mind, out of which the physical universe was projected as materiality. Throughout, these acts of creation were directed by the all-pervading conscious energy of Spirit, God's essence. This is how mysticism explains the remarkable order of our universe. Now let us see how the new physics explains this fact.

Echoes in the new physics.

Recall that both mysticism and physics hold that, contrary to appearances, existence is composed of only one substance. These monistic views diverge, however, when it comes to what that single entity is: spiritual science says consciousness; material science generally argues for matter—or energy, the two being equivalent. Each discipline agrees that the physical universe was created, and is sustained, by enormous energy. Since mysticism teaches that the force of Spirit is conscious, there is no difficulty in understanding how what that force creates comes to possess order. If an intelligent man with a good set of tools and a creative mind sets out to build a birdhouse, it surprises no one when that structure turns out to be solidly constructed, functional, and attractive.

However, the prevailing perspective of the new physics is that consciousness, or mind if you like, emerged from matter at some point. It is not known how this occurred, just as there is no consensus about how life itself emerged from inanimate substances. So from one point of view, physics does not have much to say about the nature of consciousness. As was noted previously, even material scientists who specialize in studying the workings of the mind and the brain have no solid understanding of what consciousness is, or how it comes to be. Some even deny that it exists at all. To find echoes in the new physics of this principle of Ultimate Reality, it is necessary to take an indirect approach. What is the evidence implying a conscious universe which confronts materialistically-minded physicists?

An amazingly well-ordered universe

That evidence is all around us. We live in a remarkably well-ordered universe. Our earth, along with the other planets, orbits a stable sun which provides a constant source of energy to maintain life. Laws of nature are precisely balanced to permit change and diversity within well-defined limits. The sun's surface, for example, is boiling with activity. Solar flares can leap as much as a million miles into space, and sunspots as large as one hundred earths mark areas on the sun that are one-third cooler than normal. Nuclear reactions in the sun's core emit as much energy in one second as mankind has consumed in the whole of history.[14] Yet all of this activity is controlled and seemingly purposeful. Even exploding stars—supernovas—play a constructive role in the universe, for this is how the rare and heavy elements necessary for life on earth were formed. Every chemical element once was inside an exploding star. Our bodies literally are stardust.[15]

So how likely is it that the Big Bang was an unconscious "random quantum fluctuation" which produced the raw matter and energy, and laws of nature, that resulted in this ordered universe of ours? This may appear to be an unanswerable question, but physics has found a way to pin order down numerically through the concept of *entropy*. Entropy is a measure of disorder, the flip side of order. So high entropy connotes disorder, and low entropy means order. If a helium-filled balloon is floating in the middle of your living room, this situation possesses a relatively low entropy. Normal air is outside the balloon, and all the helium is inside it. Nice and tidy. But if you pop the balloon, the entropy in the room becomes higher—for soon the air and the helium are mixed together.

Similarly, a pile of sugar mixed with salt possesses high entropy. It is disorderly. Sorting the sugar and salt into two separate piles lowers the entropy of the mixture by making things more orderly. However there is a cost: the energy you put into the sorting process. In the course of laboriously using a toothpick and a magnifying glass to pick apart the sugar and the salt, you would burn up quite a few calories. All of the heat energy radiated by your body into the room during the sorting job thus increases the overall disorder of the universe at the same time you are producing order in the smaller world of the sugar and salt pile. So physicists say that the universe as a whole is moving toward increasing entropy, or disorder, even though discrete bits of it—such as the sugar and salt pile—seem to be tending toward greater order. Generally, order on earth is being "bought" by steadily increasing disorder in the core of the sun (almost all the energy we use is converted solar energy) and eventually the sun will run out of low entropy nuclear fuel.

The point is this: the universe is known to have started out in a very orderly fashion. Ten to fifteen billion years have passed since the Big Bang. Every year, according to the laws of physics, the overall entropy—or disorder—of the physical creation has increased. This is the Second Law of Thermodynamics: "the entropy of an isolated system increases with time." Mysticism, of course, teaches that the material universe is not isolated from the rest of creation. Viewed through the eyes of the laws of physics, however, the total level of disorder in the universe must increase over time. Regarding entropy, the physical creation is something like a clock with a massive mainspring. Since it will take many tens of billions of years for the physical universe to run down, it must have been wound up very tight indeed at the instant of the Big Bang.

Unity is the highest order

This needs some explaining, however. For common sense argues that order has been increasing since the Big Bang, not decreasing. That is, now we see all kinds of separate entities: trees, people, rocks, stars, cats, and so on. Fifteen billion years ago there was no separateness at all. So we need to understand why a united mixture of helium and air is considered disorderly by physicists, while the unified Big Bang is seen as orderly. At the moment of creation, physics believes that all of the forces of nature were merged into a single "superforce." Neither matter nor energy as we know them existed then. Gerald Schroeder says that "at the earliest instant about which we can theorize, temperatures exceeded 10^{32} degrees C or 10 million billion billion times hotter than the temperature at the center of the Sun. No one knows what was present at that explosive instant but whatever was there was in an exotic state of madly rapid motion."[16]

This immensely high energy generated an immensely strong gravitational force. While we generally think of gravity affecting only things made of matter, the equivalence of matter and energy means that all forms of energy generate gravitational fields. Massless starlight has been found to be attracted, and thus bent, as it passes the sun. Therefore, at that earliest instant of the Big Bang described by Schroeder, physicists believe that gravitational forces were as strong as any other forces. This made for some strange goings-on. Steven Weinberg writes about that time: "Not only would gravitational forces have been strong and particle production by gravitational fields copious—the very idea of 'particle' would not yet have had any meaning."[17] So for a brief moment in the extremely high energies of the Big Bang, gravity—which presently is by far the weakest force in nature—was the only entity in existence.

Though material science generally avoids poetic license, it is difficult not to draw a parallel between mystical love and physical gravity. Each unites, rather than separates. Each is considered to be the force which first manifested in creation. Each has no opposite: hate is considered by mysticism to be the absence of love, not a co-equal opposing force; similarly, gravity is universally attractive, so one object in the universe feels the pull of all the other objects in the universe.[18] No anti-gravity has been discerned, even though it is a common feature in science fiction. Thus there is a certain beauty in the fact that the high degree of order in the Big Bang was a function of *gravity*.

Mathematician Roger Penrose explains that gravity is what makes the uniformity of the Big Bang possess a very low entropy (high order), while the uniformity of a gas in thermal equilibrium—such as a diffuse mixture of air and helium—has a *high* entropy (low order).[19] I hope that this does not sound too confusing, for beneath the technical language of thermodynamics is an important echo of mystical truth. *Unity is the highest form of consciousness, possessing the greatest order, energy, and truth.* The Big Bang was formless. Even as long as one hundred thousand years after creation, the universe still was composed of glowing plasma and relatively featureless, atoms not having been able to form yet.[20]

Penrose demonstrates that for a system of gravitating bodies—which describes our physical universe—high entropy, or disorder, is achieved by gravitational clumping. Thus gravity plays a key role throughout the process of creation. In the beginning, gravity is all that there is. Matter, energy, particles, forces: it is all one. After the universe has expanded and cooled, gravity begins to pull together separate entities into clumps. Gases combine to form stars, which explode and eventually form the stuff of planets, some of which come to be able to sustain life through the energy provided by gravitationally-produced stars.

Life thrives on order, not energy

The beauty of all this—from the standpoint of mysticism—is that the formlessness of the Big Bang is a reservoir of low entropy, or order, from which both we and the universe as a whole continue to draw sustenance. Actually the survival of humans, and other living things, is dependent more on low entropy than on energy. Roger Penrose writes:

> Often one hears it stated that we obtain *energy* from our intake of food and oxygen, but there is a clear sense in which that is not really correct. . . . Since energy is conserved, and since the actual energy content of our bodies remains more-or-less constant throughout our adult lives, there is no need simply to *add* to the energy content of our bodies. . . . However, we do need to replace the energy that we continually lose in the form of heat. . . . Heat is the most *disordered* form of energy that there is. . . . To keep ourselves alive, we need to keep lowering the entropy that is within ourselves. We do this by feeding on the low-entropy combination of food and atmospheric oxygen, combining them within our bodies, and discarding the energy, that we would otherwise have gained, in a high-entropy form.[21]

Once again, beneath this somewhat technical language there is a clear echo of spiritual truth. Life thrives on ordered energy. Energy alone is not enough. Otherwise we could simply stand under a heat lamp, or the rays of the sun, and not need to eat food. Mysticism teaches that this order does not come about by accident. It is the result of Spirit's all-pervading conscious energy, which created this universe. Jesus may not have been thinking of the second law of thermodynamics when he said, "Man does not live by bread alone, but by every word that proceedeth out of the mouth of God."[22] Still, this statement is in perfect accord with material science, if *word* is taken to mean the perfectly ordered energy of the Big Bang. Spirit continues to feed our universe with order, but the physical store of low entropy is becoming more and more diffuse with time.

Physics and mathematics have shown that every second which has passed since the instant of creation has resulted in an increase of entropy, or disorder, in the universe as a whole. "Entropy," of course, is largely a thermodynamic concept, so some argue that even though the creation may be getting weaker as energy diffuses, human consciousness is getting wiser. Mysticism suggests that this is a questionable assumption. Spiritual science is more in accord with the conclusion of the Second Law of Thermodynamics: the perfectly ordered energy of Spirit present at the earliest moment of the Big Bang was the highwater mark for the overall consciousness of the physical universe. From the standpoint of order, things have been going downhill since.

Why chance is a most unlikely creator

However, as has already been noted, mysticism teaches that there is a purpose behind this seeming devolution. Even though human logic and understanding cannot encompass the divine will, it is clear that chance is not responsible for the existence of our universe. Such is evident not only to perfect mystics, but is reflected in findings of physics. Roger Penrose has calculated the probability of a Big Bang producing a universe closely resembling the one in which we actually live. He imagines the Creator armed with a "pin" which is to be placed at a particular point in a volume of "phase" space (this is a mathematical sort of space, but it can be visualized as a massive dart board). According to Penrose:

> Each different positioning of the pin provides a different universe. Now the accuracy that is needed for the Creator's aim depends upon the entropy of the universe that is thereby created. It would be relatively "easy" to produce a high entropy [disorderly] universe, since then there would be a large volume of the phase space available for the pin to hit. . . . But in order to start off the universe in [a] state of low entropy—so that there will indeed be a second law of thermodynamics—the Creator must aim for a much tinier volume of the phase space.[23]

Simply put, there are many, many ways to create a disorderly universe, but the probability of the Big Bang producing the order we find all around us is astonishingly small. How small? Penrose says that the Creator's aim with that metaphorical pin must have been accurate to one part in $10^{10^{123}}$. This number is larger than all of the subatomic particles in the entire universe. Thus it would be as likely for God to blindly throw an infinitesimal dart from heaven and miraculously hit one particular electron in the immeasurably large physical creation, as for the Big Bang to have formed our universe by chance. According to Penrose, physics has no explanation for the precision that set the universe on its course. The answer, he says, lies in the "singularity" of the moment of creation. Since this lies beyond physics' understanding of space and time, material science has no way of explaining the creation's entirely unlikely degree of order.

Only feeble explanations have been put forward. One already has been noted: the "infinite number of monkeys typing for infinity eventually will write the works of Shakespeare" approach to cosmology. This notion, says

Paul Davies, is that "If the universe has enough time available then, sooner or later, one might suppose, whole stars, whole galaxies, will simply begin to form—accidentally. The fact that the time for such an absurdly improbably event is inconceivably long (at least $10^{10^{80}}$ years) is, in principle, no problem if one is prepared to believe that the universe is of infinite age."[24]

We might ask, of course: what created time and the laws of probability which permitted this chance event to take place? And where did the raw material come from which was acted upon by that random quantum fluctuation? If physicists argue that the universe has existed for infinity in various guises—disordered and ordered—then an eternal creative force is implicitly assumed. Yet material science knows no way to prove its existence, and so remains mired in theoretical hypothesizing. In Davies' quotation above there are two *ifs* and one *suppose*, not a particularly solid foundation for a Theory of Everything, or the truth about Ultimate Reality.

> *A certain stranger was hastily seeking a home,*
> *so a friend took him to a house in ruins.*
> *"If this house had a roof," he said,*
> *"you could live next to me.*
> *Your family would be comfortable here, too,*
> *if there were another room."*
> *"Yes," he said, "it's nice to be next door to friends,*
> *but my dear soul, one cannot lodge in 'if'."*
> —Rumi[25]

In fairness to Paul Davies, however, I should point out that his equivocal language probably is due to his lack of commitment to the "typing monkeys" approach to cosmology. In his book, *The Mind of God*, Davies points out that the principle of Occam's razor has served science well and should not be abandoned by the new physics. (In accord with the properties of a valid scientific theory discussed previously, Occam's razor cuts away unnecessary explanations of phenomena.) Davies says, "You pick the theory with the least number of independent assumptions . . . if one thinks of a theory as a computer program, and the facts of nature as the output of that program, then Occam's razor obliges us to pick the shortest program that can generate that particular output."[26]

Let us try out this razor. Question: What caused the universe to be so orderly, and allow consciousness to develop within it? Theory 1 says: God, who *is* consciousness. Theory 2 says: a random quantum fluctuation, which occurred within a preexisting vacuum state and already existent time, that utilized established laws of nature to form a stable universe by chance, after having created innumerable other physical universes unsuitable for human life which never can be known to us. Even this is an over-simplification of Theory 2, but you get the picture. Count up the number of separate assumptions: Theory 1 (God)—one assumption; Theory 2 (randomness) — six assumptions.

It's simple: Consciousness produces consciousness

So faced with a choice between mindless accident, and purposeful deity, Paul Davies is a physicist who leans toward God: "My conclusion is that the many-universes theory can at best explain only a limited range of features, and then only if one appends some metaphysical assumptions that seem no less extravagant than design. In the end, Occam's razor compels me to put my money on design."[27] This is a good bet. The simplest way of explaining how consciousness, and order, came to exist in the physical creation is to accept that it was there from the beginning. Sawan Singh writes, "If there had been no consciousness in the original Cause, then there would be no consciousness in the universe."[28]

Many physicists would agree with these sentiments if consciousness is equated with the mind and mathematics. As we have seen, mysticism considers the mind to be unconscious and mechanical, yet also knows that mind is a higher form of Spirit than completely inert matter. Universal mind, which is the form of consciousness dominant on Levels 2 and 3 of the five levels of existence, was the mechanism which created the physical universe. According to mysticism, this is what makes it possible for the limited human mind to unravel many of the workings of the laws of nature. To some extent, the thinking of material scientists can come in tune with the thought of Universal Intellect, just as a drop knows firsthand something of what the ocean is like—albeit in an imperfect and incomplete sense.

It already has been noted how accurately the laws of nature understood by the new physics can be mirrored by mathematical equations. Even if one does not accept all of the tenets of mysticism, this fact has been a source of inspiration to many metaphysically-minded physicists. Sir James Jeans, a mathematician, physicist and astronomer (scientists tended to be more eclectic in the early twentieth century) went so far as to write: ". . . the universe appears to have been designed by a pure mathematician. . . . We may think of the laws to which phenomena conform in our waking hours, the laws of nature, as the laws of thought of a universal mind. The uniformity of nature proclaims the self-consistency of this mind. . . . If the universe is a universe of thought, then its creation must have been an act of thought."[29]

So we return to the words of Rumi which began this section: *This world is a single thought of the Universal Intellect.*

Thinking is intentional, however. We do not think without a purpose. The purpose of a thought may be as diffuse as to avoid boredom or an empty feeling in our minds, or as focused as to design an improved computer chip. As shall be discussed under a succeeding principle, both spiritual and material science teach that in the domain of everyday life there is no effect without a cause. Assuming, then, that physics agrees that consciousness is woven into the fabric of the universe (mysticism goes further and says that it *is* the fabric), have physicists been able to suggest any reason why this should be so?

Weaknesses of the Strong Anthropic Principle

Yes, they have. And even though this reason—the Strong Anthropic Principle—is scoffed at by many material scientists, who consider it to be a nonsensical bit of metaphysics that has no place in "hard" science, in some respects it bears a close resemblance to Rumi's words:

> God says, "I was a Hidden Treasure, so I wanted to be known." In other words, "I created the whole of the universe, and the goal in all of it is to make Myself manifest. . . . "

The Strong Anthropic Principle has been stated by Steven Weinberg as: "the laws of nature should allow the existence of intelligent beings that can ask about the laws of nature."[30] Or as Willem Drees puts it, "any possible universe must have the properties for life (or intelligent and observing life) . . . it [the Strong Anthropic Principle] leads to an explanation of properties of the Universe in terms of purpose: a property that is necessary for life is necessary for the Universe."[31] However, before jumping to the conclusion that physicists have become mystics, it is necessary to point out that the Strong Anthropic Principle generally is linked with a many universes theory: all kinds of universes spring out of quantum fluctuations, each with varying laws of nature. Finally one comes along which is just right to support life, and fifteen billion years later humans are able to ask how the material creation came to be.

Thus even though the Strong Anthropic Principle comes close to the truth—seekers of Ultimate Reality exist in this universe because God wants to be sought—it obscures that knowledge by denying the existence of a conscious Creator. Once again, physics ends up in the endless loop that comes about when the human mind tries to explain the force that created the human mind. In the words of Drees, "Anthropic arguments employ something that we do not yet understand, complex life and consciousness, to explain something else."[32] Similarly, David Lindley describes the problem faced by proponents of the Strong Anthropic Principle:

We live in but a single universe, and we are asked to decide whether we live in the one and only universe that has been created, in which case we have to explain why that universe should have properties congenial to our existence, or [we live] in one of an infinite variety of universes, in which case the fact that we live in a congenial one becomes a tautology.[33]

To illustrate the nature of this tautology—or redundancy—Bryan Appleyard puts forward the Fridge Light Hypothesis: "Every time we open the fridge door there is a light on inside, so we speculate that there is always a light on inside. In reality, opening the door turns on the light—what we see is determined by our presence and our method of observation."[34] Many consider this to be a weakness of any philosophy which views consciousness as either the purpose, or inevitable product, of creation. From that skeptical perspective we see the light of consciousness, or mental cognition, only because we happen to live in a universe which, metaphorically speaking, has both a door that can be opened and a light that goes on when you do so.

This view holds that a fortuitous chance event caused the laws of nature to come together in just such a way as to permit both an ordered universe and the development of conscious beings. These beings then look outward at the universe, and are amazed to find order and consciousness. "Of course they do," goes the materialistic argument. "If things were otherwise, it would not be possible for matter to evolve into 'mind,' which actually is only highly complex matter. These deists and Aristotelians—who believe that the universe is designed to form the end product of consciousness—fail to understand that there are an infinite number of universes, most of which are not conducive to life."

In other words, lined up beside the refrigerator whose door we are able to open and perceive a light on inside, are an endless array of misshapen and malfunctioning refrigerators. Most do not even have a door that opens, and of those which do, only a tiny percentage have a light that works. Thus the consciousness which seems to permeate our universe is not an attribute of Ultimate Reality, but merely a happenstance event which once in a great while chances to make its appearance on the stage of physical existence.

This, of course, is purely a philosophical position not grounded in science. It is a last-gasp attempt to explain away the obvious fact of consciousness in the material creation. As David Lindley points out,

. . . proponents of the strong anthropic principle are multiplying whole universes in order to provide a home for us, and then, moreover, concealing those universes from us so that we can never see the extravagances enacted for our benefit.[35]

These materialists are unnecessarily complicating a truth that is as simple as it seems: consciousness is the very fabric of creation. We see the shining of that light because it always is on. Unlike the Fridge Light Hypothesis, the illumination of Spirit never burns out, and is there whether or not a human being opens his mind to look for it. Mysticism teaches that it is the objective essence of everything in existence, so there is no question of consciousness sometimes being present and sometimes being absent.

Parts do not make the whole

A similar conclusion was reached by physicist Menas Kafatos and social scientist Robert Nadeau in their book, *The Conscious Universe*. Since quantum physics has proven that reality is "non-local," as demonstrated by Bell's Theorem, this implies that whatever is found in a part of the universe, is contained in the whole. However, Kafatos and Nadeau point out in the quotation below that it is impossible for material science to prove conclusively that "reality-in-itself" is conscious by studying parts of that undivided wholeness. This is the province of mysticism, where understanding of the whole is accomplished by becoming it.

> If consciousness is an emergent property of the universe in the case of human beings, would not this also imply, given the underlying wholeness of the cosmos, that the universe is itself conscious? . . . We believe, providing of course that we refrain from viewing a conscious universe in anthropomorphic terms, that it does. Can we then leap to the conclusion that science has now proven that the universe is conscious? We cannot for the obvious reason that the quality of consciousness would be a property of reality-in-itself, or a property of the undivided wholeness which is "inferred" by our scientific description of the character of the universe but which is in no sense "disclosed" or "proven" to exist by that description.[36]

So again we find the new physics being able to hear an echo of Ultimate Reality, but unable to understand the nature of its source. Spiritual science is the sole means of knowing directly the truth which lies behind physical phenomena. For that truth is the all-pervading conscious energy of Spirit, which unites all of creation—material, mental, spiritual— into an unbroken wholeness. The general inability of mankind to come into contact with Spirit in no way diminishes the fact of its existence. After all, few people have the technical expertise to understand the mathematics of quantum mechanics, but this does not mean that those laws are an illusion.

Why, then, is it so difficult and rare for a person to merge his consciousness with Spirit, and thereby uncover the truth about Ultimate Reality? If God wants to be known, why does He keep himself so well hidden? Since mysticism teaches that spiritual domains of creation exist, what prevents this fact from being as well accepted and proven as the laws of physics? These are reasonable questions. Even though they have been addressed previously, the next principle of spiritual science will add to our understanding.

Principle 4.
Barriers and veils prevent easy passage between levels of creation, or states of higher consciousness.

Man is a mighty volume; within him all things are written,
but veils and darknesses do not allow him
to read that knowledge within himself.
The veils and darknesses are these various preoccupations
and diverse worldly plans and desires of every kind.
—Rumi[1]

Perfect mystics say that since it is impossible for our limited minds to fully understand why God created this universe, it is equally impossible to fathom why we are being kept in this lowest domain of existence. The two questions, after all, point in the same direction: the purpose of creation. However, even though the "why" of the physical universe is most difficult to comprehend, the plain fact is that it—and we— are here, separated from the primal divine unity. Since God-as-God is formless and nameless, complete without a second, that unbroken wholeness had to be divided into parts to reveal His qualities. Rumi says:

> So God compounded animality and humanity together so that both might be made manifest. "Things are made clear by their opposites." It is impossible to make anything known without its opposite. Now God most High possessed no opposite. He says, "I was a hidden treasure, and I wanted to be known." So he created this world, which is of darkness, in order that His Light might become manifest.[2]

Bound by chains of mind and matter

As the opposite of light is darkness, so does mysticism consider the opposite of freedom to be justice—or determinism. God possesses perfect freedom to act, or not to act, as He wills. After all, there is nothing other than Him, so who is there to limit His freedom? We, on the other hand, find barriers in our path at every step. As shall be discussed more fully under the next principle, our freedom is limited by the laws of cause and effect which prevail in the lower levels of creation. Call these laws "determinism," "cause and effect," "karma," "action and reaction," or "justice" (as in the Biblical injunction "eye for eye, tooth for tooth"[3]), the power they hold over us is transparently evident and undeniable.

Who in the world can say that he is free to do whatever he wishes? Create a universe. Destroy a galaxy. Establish different laws of nature. Become pure consciousness. Our will appears free to us only because we are so accustomed to being enslaved. Given the opportunity to look through the bars of our cell, or pace a few steps in that tiny enclosure, we are grateful for even those slightest tastes of freedom. Perhaps a memory lingers of when we had no eyes or feet at all, and were chained even more tightly to materiality.

Mysticism teaches that there are only two veils between our soul and God. These are mind and matter. As was noted previously, the essence of everything is Spirit, God-in-Action, but some things are purer manifestations of Spirit than others. Matter is by far the crudest form, and mind somewhat less crude. In our present state as human beings, these two entities surround the luminous consciousness of the soul as several layers of thick cloth cover a lamp. Only dim light shows through in either case. This is what allows the illusion of physical existence to continue. The darkness and ignorance in which we find ourselves is not an accident, or a veil which can be removed by an act of human will. They exist for a divine purpose, and so are not easily rent. When asked why we are imperfect, Charan Singh said:

> If we were not imperfect, this world would not exist today, we all would have returned to Him. Because the Lord wants this world to go on, He wants this creation, so imperfection is bound to be here. This world is the region of the *combination* of good and bad. . . . When He would like us to be perfect, we will become perfect, and we will go back and merge into Him.[4]

The physical universe, then, is more akin to a benevolent reform school than a punitive prison. The barriers of mind and matter which keep us here certainly are real and difficult to remove. But they exist so that an intense longing for freedom may develop. The more onerous the conditions of imprisonment, the greater becomes our desire to return Home. We are, say the perfect mystics, strangers in a strange land. Visitors to a foreign country. Birds roosting in a treetop for a single night before flying on. Those who are happy in the material creation are not motivated to search for God. Like Rumi's chickpea, we have to boil for a time in physical existence so as to become subtle and tender enough to merge in the ethereal Spirit—and be carried by that wave back to the ocean of consciousness which we left long ago.

Many people have a desire to sojourn, now or after death, in one of the regions of creation above this universe, but below God-as-God. These commonly are referred to as "astral" or "mind" planes, where matter as we know it is absent. This desire is a mistake if one's goal is Ultimate Reality, the region of pure Spirit. Charan Singh notes in this regard that "we make much better spiritual progress by being in this human form than from inside these astral planes, because here we see so much misery around us that we want to escape from it. . . . We do not want to remain here, so we try to work hard to go back to Him. Without intense longing we cannot merge back into Him. And there [astral plane] since there is no misery, no birth nor death, the yearning to go back is very little."[5]

Rehabilitation before freedom

The barriers which prevent us from reaching higher domains of consciousness exist for our ultimate good. Here in the United States many prisons are terribly overcrowded. Felons often are set free before their sentence is up to make room for someone else, or because a judge has ordered that the crowded conditions in prison are inhumane. Frequently one reads in the newspaper that within a few days after a prisoner was released, he committed another crime, sometimes worse than the one for which he was first sent to jail. This is what happens when freedom is granted before rehabilitation is complete. For a like reason, removing the coverings of mind and matter which veil our soul cannot be rushed. Perfect mystics liken this process to separating a fine silk cloth from the thorny bush in which it has become enmeshed. To avoid ripping the cloth to pieces, it must be gently and carefully disentangled from the thorns. This takes time.

> *God has created these veils for a purpose. For if God's beauty should display itself without a veil, we would not have the power to endure and would not enjoy it. Through the intermediary of these veils we derive succor and benefit.*
> —Rumi[6]

But mysticism teaches that the intended purpose of mind and matter— to serve as intermediaries between our ethereal soul and the material creation—has become subverted. It is meaningless to ask whether this subversion is part of a divine plan, or the result of human ignorance. That question cannot be answered with our current limited understanding, even though the source of evil is a quandary which has occupied philosophers and ethicists for millennia. Mysticism takes a practical approach to this problem: whatever the reason for man's descent into the darkness of mind and matter, the fact is that we are here. To get out of this mess, we need to understand only what keeps us stuck in the mire—not how we became embedded in the first place.

Rumi says, "By the time intellect has deliberated and reflected, love has flown to the seventh heaven."[7] This statement describes in a simple and poetic manner both the barrier that prevents the seekers of Ultimate Reality from accomplishing their goal, and the means by which that obstacle is removed. "Intellect"—with a small "i"—is Rumi's term for the limited human mind. "Love" is a synonym for unity. Our minds are what veil us from final truth. Every thought, image, desire, concept or emotion that enters our mind stands between us and the unbroken wholeness of reality. Perfect mystics tell us that this fact derives from the objective structure of creation, and is not a subjective philosophical position which simply can be wished away. This structure came about through God projecting Spirit, Spirit forming universal mind, and that mind projecting itself as the physical universe.

Mind—a mechanical force of nature

Hence, mind—both in its limited and universal forms—is the only barrier between our soul and God, the only force that prevents the drop from returning to the ocean. Yet the barrier of the mind initially appears to be the means of knowing Ultimate Reality. This is the primary source of confusion for both spiritual and material scientists who seek final knowledge about existence. When we use our mind to try to penetrate through the veil of our mind, the altogether predictable result is circularity. Dog-chasing-tail. Tautology. Gödel's Theorem. Uncertainty principle. Baby universes that are their own parents. Narcissus staring at his own reflection. Perfect mystics veritably scream this fact at us from the rooftops: *mind is the primary barrier to truth, not the means to it.*

Unfortunately, what hears this warning about the mind is—what else?—the mind. It may seem to agree with the logic of the argument I have just stated, but strenuously resists being demoted from its dominance over the soul. Mysticism advises: do not listen to it; fight it. The mind is mechanical, not conscious. It has become habituated to sensual pleasures and intellectual reasoning rather than the pursuit of truth. The soul and Spirit are the only forces which are more powerful than mind. They *can* subdue it, and thereby eliminate the primary barrier to knowing Ultimate Reality. But this requires a battle. Yes, God wishes to be known. However, the only way to know Him is to remove the coverings from the soul, which is not easy. If it were so simple to do, our immersion in this vast creation would not have been necessary.

> *The prophets and saints do not avoid spiritual combat. The first spiritual combat they undertake in their quest is the killing of the ego and the abandonment of personal wishes and sensual desires. This is the Greater Holy War. . . . All eyes*

and ears are shut, except for the eyes and ears of those who have escaped from themselves.

—Rumi[8]

It is important to understand that spiritual science views the mind as another force of nature. We need to think about this carefully. Material scientists use their minds to learn about the workings of physical forces: electromagnetism, gravity, and so on. Physicists believe that one day they will arrive at a Theory of Everything that unifies those forces into a single "superforce." Since this superforce was responsible for producing everything that we now find in the universe, which naturally includes the minds of physicists and other humans, those minds themselves must be an offshoot of the original unified force. The only alternative is to assume that our limited human mind is identical to whatever was present at the instant of the Big Bang. Since mankind now finds itself neither able to create another universe, nor even to understand how and why creation took place, it appears evident that the human mind is a product of some higher force—and not that force itself.

So to arrive at a true Theory of Everything, or God, mysticism teaches that it is necessary to learn how to unify the force of the mind of man into a more inclusive force. Physics, as shall be discussed in a following section, has taken the same approach with material entities: electromagnetism and the weak nuclear force once were thought to be separate and distinct, but now have been shown to be part of a single electroweak force which is apparent only at high energies. In other words, when spiritual science says that mind is the only barrier to knowing God, one interpretation of this statement is that the human mind is a low energy manifestation of the unity that is Spirit, God-in-Action.

Just as the unified electroweak force is not evident at normal energies—which permits electromagnetism and the weak force to exist as separate forces—neither does the all-pervading conscious energy of Spirit manifest on the same plane of consciousness as mind and matter. This would be like using an electric current to try to reproduce the force of the Big Bang. No matter how many generators are hooked together, their electrical output never would equal that original creative energy. And even if they *could*, at that point their combined energy would not be electricity, but something else—the superforce. This analogy points to why the human mind cannot be used to know the Universal Mind, or why Universal Mind is inadequate to comprehend Spirit. In either case, a lower-order of consciousness must be raised to a higher-order—at which point it no longer is what it used to be.

God most High is not contained within this world of ideas, nor in any world whatsoever. For if He were contained within the world of ideas, it would necessarily follow that he who formed the ideas would comprehend God, so that God would then not be the Creator of the ideas. Thus it is realized that God is beyond all worlds. —Rumi[9]

Contemplative meditation leads to freedom and truth

Mysticism uses the research tool of contemplative meditation to eliminate the barriers which stand between our normal state of consciousness and knowledge of Ultimate Reality. The four states of being which we have previously discussed will help us understand how this is accomplished. As Rumi implies, the world of sensory observation, the world of thinking, and the world of imagination all must be left behind in order to enter the domain of God. Recall that our model of consciousness looks like this:

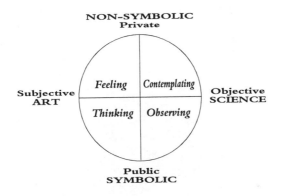

Figure 4. Consciousness divided into the four states of being

I noted earlier that the state of being known as *contemplating* is unknown to most people. This is the private and objective domain of consciousness realized through contemplative meditation. Hence this is a more accurate depiction of the states available in everyday human awareness:

This is the point from which the beginning student of spiritual science starts his search for Ultimate Reality. His research hypothesis is that this truth can be found only in a state of contemplation. However, at present his consciousness is occupied in experiencing the other states of being. As has been noted, mysticism teaches that contemplation is a unification of those lower forms of consciousness. Thus it is necessary to merge these subsidiary forces of observing, thinking, and feeling by changing one's focus of attention.

Stop observing

The first to be transcended in contemplative meditation is "observing." This refers to observations of the outside world made by the five senses. By shutting the eyes, sitting in a quiet place, and ceasing to pay attention to either the external environment or one's own body, this state of being is temporarily deactivated. Just as flipping a breaker switch causes all of the electricity in one area of a house to be shut off, so is the flow of consciousness which used to be directed outward through sensory observations. This results in:

Keep in mind, of course, that in this discussion I am speaking of the ideal, or goal, of the student of spiritual science. Those who have attempted any form of meditation know how simple it is in theory, and how difficult it is in practice. Just as it is rare for any sophisticated experiment in material science to work perfectly the first time it is attempted, so does early progress in meditation generally proceed in fits and starts, three steps forward for every two back. At some point, however, all of the student's attention will be inside, as opposed to being spread out in the outside world via the senses. Each of us, of course, accomplishes this feat every time we fall asleep—the only difference being that in sleep we lose awareness of the external world and fall to a lower state of consciousness, while in meditation the aim is to remain awake and completely aware.

Many people do experience being inward and awake, but usually only when thinking intensely about something. Rodin's famous statue of *The*

Thinker beautifully captures the mood of this state. Nothing exists in the mind of the deep thinker except what is being thought about. (This is the condition of the "absent-minded professor" who forgets what day it is, or wears his bedroom slippers to class instead of his shoes because his attention is absorbed on a particular problem.)

Stop thinking

Mysticism teaches that this degree of concentration is admirable, but in contemplative meditation it is necessary to focus on only one thing—not many. For our flow of thought invariably is symbolic, hence subjective. It does not lead to objective knowledge. In addition, there is nothing to think about other than what has come to us through the senses over the course of a lifetime. Even abstract concepts ultimately are the product of reading and hearing the words of others with our eyes and ears. Yet if thoughts could come to us directly from some higher domain of existence, and perhaps they can, these too would be merely a symbolic reflection of that realm. Sitting in a cold auditorium in the dead of winter and watching a narrated slideshow about tropical islands is not at all the same as actually lying on a warm beach with palm trees swaying overhead.

So thinking also has to cease. As was noted earlier, this generally is accomplished by repeating a word, or words, which have no connection with anything in the physical universe. This minimizes the opportunity for the mind to produce thoughts regarding those words, while occupying the thinking faculty in a focused manner. Perfect mystics tell us that without such repetition it is impossible to cease thinking for more than a brief time. Without some mental activity to occupy it, the mind—being a well-functioning machine—will continue thinking about what it habitually ponders: work, family, religion, the opposite sex, sports, politics, philosophy, science. You name it, at this moment someone in the world is thinking about it. Repetition of a word or words is the only way to stop this ceaseless, and largely purposeless, chatter in our minds. This produces:

Stop feeling

Feeling, being subjective and non-symbolic, is difficult to describe and can take a number of forms. Emotions are perhaps the most obvious manifestation of this state of being. Eliminating these from one's consciousness serves to differentiate contemplative meditation from what might be termed ecstatic religion. Many people claim to have felt the presence of God, or one of His messengers, during some sort of ineffable "mystical" experience. Almost always, this feeling has nothing to do with a genuine state of higher consciousness, as such is beyond human emotion. This is not to say that true contemplative meditation is devoid of sensation, for mysticism teaches that perfect bliss and love is an objective attribute of the purely spiritual domains of existence. But this is a different state from feeling "tears of joy," or "waves of happiness," or whatever other purely human emotions are felt in a lower form of spiritual experience.

A more subtle form of feeling is simply being aware of oneself as an independent and self-willed entity. This is closely related to "ego," or sense of "I-ness," which mysticism regards as being a great barrier to reaching higher domains of consciousness. Charan Singh writes:

> It is our ego that gives rise to our ills, and leads to our continued return to the world of phenomena, and to our recurring suffering and misery. Once we get rid of this ego, which stands as a separating wall between us and the Lord, we go beyond the domain of matter and mind; we are entered into the refulgence of the Lord. The drop merges into the ocean and itself becomes the ocean.[10]

Deflate the puffed-up ego

Volumes could be written about the workings of the *ego*—which in spiritual science has a different connotation than the Freudian, or psychological, use of this term. This subject cannot be adequately addressed here. However, everyday language comes close to explaining how mysticism regards this attribute of the mind. In talking about someone else, people may say, "He's got a healthy ego." Usually this is meant more as a compliment than as an insult. It means that the person referred to has a sense of self-confidence, of knowing his capabilities and using them well in daily life. On the other hand, saying "He's an insufferable egotist" has a different meaning. That person is considered to have an exaggerated self-importance, believing that he is better than other people and deserves all the credit for his marvelous qualities and accomplishments.

Similarly, mysticism recognizes that some feeling of personal identity is necessary to function in this world—and indeed any world below complete union with God. After all, if unity is not present, separateness is. And this requires that conscious beings be aware of their individuality. Otherwise there would be no longing to assuage the feeling of loneliness, which we all have to some degree or another, by merging our identity with God's unbroken wholeness. That is the final cure for loneliness. Otherwise, individuality is inevitable on lower planes of existence.

> *Forgetfulness of God, beloved,*
> *is the support of this world;*
> *spiritual intelligence its ruin.*
> *For Intelligence belongs to that other world,*
> *and when it prevails, this material world is overthrown.*
> —*Rumi*[11]

It is only when our ego becomes unnaturally dominant over other faculties of the mind that a problem arises. Julian Johnson writes: "The normal ego is all right, but when it begins to swell up out of all proportion, then it takes on the nature of a disease. . . . That faculty, which is quite necessary for the preservation of the individual in this life and for the proper placement of that person in relation to all others, becomes so overgrown that the normal-self becomes for him the center of the universe."[12]

. . . *the normal-self becomes for him the center of the universe.* This description shows how widespread is the disease of ego. For virtually everyone in this world has a sense that life revolves around them, or at least it *should.* "Egotistical" people certainly are at the far end of the spectrum, and stand out for that reason. But compared to the ideal of spiritual science, almost all of us have far too much ego for our own good. In mysticism this is less a moral judgment and more a reflection of scientific fact. Ego is ignorance, not evil. The plain truth is that no one has anything to be proud of, other than God, who is beyond pride. If we had created this universe, there might be a reason to puff up our chest as we admire the wonders that we brought into being. In actuality, our individual existence is entirely due to causes beyond our control, and matters not a whit in the overall order of things. In the words of Charan Singh:

> Our ego and pride stand in our way and make the mind powerful and strong. It is humility which will rid us of our ego and self-importance. All Saints teach us this lesson in their writings. What is our value after all? What is the individual's value in this vast creation that the Lord has created? Billions of universes like ours lie within man, the microcosm in this macrocosm. Our existence, that of the individual, has absolutely no value and we should not attach much importance to ourselves.[13]

In the course of discussing the next principle of spiritual science we shall find that ego and free will are closely related. As commonly understood—or rather misunderstood— by the general public and material scientists alike, these apparent properties of human consciousness stand as the greatest barriers to knowledge of Ultimate Reality. In essence, they are tricks played by the Master Magician, God, in order to keep this physical universe running smoothly as a reform school for our wayward souls. Ego is one of the forces which serves to forestall a premature, or accidental, escape from the material creation. Just as inmates in a penitentiary are allowed to know the rules which govern the inner workings of prison life—when meals are served, how privileges are earned by good behavior, visitation procedures—the laws of nature which control the *internal* operation of our universe are relatively open to discovery.

However, no inmate is told how the guards patrol the outer walls of the prison, or all of the security measures which prevent escape. Neither is he given a key to the cellblock, or issued a pass that enables him to walk out of the front gate—at least until his time has been served. In the same manner, mysticism teaches that the forces which keep human consciousness limited to knowledge of this physical universe (Level 1 of a five-level creation) are not within our power to overcome unless the gatekeeper, God, assents to our release. This is a matter of grace.

Mysticism has more to say on this subject, but the main point is that some outside force is necessary to break us out of the domain of mind and matter. Scientifically speaking, this is the message of Gödel's Theorem: truths exist which cannot be proven to be true within any organized system of logic. Complete truth only can be attained by transcending the limits of that system. Similarly, Spirit or God-in-Action is the only force which can free us from the confines of our limited consciousness. How this occurs in contemplative meditation will be the focus of the final principle.

Come alive by dying

For now, we need to return to the point at which the student of spiritual science has succeeded in reducing his sense of ego. Having ceased observing the outside world, thinking about his inside world, and feeling that he is an independent entity capable of self-willed action, there comes a point when all of those three states of being—observing, thinking, feeling—are essentially inoperative. This occurs when the energy of the mind is completely focused on repeating the words which occupy one's attention in the initial stage of contemplative meditation. In this way consciousness is withdrawn from the material world, and becomes poised to enter higher domains of existence. By merging into the repetition of those words which have no connection with the physical universe, the limited human mind takes a large step toward the much vaster power of Universal Mind.

Charan Singh says in this regard, "Your mind should merge into the words. . . . It must merge along with the words. You should be *in* those words, not somewhere away from them."[14] This eliminates the subject/object dichotomy which we shall find in the following section to be at the root of quantum physics' confusion about the ultimate nature of material reality. That is, from Principle 1 we know that both mysticism and physics recognize that an unbroken wholeness is the subtle essence of everything in existence. Yet our sense of ego leads us to feel that we are a part, not a whole. Clearly there is a problem here for the seeker of final truth. If reality is a whole, and I am a part, then there are only two possible ways to bridge that gap: the whole becomes me, or I become the whole. However, Rumi notes that one of these ways is closed:

> *With God there is no room for two egos. You say "I," and He says "I." In order for this duality to disappear, either you must die for Him or He for you. It is not possible, however, for Him to die—either phenomenally or conceptually—because "He is the Ever-living who dieth not." He is so gracious, however, that if it were possible He would die for you in order that the duality might disappear. Since it is not possible for Him to die, you must die that He may be manifested to you, thus eliminating the duality.*[15]

Thus it is necessary for the student of spiritual science to die to himself in contemplative meditation. The result would appear to be:

Nothing. Nonexistence. Emptiness. What possibly could remain in consciousness after observing, thinking, and feeling as we know them have been removed? Yes, those words which are the initial focus of concentration are being repeated by the mind, which has merged itself so completely into them that essentially it has become them. But what kind of existence is this to be mere words? Thoughts such as these discourage some people from taking up the practice of deep contemplative meditation. Others meditate only half-heartedly, consciously or unconsciously fearing this apparent state of non-being. The happy truth, however, is that this takes place:

Contemplating spiritual realities

As soon as one's attention is diverted completely from those states of being which have a connection with the external material world, or subjective imagination, the higher faculty of *contemplating* becomes fully operative. In this state, access to "inner" domains of creation is possible. These realms are objectively real, but non-symbolic in regards to human language or mathematics. Since the essence of Ultimate Reality is consciousness, knowledge of the all-pervading conscious energy which sustains and enlivens the physical universe is attained only through contemplation. We must become non-material consciousness to understand the force of Spirit which preceded and created matter.

A central tenet of mysticism is reflected in the First Law of Thermodynamics: energy cannot be created or destroyed, only shifted from one type to another.[16] Similarly, consciousness cannot be created or destroyed, only shifted from one state of being to another. Since the essence of everything is consciousness—God-in-Action—being able to create or destroy consciousness would be tantamount to being God. This is why spiritual science says that suicide is not only wrong, but senseless. Killing one's body does not remove the true source of suffering—the mind—which continues to exist after physical death. The materialistic world view which holds that mind is produced by matter, and soul does not exist at all, is untrue. Each person, of course, will be able to test this hypothesis for himself when his body dies.

The only other means of determining whether consciousness exists independently of the brain and other matter is through contemplative meditation while one is alive. Mysticism calls this "dying while living." This research approach of spiritual science is eminently logical and practical: shut down those states of being which relate either to the physical world or one's subjective self, and see what remains. If it is true that consciousness simply changes form when released from its attachment to mind and matter, the student of mysticism only has to be quietly aware of that new state in which he finds himself. This will be an immediate and unmitigated awareness, not requiring the transfer of symbolic knowledge, since it is *being* itself which has been raised to a higher form. Julian Johnson describes what occurs when the state of contemplation has been reached:

> When such concentration has been gained, the attention is all inside. The whole of the mind and soul have left the outer world and gone inside. Only the inner worlds exist for us, the outer world having been completely shut out from our consciousness. Then we go on holding our attention at the inner center. Slowly and gradually the soul and mind gather all their forces at that inner center and, finally leaving the physical world entirely, penetrate through some inner aperture and enter a higher region. We may call it a higher dimension, at that moment the soul passes through the inner "gates of light" and steps out into a new world.[17]

Mysticism is less "mystical" than physics

To some this undoubtedly sounds fantastic, unbelievable, wishful thinking—a mental deception that occurs entirely within some sort of personal dream state. But the well-defined character of the "experiment" of contemplative meditation by which mystical knowledge is acquired, and the fact that whoever diligently undertakes that experiment arrives at the same "findings," is as much proof as any other science is expected to provide a skeptic.

Further, it has been noted that the new physics—the pinnacle of material science—is not at all loath to theorize about domains of existence other than our material universe. In quantum cosmology, physicists raise the possibility that countless physical universes exist beyond the confines of ours. Superstrings, the leading candidate for a Theory of Everything in physics, requires that reality be composed of ten dimensions—six in addition to the three of space and one of time in which we live. A popular "many worlds" interpretation of quantum physics assumes that every instant we, and everything else, branch off into innumerable parallel universes—all but one of which is inaccessible to us. Other mathematical theories in physics posit features which cannot be perceived in everyday life, such as Steven Hawking's "imaginary time" alluded to previously.

Does it not seem strange that physicists have no problem with seriously considering theories that entail planes of existence which supposedly are real, but cannot be known directly—while criticizing spiritual science for being "mystical" because it teaches that other domains of reality exist which can be known through contemplative meditation? Which discipline is being more other-worldly: the one which theorizes without offering any means of determining whether those suppositions are true, or the one which puts forward a concrete method by which the validity of its assertions can be tested?

Believe it or not, the barriers which prevent us from knowing Ultimate Reality are not immutable. The primary obstacle in our path is our own mind. We have become so accustomed to relying on this mechanical means of obtaining limited knowledge about a small part of the physical world, that we believe the human intellect is the highest possible form of consciousness. Mysticism says, "This is wrong. Your thoughts about reality are what prevent you from knowing the truth about it. Pure consciousness is the essence of Ultimate Reality. Become that essence, and you have attained everything—including God. Otherwise, you and all that you know through your limited mind are nothing."

> *Fear the existence in which you are now!*
> *Your imagination is nothing, and you are nothing.*
> *A nothing has fallen in love with a nothing,*
> *a nothing-at-all has waylaid a nothing-at-all.*
> *When these images have departed,*
> *your misunderstanding will become clear to you.*
> —Rumi[18]

Echoes in the new physics

What echoes can be heard in physics of this principle of spiritual science? *Barriers and veils prevent easy passage between levels of creation, or states of higher consciousness.* Material science is not concerned with any level of existence other than this physical universe. Likewise, it views the everyday mental activity of observing, thinking, and feeling as the highest state of consciousness mankind can attain—assuming that even this can be called "consciousness" and not simply a byproduct of complex neural connections in the physical brain. So at first glance this principle would not seem to be reflected in physics. Or if it is, it would have to be restated to refer to barriers which material scientists face in using their minds to lay bare the truth about the physical universe.

Only two small clouds . . .

Up to the twentieth century, the existence of even this sort of barrier would have been denied by physicists. As has been noted by many historians of science, the late nineteenth century was marked by an unbounded—and unfounded—optimism that physics was on the verge of knowing all there was to be known about the universe. In the words of Kafatos and Nadeau:

Toward the end of the nineteenth century, Lord Kelvin, one of the best known and most respected physicists at that time, commented that "only two small clouds" remained on the horizon of knowledge in physics. In

other words, there were, in Kelvin's view, only two sources of confusion
in our otherwise complete understanding of material reality. . . . What
Lord Kelvin could not have anticipated was that efforts to resolve these
two anomalies would lead to relativity theory and quantum theory, or to
what came to be called the "new" physics.[19]

What a difference a century makes. In the late 1800s, the old physics
seemed to be two steps away from complete knowledge of reality. By the
late 1900s, the new physics essentially has admitted that the methods of
material science *never* will be able to reach the complete truth about
physical existence. Kafatos and Nadeau note that in Lord Kelvin's time,
"some established physicists were encouraging those contemplating
graduate study in physics to select other fields of scientific study where there
was better opportunity to make original contributions to scientific
knowledge."[20] In other words, there was so little left for physics to discover
that the physicists already on duty could tie up the few theoretical loose
ends without any help from the next shift.

Physics throws in the towel

Presently physicists are being quoted as saying that their discipline
appears to be barred from generating experimental evidence that could
confirm the penultimate Theory of Everything. Howard Georgi observes,
"Now I'm afraid we're going to have to be satisfied with data that convince
only those who were already convinced. . . . It's a permanent problem;
we're running up against a fundamental limit of nature."[21] Just as in
Kelvin's era, some undergraduate science students reportedly are being
advised to shun physics, but for the opposite reason: brain science, chaos
theory, computer science, or even Wall Street may be a better career
choice. This is not because there is so little left to learn in physics, but—as
John Horgan says—" . . . physicists face the possibility that their journey [to
a Theory of Everything] might end short of its destination."[22]

Kafatos and Nadeau conclude that the contest is over between material
science and other approaches, such as mysticism, to knowing the truth
about Ultimate Reality. Physics is being forced to throw in the towel:
"What seems clear at this point in time is that science must drop out of this
competition—it cannot, in principle—provide a description of reality-in-
itself."[23] During the past one hundred years, what caused physicists to
become so pessimistic about their chosen discipline's ability to reach final
knowledge? In previous sections we explored some reasons for this
turnabout, such as Gödel's Theorem and the fading away of solid matter
into ephemeral quantum phenomena. Here this question will be addressed
from a more fundamental perspective: what is the central barrier which

prevents the human mind from knowing the complete truth about *any* aspect of physical existence?

One barrier to complete truth

Atoms, the brain, biology, the Big Bang—only one veil prevents us from learning the complete truth about each. As was discussed in the preceding section, we shall find that this is the same barrier which confronts the spiritual scientist in contemplative meditation. I will argue that just as mysticism and physics agree that Ultimate Reality is an unbroken wholeness, so do these branches of science agree on what prevents us from knowing that unified truth. However, the thread of this argument will take a few pages—and patience on the reader's part—to unravel. We will be moving into some areas of physics which confuse even professional physicists, and I want to proceed at a pace which will be manageable to a reader who has no previous acquaintance with quantum mechanics.

To those who are tempted to skip this section, I offer both a piece of advice and an "out." The advice is to stick with it, because even those whose bent is more toward spirituality than physics should find that their understanding of mysticism is furthered by the discussion that follows. After all, the same reality underlies both spiritual and material science. A deeper knowledge of one can only strengthen faith and confidence in the other. However, if you become frustrated with the analysis which follows, this is the "out"—a two-sentence synopsis of the rest of this section: *a part can know only about other parts, never the entirety of the whole; to know the whole, become without parts.* This sounds entirely mystical, which it is.

These sentiments also are in accord with the traditional interpretation of quantum physics, particularly the first sentence. The second sentence is, in my opinion, a logical extension of the first—and only fails to be an accepted principle of the new physics because that discipline does not believe that it is possible for the human mind to be without parts. My case is bolstered by the conclusions of a book, already cited, written by a physicist and social scientist: *The Conscious Universe: Part and Whole in Modern Physical Theory.* I will draw heavily upon the thoughts of those authors, Menas Kafatos and Robert Nadeau, in the step-by-step argument which follows. They are noteworthy exceptions to an all-too-common tendency among writers on the new physics to either avoid evident metaphysical questions, or to address them in a shallow manner.

A symbol for every part

The old physics, and material science in general, used to believe that every part of the universe could be described by a symbol, or a representation of reality. Physics used to be based on a one-to-one correspondence between observation and mathematical theory. Sense something and assign a number to it. Or assign a number to it, and then sense it. Progress in material science occurred in both ways. Sometimes experiments would suggest a theory to explain observations, and sometimes theories would suggest experiments to sense previously unknown phenomena. Advances in modern particle theory have come about through the latter approach: high-energy particle accelerators often find new, or rather undiscovered, subatomic particles after some purely theoretical breakthrough suggests where they should be found.

Yet regardless of whether observation preceded theory, or the other way around, physicists assumed that—like in an orderly workshop— everything in the universe had a place, and there was a place for everything. By "place" was meant a nice tidy symbol that represented a corresponding physical reality. A dining room table has all kinds of things on it: plates, cups, forks, spoons, salt and pepper shakers, candlesticks. Each and every one can be described by a word (I've just done that), or if you like—a number (first plate = P1, second plate = P2, first cup = C1, and so on). Nothing on the table lacks a symbol. If a dinner guest points at something and asks, "What's that?", the host never says, "I'm sorry, I wish I could tell you, but there's no way to express what it is."

Kafatos and Nadeau note that at the root of this belief of the old physics was a dualistic assumption about mind and matter: "Observed systems in classical physics were understood as separate and distinct from the mind that investigates them, and physical theory was assumed to bridge the gap between these two domains of reality with ultimate completeness and certainty. The measuring instruments were simply more refined means of gathering sensory input, and the expectation was that carefully controlled experiments would, in principle if not always in fact, confirm a one-to-one correspondence between every element of the physical theory and the observed reality."[24]

Round up all the symbols to corral the whole

This belief—that everything in the universe could be represented by a symbol— led physicists to be confident that eventually the whole universe could be known by adding up its separate parts. This confidence was understandable, and entirely justified, given the scientific facts that had been accumulated up to the early twentieth century. Until that time, physical theories had been remarkably successful in explaining phenomena. For example, even though in the 1800s no one could see electricity or knew exactly what it was (the same is true today), equations had been developed which enabled scientists to understand how electricity was generated and moved through electric currents. Even if you could not see it, you could label it. And since all physical phenomena could, in theory, be "branded" with a symbol, at some point there would be no stray parts of reality running around outside of human knowledge. The human mind would have succeeded in completely corralling material existence within the theoretical fences of physics.

Recall this previously-cited quotation from Stephen Hawking: "It turns out to be very difficult to devise a theory to describe the universe all in one go. Instead, we break the problem up into bits and invent a number of partial theories." This is the approach which held sway in physics during the nineteenth century, and continues to be accepted by many physicists today despite conclusive evidence that it cannot succeed in knowing the truth about Ultimate Reality—only limited reality. Nevertheless, it is an appealing notion. If you are dining at a table so large that all of it cannot be seen from your seat (perhaps you are in the vast banquet hall of a king), then one would think that you still could make a list of everything on the table by starting at one end, and systematically writing down what you observe while making your way to the other end.

Or to make things easier, you and two companions could decide to divide the table up into thirds, with each person enumerating what they find on their own section. Combining the three lists, according to the classical view of physics, would result in a complete and accurate description of the entire dinner table. Since each person studied a different section of the table, and those sections were distinct from each other, adding up the contents of each would result in neither any overlaps nor any omissions. "One hundred plates in the first section, plus ninety plates in the second section, plus one hundred ten plates in the third section equals three hundred total plates." None of the plates have been double-counted, because obviously it is impossible for one plate to be in two places at the same time. And every plate *has* been counted, because each is open to the view of a careful observer.

A belief that distinct regions exist in space is central to both plate-counting and classical physics. Kafatos and Nadeau write: "That belief allowed one to assume that the whole could be described by the sum of its parts, and thus the ultimate extension of theory in the description of all the parts would 'be' the whole. To put it differently, it was presumed that reductionism was valid, and, therefore, that one could analyze the whole into parts and deduce the nature of the whole from the parts."[25] Thus one support of classical physics was that every part of physical reality could be assigned a symbol, and another support was that those parts were separate and distinct from one another. Without that second assumption, of course, parts do not really exist. It would have been difficult for physicists to believe that a whole was the sum of its parts if those parts could not be clearly identified. Labeling, assigning a symbol to something, presupposes that the "thing" actually exists as a separate entity.

Whoops! Physics' grand plan collapses

Quantum physics demolished both beliefs: that everything in the universe could be represented by a symbol, and that reality consisted of separate parts. Previously it was noted that discovering the laws of quantum physics solved some perplexing problems in material science, while raising new questions. Solved, for example, was the problem of what kept orbiting electrons from rapidly radiating away electromagnetic energy and spiraling down into the atomic nucleus. Subatomic particles gain and lose energy not in a continuous manner, but in discrete steps—quanta—like a person walking up and down a staircase. Just as he must pause on one step or another, since it is not possible to stand on the vertical area in-between steps, so are electrons forced by the laws of quantum mechanics to orbit only in certain energy levels. When the lowest "step" is reached, an electron can go no further. This makes possible the stable existence of the physical universe.

Raised, on the other hand, were questions such as: how does an electron move from one orbit to another without passing through the space in-between? When a person walks up a flight of stairs he can pause only on the steps, but his feet move continuously from step to step, passing through the vertical space that separates each step without stopping there. Electrons, though, were found to take a "quantum leap" from orbit to orbit. Now here, now there, without any discernible movement in time or space that linked the here and there. This behavior was unlike anything that had been observed in classical physics.

Some of the most brilliant minds in physics—Planck, Bohr, Heisenberg, de Broglie, Schrödinger—eventually pieced together the answer to this puzzle. The key is the distinction between a wave and a particle. In physics a *wave* of matter or energy or information, just like an ocean wave, flows continuously in time and space. It is an unbroken whole. A *particle*, however, is like a fleck of foam cast off by a breaking ocean wave. It is localized in time and space, existing as a separate entity. Quantum physics found that every subatomic particle, such as a photon of light, sometimes acts like a wave and sometimes like a particle. This is what produces the puzzling behavior of the quantum leap.

Ghostly whales and real wasps

Physicist Nick Herbert likens the situation to a whale and a wasp—understanding that while a whale and a wasp both are animals, a wave and a particle are very different creatures. He writes:

> Quantum theory suggests that before we measure the particle (wasp), it's not a particle at all but something as big as a whale (wave) . . . . Imagine that the whale dwells not in the real world but on the spirit plane; the wasp is real. . . . Whenever a measurement occurs anywhere in the world,

something like a ghostly whale (immense, insubstantial, permeable, and wavelike) turns into something like a real wasp (minute, substantial, and particle-like). . . . In terms of the whale/wasp analogy, the quantum reality question divides into two parts: 1. what is the nature of the whale? 2. how does the whale change into a wasp? The quest to describe the whale is called the "quantum interpretation question". . . . The matter of how whale becomes wasp is called the quantum measurement problem. . . . [26]

Both of these questions—quantum interpretation and quantum measurement—sounded the death knell for the belief that a one-to-one correspondence could be made between everything in the universe, and an associated symbol. Even though these questions remain unanswered—despite an abundance of conflicting theories that have been put forward—their very existence is what is important. No matter what the answers to them are, they will not involve the sort of tidy solid reality which the old physics assumed to exist.

For the quantum wave generally is considered to be a wave of information, rather than a physical entity. Physicist Fritjof Capra says, "The waves associated with particles . . . are not 'real' three-dimensional waves, like water waves or sound waves, but are 'probability waves'; abstract mathematical quantities which are related to the probabilities of finding the particles in various places and with various properties."[27] Capra notes the upshot is that a subatomic particle "manifests a strange kind of reality between existence and nonexistence. . . . It is not present at a definite place, nor is it absent. It does not change its position, nor does it remain at rest."[28] This is due to the wave-particle duality evident in quantum physics. When an entity is not being observed, or measured, it has the qualities of a wave. When it *is* being measured, that entity appears to be a particle.

Thus the classical one-to-one correspondence between reality and its symbolic representation breaks down in this way: the behavior of the quantum "whale"—or wave function—can be determined precisely. Kafatos and Nadeau write, "If, then, we look at the quantum reality from the perspective of wave mechanics, we have a mathematical formalism describing a wave propagating deterministically with well-defined characteristics. In the absence of observation, there is a complete correspondence between every element of the physical theory and the physical reality in the classical sense. When, however, we make a measurement to determine what is actually there, some of the possibilities appear, and others disappear."[29]

In other words, quantum theory holds that subatomic particles—and indeed everything in the universe, to some degree—only have tendencies to exist in a particular state, such as in a certain place and moving at a certain momentum. The wave function is considered to describe those probabilistic tendencies completely and exactly, as in "that particle has a sixty percent chance of being there, and a forty percent chance of being here." Thus the particle supposedly is neither here nor there until a measurement is made, at which time it is found definitely to be in one place or the other.

From the standpoint of knowing Ultimate Reality, then, the problem is obvious. Returning to the dinner table analogy, the quantum-level "plate counter" still is able to see what is on the table when he looks at it. But his knowledge of the table's contents now is less secure than it was in the classical state of affairs. Yes, he can still count the number of plates. However, he cannot assume that he has reached the rock-bottom level of dinnerware reality.

For the plate-counter now knows that some other force which he cannot observe directly, the quantum wave function, is responsible for giving the plates their seeming solidity. And because the position of the plates only can be determined with a certain probability in the absence of observation, the quantum plate-counter has no guarantee that the next time he looks at the table, its contents will be in the same places as they were before. There even is a small possibility that one of the plates will have vanished completely and found its way to underneath the table—a phenomenon known as quantum tunneling, which is produced by quantum leaps across a barrier. If the plates were quantum objects, you could count them once and get a total of 300. Count them again, and there might be 299 or 301, due to the tunneling effect. This is not the way to Ultimate Reality, where truth is permanent.

And if one wanted to go beyond mere quantum plate-counting, and tally the attributes of those plates, an additional difficulty for the classically-inclined dinner guest would arise. Say the plates come in two colors, white and black. In the old physics, tableware had the good manners to remain the same color while a person tallied up the count for each hue. If you wanted to be sure that no one changed the color of a plate after it had been counted, a guard could be placed in front of each place setting to prevent anyone with a can of paint from drawing near. However, if we view those plates as akin to subatomic particles, quantum theory says that it is impossible to guarantee that their color will not be changed by an influence that could be anywhere in the universe—just as we found in Principle 1 to be the case with Bruce and Bob's hats. There is no way to put either a physical or conceptual boundary around the table and say, "*This* is the absolute limit of what I am studying."

This reflects the previously-discussed finding of Bell's Theorem: non-locality is an undeniable fact of the universe. Ultimately, everything is connected to everything else, which makes it impossible to view reality as being composed of distinct parts. You may think you are examining only the top of a banquet table, but in a certain sense you are studying the entire creation. Kafatos and Nadeau say, "The results of the experiments testing Bell's Theorem suggest that all the parts, or any manifestation of 'being' in this vast cosmos, are seamlessly interconnected in the unity of 'Being.'"[30]

However, for now I want to lay aside the issue of wholeness, and look more closely at why we cannot know all the truth about even a part. This, of course, strikes another blow at the reductionist approach of the old physics. For even if it could be assumed that piecing together all the parts *would* result in a whole, the entire truth about the whole never can be attained if knowledge about each part inevitably is limited. Quantum physics has found that the uncertainty principle is yet another barrier to understanding the underlying nature of reality.

Planck's Constant—the tiniest part of the whole

The vanishingly small Planck's Constant prevents physicists from knowing everything about anything—whether it be the Big Bang or the tiniest subatomic particle. It is interesting, but not all that surprising, that the same law of nature which permits the physical universe to exist also keeps us from knowing the deepest truth about the reality in which we live. This law is known as Planck's Constant and was referred to earlier in our discussion of what prevents cosmologists from tracing the energy of the Big Bang all the way back to the moment of creation. It appears in the new physics in various guises: as a unit of space, as a unit of time (how long it takes light to travel across that space), and as a unit of energy (when all the forces of nature merge into a single superforce).

Recall that prior to the Planck time—10^{-43} of a second after the instant of the Big Bang—is when physicists believe time and space were "smeared out" or blurred together. Similarly, the Planck length, 10^{-33} of a centimeter, is considered to be the smallest possible unit of space. Go smaller, and space is thought to break up into some sort of quantum foam. So Planck's Constant essentially serves as the dividing line between material and non-material existence. It makes matter possible. We have seen that physical reality is not continuous, because if it were, atoms could not exist: in an instant electrons would spiral down continuously right into nonexistence, thereby dissolving the universe as we know it. Instead, materiality is composed of innumerable quantum events, each possessing a size, time, and energy in accord with Planck's Constant.

Physical reality seems to be so solid and whole only because we look upon it with such crude organs of perception. Heinz Pagels writes, " . . . if we look at a pile of wheat from a distance, it appears to be a continuous smooth hill. But up close, we recognize the illusion and see that in fact it is made of tiny grains. The discrete grains are the quanta of the pile of wheat. Another example of this 'quantization' of continuous objects is the reproduction of photographs in newspapers. If you look closely at a newspaper photo it consists of lots of tiny dots; the image has been 'quantized'—something you do not notice if you view the photo from afar."[31] Planck's Constant is what makes parts out of the continuous whole which, in the view of both mysticism and the new physics, underlies the world's phenomena.

Pagels notes that Planck's Constant "specified, if you like, the size of a single grain in the pile of wheat. If Planck's constant could be set to zero, the grain reduced to zero size, then the continuous nature of the world would reappear. The experimental fact that Planck's constant is not zero came to mean the world is in fact discrete."[32] Well, how discrete is it? Not very. In discussing the Planck time, we saw that if a second could be divided into 10^{43} parts, those parts are about equal to how many electrons laid side-by-side would fit across the span of the entire observable universe. The Planck length can be visualized as a similar vanishingly small distance. Yet the amazingly thin veil of Planck's constant serves as an impenetrable barrier between the wave-like and particle-like aspects of physical existence. No Planck's constant, no parts. Also, no material universe. So in both physics and mysticism, the tradeoff for being separated from unbroken wholeness is existing as a distinct entity.

Uncertainty is certain

And there is an additional price to pay for being a part rather than the whole: ignorance of the complete truth about even other parts. This is another feature of the universe which is produced by Planck's constant— uncertainty. One way of looking at the uncertainty principle in quantum physics is that attempts to measure one attribute of a subatomic particle inevitably disturb it, and this makes it impossible to know the undisturbed value of other attributes. If you want to know the position and velocity of a single subatomic particle, such as an electron, Hawking says:

> The obvious way to do this is to shine light on the particle. . . . However, one will not be able to determine the position of the particle more accurately than the distance between the wave crests of light, so one needs to use light of a short wavelength in order to measure the position of the particle precisely. Now, by Planck's quantum hypothesis, one cannot use an arbitrarily small amount of light; one has to use at least one quantum. This quantum will disturb the particle and change its velocity in a way that cannot be predicted.[33]

Thus, notes Hawking, the more accurately you try to determine the particle's position, the shorter the wavelength of light that must be used to observe it, and the more energy is transmitted to the particle—which disturbs its velocity by a greater amount. "In other words," he says, "the more accurately you try to measure the position of the particle, the less accurately you can measure its speed, and vice versa."[34] It turns out that the uncertainty in the particle's position times the uncertainty in the particle's momentum is greater than, or equal to, Planck's constant.

The reason for this is difficult to explain in a non-technical manner, but Kafatos and Nadeau say that it involves the distinction in quantum physics between a wave and a particle: "If Planck's constant were zero, there would be no indeterminacy because we could predict both momentum and position with the utmost accuracy. A particle would have no wave properties, and a wave no particle properties—mathematical map and the corresponding physical landscape would be in perfect accord."[35] Do not worry about trying to understand why this is so. The point to remember is this: while Planck's constant makes it possible for the physical universe to exist as a collection of discrete quantum parts, rather than a continuous wave like whole, that constant also puts an irreducible limit on what can be known about quantum phenomena.

For another rule associated with the uncertainty principle is that the dynamic attributes of quantum phenomena always come in pairs. Position and momentum are one such pair. Energy and time are another pair. In a single measurement you can know the exact value of any one attribute in a pair, but not both. This way of looking at the mathematics underlying quantum uncertainty has nothing to do with how much a particle is "disturbed" while observing it. Thus it would be possible to determine the precise position and energy of an electron, while having no clue to its momentum or the length of time it has existed in that energy state. As Nick Herbert puts it, this means that half of a particle's attributes "are always hidden from view. This restriction on the mutual measurement precision of certain attributes is just sufficient to prevent you from devising an experiment that would show you what's really there. . . . "[36]

Illusory knowledge of the everyday world

If this is true, why does the everyday world appear to be so predictable and open to view? After all, even a beginning tennis player frequently can hit the ball over the net. This requires being able to track with one's eyes the ever-changing position and velocity of the tennis ball hit by your opponent, and to intercept it with your racquet at just the right moment. The uncertainty principle implies that this should be virtually impossible to do, since the more you know about the ball's position, the less you will know about its velocity. When you sense that the ball is in the right position to hit it, ignorance of its speed means that often it will be past you before you are able to swing your racquet.

Fortunately for tennis players, the importance of quantum effects is related to the size of an object times its momentum, as compared to Planck's constant. So large objects turn out to have a very small uncertainty. Pagels says that "For a flying tennis ball, the uncertainties due to quantum theory are only one part in about ten million billion billion billion (10^{-34}). Hence a tennis ball, to a high degree of accuracy, obeys the deterministic rules of classical physics. . . . For atoms in a crystal we are getting down to the quantum world, and the uncertainties are one part in a hundred (10^{-2}). Finally, for electrons moving in an atom the quantum uncertainties completely dominate. . . . "[37]

In the everyday world, nothing appears uncertain or unknowable— theoretically, at least. It seems that our ignorance about the nature of something is just that, insufficient knowledge, rather than an irreducible limit on what can be known. Only in the subatomic domain are the laws of quantum physics, including the uncertainty principle, plainly evident. Only there, we might like to think, does that principle serve as a barrier to knowing any phenomenon completely. Yet it is not true that quantum uncertainty affects only those studying the minute building blocks of physical existence. According to Kafatos and Nadeau, "quantum indefiniteness or quantum uncertainties are not confined to the microscopic realm: They spill over to the macroscopic realm. This we believe, has important consequences for understanding the character of physical reality at all scales. . . . "[38]

They note that hitting a tennis ball amounts to being subjected to a "macro-level illusion." That is, because we deal in everyday life with large material objects—the crudest form of existence from the perspective of mysticism—we fail to realize the ubiquitous nature of the laws of quantum physics. Toasters, cars, coffee cups, lettuce, clouds, our own bodies—all of the things we perceive apparently come into existence in accord with quantum laws. This is what allows Kafatos and Nadeau to argue that since *everything* obeys the rules of quantum mechanics, the uncertainty principle holds sway over more than just subatomic particles. Spiritual science agrees with them. Whatever prevents us from knowing simultaneously the position and momentum of an electron, also keeps us from knowing the entire truth about ourselves, and also bars our consciousness from knowing the unbroken wholeness of Ultimate Reality.

The mind's Midas touch: everything turns to matter

Normally, human consciousness only is able to know something about parts, rather than everything about the whole. If the human mind is viewed as a calculator, it seems to be designed for division, rather than addition. Whatever the mind perceives, it divides into parts. This occurs on every level of physical reality, from the quantum realm on up. In quantum theory we have seen that the wave function of a subatomic particle is considered to contain the possible attributes of that particle—such as its position and momentum—in the absence of observation. Like any other kind of wave, that wave function is a continuous whole. Nick Herbert says:

> For the electron considered as a particle, the hydrogen atom is mostly empty space. However, considered as a wave (realm of possibilities), the electron fills its atom brim to brim. It is hard to believe that the electron is physically smeared out across its realm of positional possibilities, because every time we measure it we never see a smeared electron, always a point particle. In each atom, however, something seems smeared out to fill the atom, an indescribable something we call the "probability cloud," "realm of positional possibilities," "electron wave function," or "quantumstuff" without really being very sure what we're talking about. Whatever it is, though, the whole world is made of it.[39]

As was noted previously, the question of what that ethereal "wave function" actually is constitutes the *quantum interpretation problem*. The quantum waves cannot be seen, or perceived directly in any fashion whatsoever. Only their effects are discernible as they provide the underpinning for all of physical existence. In this sense the quantum wave function is to material science as Spirit is to spiritual science: everpresent, filling all space, guiding each part of creation, imperceptible by the senses, the unmoved mover. Just as Spirit, or God-in-Action, is the first manifestation of apparent separateness from the absolute unity that is God-as-God, so does the wave function appear to be the entity which produces matter in the physical creation.

How it does this is known as the *quantum measurement problem*. Physics does not begin to know the answer to it. But in one way or another, observation or measurement of the wave function seems to be key in producing a discrete particle from a continuous whole. Nick Herbert writes, "One of the main quantum facts of life is that we radically change whatever we observe. Legendary King Midas never knew the feel of silk or a human hand after everything he touched turned to gold. Humans are stuck in a similar Midas-like predicament: we can't directly experience the true nature of reality because *everything we touch turns to matter*."[40] In other words, whenever a measurement of some aspect of the quantum realm is made by physicists, a material entity with specific attributes always is observed—never the underlying wave function itself, which somehow becomes transformed into a point-like bit of matter.

Under the next principle the question will be raised of whether it is observation by a limited mind which forms matter out of "quantumstuff," or if a more universal force is at work here. For now, the point I am making is that the wave function is another reflection of the unbroken wholeness which the new physics recognizes as lying beneath the seeming separateness of physical reality. Bell's Theorem is one demonstration of this fact, and the wave function is another. Whenever one penetrates beneath appearances in quantum theory, unity is evident—though unseen. Acts of observation divide that unity. This division is perceived, rather than actual, since the underlying wholeness of reality remains untouched. Still, the illusory parts of existence created by observation seem real enough to the one doing the observing.

To measure is to divide

Imagine someone about to walk into a room filled with people at a party. There they are, all sorts of guests, eating, drinking, and talking animatedly with each other. The newcomer pauses in the doorway and takes in what he sees. Suppose he is a young man, single and without a girlfriend at the moment. More than likely, his mind begins dividing the visual data being received by his eyes into two categories: men and women. Then another mental division occurs: "attractive" and "unattractive" women. And the process continues, depending upon his personality and predilections: young/old, blond/brunette, tall/short, and so on. Within a few seconds the computer-like human mind has sorted through a complex array of information and presented the young man with a single summary thought: "walk over and introduce yourself to the woman in the red dress standing by the punchbowl."

From our newcomer's perspective, in just a few seconds his view of the crowded room changed dramatically. At first he saw the guests as a teeming whole, something like a milling herd of cattle holding cocktail glasses. But his frame of mind—call it desire, if you like—rapidly, and essentially uncontrollably, divided that unity into parts until a sole focus of attention remained. Someone else, of course, would have seen the guests in quite a different light. A stockbroker might have zeroed in on those wearing the most expensive clothes and jewelry; a woman looking for her lost son would ignore anyone who did not match his appearance; a hungry person would be attracted to whoever is holding the snack tray.

While this is a commonplace example of observation in everyday life, it illustrates how measurements divide rather than unite phenomena. Interestingly, the Sanskrit term *maya*—which means the illusory world of appearances—is derived from the root *ma*, "to measure."[41] This reflects the fact that assigning symbolic attributes to anything through an act of measurement only leads one further away from the deeper non-symbolic, and unbroken, reality that lies beyond those attributes. Limited reality certainly can be known through observation and logical analysis, but not Ultimate Reality. Both the new physics and mysticism agree on this point.

The fourteenth century Christian mystic who wrote *The Cloud of Unknowing* says, "For though we through the grace of God can know fully about all other matters, and think about them—yes, even the very works of God himself—yet of God himself can no man think. Therefore I will leave on one side everything I can think and choose for my love that thing which I cannot think!"[42] Similarly, Kafatos and Nadeau, have this to say about the limits of material science:

> Science is a dialogue with nature in which we can only correlate relations between particulars, and thus any proof that the parts constitute the whole is not and cannot be the subject of science. How, then, do we even grasp the notion of a whole? The answer is, we think, quite obvious—we do so because that sense of wholeness is a "given" in consciousness. But science, again by definition, does not define wholeness any more than mathematicians can define mathematically an empty set, or cosmologists can define the universe before its origins. Definition requires opposition between at least two points of reference. Although we may, on the deepest level of awareness, sense or "feel" the underlying unity of the whole, science can only deal in correlations between the behavior of parts. It can say nothing about the actual character of the undissectable whole from which the parts are emergent phenomena. This whole is literally indescribable in the sense that any description, including those of ordinary language, divides the indivisible.[43]

Remember, these thoughts are from a physicist and social scientist writing about the implications of quantum theory. While the quotation above is in perfect accord with the teachings of mysticism, the conclusions of Kafatos and Nadeau are founded on findings of material science—which is the "science" being referred to in the lines above. I say this because spiritual science is capable of knowing the whole of reality in an immediate and non-symbolic manner through contemplative meditation. A perfect mystic has learned how to avoid separating unbroken truth into parts. He can do this because his own consciousness is whole and without divisions. Higher planes of reality are contemplated in the absence of ego or desire, so what is there is seen as it *is*. This is impossible when one's limited mind is the research tool for attempting to learn about reality.

Mind, logic, measurement, analysis, observation, mathematics, theory. All of these words amount to the same thing: dividing a whole into parts. This is their function, and they do their work well. But as Kafatos and Nadeau imply, they are not only useless in knowing the unbroken wholeness beneath physical phenomena, but actually lead one further away from that deep reality. Consider this: an ethereal quantum wave function contains the potential attributes of a material electron. One wave, one electron. The mind of a physicist, using appropriate technology as an extension of the senses, observes that particle. A series of measurements reveal it to be in a particular position, with a particular momentum, and so on. One mind, many parts—or attributes—of an electron.

What you desire is what you get

How did this come about? Because the physicist's mind *desired* those attributes to be manifested, his willful acts of measurement brought those parts into his awareness. A single electron became a collection of descriptions, in much the same way as our party guest turned a roomful of people into dualistic clumps: men/women, young/old, beautiful/plain, and so on. In neither case does the observer know a part of existence in its essence. The physicist and guest each obeyed the rules of quantum physics, which state that it is not possible to separate what is being measured from who, or what, is doing the measuring. They form a single entity. Thus, say Kafatos and Nadeau, not only does the uncertainty caused by Planck's constant "prevent us from knowing precisely all the properties of that electron which we might presume to be there in the absence of measurement . . . our presence as observers and what we choose to measure or observe are inextricably linked to the results we get."[44]

So a physicist is able to know only about parts, and worse, just about a portion of those parts. And the same is true for everyone who uses the human mind to understand existence. Every person has to decide whether this state of affairs is acceptable to them. This depends upon one's goal: limited or Ultimate Reality? parts or the whole? the creation or God? The mind is a valuable research tool for realizing the nature of what lies within the walls of physical existence. But we have seen that it is useless if we want to know what lies on the other side of the barriers which keep us confined in this realm of time and space. We have arrived, then, at three conclusions:

Ultimate Reality is undivided

First, the plain fact is that the universe is One. Both mysticism and physics understand this. Kafatos and Nadeau note that from the perspective of material science, "The cosmos is a dynamic sea of energy manifesting itself in entangled quanta which results in a seamless wholeness on the most primary level, and in seamlessly interconnected events on any level."[45] However, they point out, we are unable to perceive, or even prove the existence of, this unity—which is on a level completely beyond our limited minds and measurements:

> What is determined by the act of observation, as we understand it, is a "view" of the universe in our conscious construction of the reality of the universe, and the "existence" of the reality-in-itself is not legislated by this

view. In other words, although our views of reality are clearly conditioned by acts of observation, the existence of the reality itself is not in question. . . . What comes into existence as an object of knowledge was not, as we understand it, created or caused by us for the simple reason that it was always there—and the "it" in this instance is a universe which can be "inferred" to exist in an undivided wholeness.[46]

Mind and senses divide reality into parts

Second, then, what we are able to observe with the mind and senses is not that One, but our "conscious construction" of reality. Where does this leave us? The traditional interpretation of quantum physics attempts to go no further. What is observed is what there is, goes the reasoning. If it is impossible to make direct contact with the unbroken wholeness that exists beneath appearances, then material science has nothing to do with that deeper reality. All right, fair enough. Where the new physics leaves off, mysticism takes up the search for final truth. Kafatos and Nadeau echo the distinction made in this book between symbolic and non-symbolic domains of objective reality. This distinction accounts for the varying research methods of material and spiritual science: one constructs abstract mathematical or verbal "models" of existence, the other uses pure consciousness to perceive directly higher planes of reality.

> . . . let us return to the earlier distinction between the content of consciousness, in which we construct in both ordinary and mathematical language conceptions of reality, and the background of consciousness, in which we apprehend our existence prior to any conscious constructions. . . the fact that we cannot disclose this undivided wholeness [of the universe] in our conscious constructions of this reality as parts does not mean that science invalidates the prospect that we can apprehend this wholeness on a level that is prior to the conscious contents. . . . Thus any direct experience we have of this whole is necessarily in the background of consciousness, and must be devoid of conscious content.[47]

Only an undivided consciousness knows the unity of reality

Third, it is consciousness without conscious content which reveals the deep unbroken wholeness of reality. After following a winding trail of arguments based on findings of the new physics we find ourselves at the same conclusion reached by spiritual science: only by unifying our divided personal consciousness can the hidden unity of reality be revealed. What bars us from knowing final truth is the mind and senses, for they invariably divide reality into parts—leading us ever further from our goal. Every observation, every thought, every measurement, every concept produces an additional division. The mind thus is the barrier we must overcome to know absolute Truth, not the means for finding it.

> *In union with God, of what value are signs?*
> *The one who is blind to Essence*
> *sees Divine action through the attributes:*
> *having lost the Essence he is limited to evidences.*
> *Those who are united with God*
> *are absorbed in the Essence.*
> *How should they focus on His qualities?*
> *When your head is submerged in the sea,*
> *how will your eye fall on the color of the water?*
> —Rumi[48]

This is not only a mystical conclusion, but a logical one. The primary function of the human mind and senses is to make contact with physical existence. If everything that exists could be perceived by the five senses, then symbols would be superfluous for understanding the world. We could see, hear, smell, taste, or touch reality directly. However, physics has found that there are levels of existence—such as the subatomic realm—which cannot be perceived by the senses. In these areas symbols, mathematical or otherwise, must be substituted for direct sensory perception. The wave function, for example, never is observed yet can be precisely expressed by equations of quantum mechanics.

Yet the new physics also has found that it is not possible to make a one-to-one correspondence between every aspect of that unperceived reality and an associated symbol. Planck's constant, the uncertainty principle, and Bell's Theorem demonstrate this fact. In each case, there is some sort of gap between "reality-in-itself" and physicists' ability to describe that reality symbolically. There is a simple reason why the refrain, "We cannot say why this is so," often is heard in writings about the new physics: as Kafatos and Nadeau put it, "Holism in this new sense means that the universe on the most fundamental level is an undissectable whole, and that discreteness of objects must be, in some sense, a macro-level illusion."[49]

Any sort of symbolic representation must, by definition, fail to describe the unbroken wholeness of Ultimate Reality. No part can contain the whole, unless the distinction between the two is completely erased. Then the part is no longer a part, but the whole. Mysticism knows, and the new physics is coming to believe, that consciousness is the essence of final truth. Yet any contents of consciousness—words, numbers, concepts, images, thoughts, desires, emotions—inevitably separate us from the unbroken unity of that fundamental background of consciousness. One word, one letter of the alphabet, one stroke of a single letter, any mark at all on a sheet of white paper divides that sheet into parts: what has been written on, and what is blank. When our consciousness is divided, it sees division in the universe. When our consciousness is united, it sees the wholeness of God.

Rising above the Cloud of Unknowing

I have used many words in this section in an attempt to convince the reader that words are useless for the seeker of final truth. If this seems a bit incongruous to you, rest assured that I feel the same way. But there is a difference between words intended to guide us out of a maze, and words which lead us deeper into it. So I want to share some thoughts from the anonymous fourteenth century Christian mystic quoted earlier, the author of *The Cloud of Unknowing*. His words—in italics—beautifully convey the spirit of contemplative meditation. My comments are in regular type.

> *Work hard and with all speed in this nothing and this nowhere, and put on one side your outward physical ways of knowing and going about things, for I can truly tell you that this sort of work cannot be understood by such means.*

Different laws pertain in non-material spheres of existence. Forget all that you have learned about how to know physical reality.

> *With your eyes you can only understand a thing by its appearance; whether it is long or broad or small or large or round or square or coloured. With your ears you understand by noise or sound; with your nose by the stench or scent; with your taste whether it is sour or sweet, salty or fresh, bitter or pleasant; with your touch whether it is hot or cold, hard or soft, blunt or sharp.*

The five senses measure only the attributes of things. Extending the senses via scientific technology, or the laws of quantum physics, merely measures them more finely.

> *But God and spiritual things have none of these varied attributes. Therefore, leave all outer knowledge gained through the senses; do not work with the senses at all, either objectively or subjectively. . . . For the natural order is that by the senses we should gain our knowledge of the outward, material world, but not thereby gain our knowledge of things spiritual.*[50]

Every measurement or observation of physical existence keeps us bound to the mind and senses, which cannot know other domains of reality. You must discard what prevents you from realizing higher spiritual truths.

> *So crush all knowledge and experience of created things, and of yourself above all. For it is on your own self-knowledge and experience*

that the knowledge and experience of everything else depend.
Alongside this self-regard everything else is quickly forgotten. For if you
will take the trouble to test it, you will find that when all other things
and activities have been forgotten (even your own) there still remains
between you and God the stark awareness of your own existence. And
this awareness, too, must go, before you experience contemplation in its
perfection.[51]

Any content of consciousness whatsoever bars us from seeing Ultimate
Reality, or God, reflected within us. This includes our own ego, or sense
of existing as a separate entity. Wholeness cannot be experienced by a part.
Give up being a part, and know the whole.

And though your natural mind can now find "nothing" to feed
on, for it thinks you are doing no thing, go on doing this no thing, and
do it for the love of God. . . . Let go this "everywhere" and this
"everything" in exchange for this "nowhere" and this "nothing". . .
Who is it then who is calling it "nothing"? Our outer self, to be sure,
not our inner. Our inner self calls it "All," for through it he is
learning the secret of all things, physical and spiritual alike, without
having to consider every single one separately on its own.[52]

What is this "secret of all things, physical and spiritual alike" to which
the author of *The Cloud of Unknowing* is referring? Spirit, the all-pervading
conscious energy of Ultimate Reality. Where is Spirit perceived? In the
depths of contemplative meditation. How? By eliminating the contents of
normal human consciousness—which are a veil between our soul and
Spirit, the drop and the ocean. The soul is conscious, and Spirit is pure
consciousness. Thus there is a natural affinity between the two. The barrier
which separates them is mind and matter, the crudest aspects of creation. It
is not within our power to eliminate mind and matter from this physical
universe, for then it would not exist. But human beings have a God-given
capacity to "go within" their own consciousness and leave aside all traces of
this material creation, including the limited personal self.

This is the ultimate scientific experiment. For it reveals the essence of
every domain of reality, physical, mental, and spiritual: Spirit, audible as
sound and visible as light. The discussion of Principle 7 will describe what
happens when the soul comes into contact with Spirit. Briefly, that
"nothing" of consciousness without content is realized to be the "All"
referred to in the quotation above. This is the secret which the new physics
has not yet fully realized. The mystery of an electron's quantum leap is not
so much that jump itself, but the supposed nothingness through which the
leap passes. In the same manner, the immediate connection between two
particles that was revealed by Bell's Theorem is less important than what is
responsible for the connecting. Unbroken wholeness is not an abstraction.

Quite the opposite, for everything in existence is abstracted—or projected—from that fundamental non-material reality: Spirit, the ground of being.

Face toward parts, or the whole?

Generally physicists, like almost everyone else in the world, have a fixation on parts. This is understandable. Parts are easy to recognize. They are entertaining. Whether they be the attributes of subatomic particles, or aspects of our own personality, the mind is endlessly fascinated with sorting and arranging parts of creation in various ways, trying this way and that way of making sense out of them. There is nothing wrong with this activity, so long as we are not distracted from the wholeness which is recognized by both quantum physics and mysticism to be the source of parts. In the words of Kafatos and Nadeau,

> . . . the ground of Being, of reality-in-itself . . . is obviously reality as a whole, which cannot be known, comprehended or defined as the sum of its parts for the simple reason that the whole serves as the ground for the existence of the parts. If the whole did not participate in and serve as the ground for the existence of the parts, the parts would not exist. If one, therefore, seeks to disclose the whole by summing the parts, one is seeking to explain their existence in the absence of the ground for their existence. . . it is the whole which discloses ultimately the identity of the parts.[53]

Part and whole, part and whole. These two words are central to both material and spiritual science. Far from being mere words, "part" and "whole" represent radically different ways of being. One points away from God and the Theory of Everything, the other towards that Ultimate Reality. The clear distinction between them creates an existential, as well as an intellectual, dilemma. Each person makes a choice—conscious or unconscious—every moment of every day of his life. Which direction to face: toward parts, or the whole? Shall I take one step closer to Truth, or move one step further away? A single step may not seem to account for much, but over a lifetime they add up. Where will we be at the moment of our death: more deeply mired in separateness, or closer to unbroken wholeness?

One's journey from limited to Ultimate Reality can take a long time. It is the work of a lifetime, or even several. A part does not become the whole overnight. There are many degrees of partness and wholeness through which one must pass. Thus I offer the following table of attributes of parts and wholes as a rudimentary traveler's guide, recognizing that ultimately no description of the whole is possible. Nevertheless, while still existing as an ego-centered part of creation, we need some signposts that indicate in which direction we are moving: toward increased partness, or greater wholeness? Virtually every subject that has been discussed in these pages can be looked at in this light.

Mysticism teaches that when the essence of our consciousness, the soul, merges with the all-pervading conscious energy of Spirit which sustains the creation, parts cease to exist. They are recognized to be what they are: illusory divisions of an unbroken wholeness. The pieces of the jigsaw puzzle of life remain only so long as we ourselves are a part of that puzzle. If you have done your best to fit together that puzzle, and find—as has the new physics—that some pieces needed to arrive at a complete understanding of existence seem to be missing, consider this: is it possible that the very consciousness which is trying to solve the puzzle is the only missing piece? Is there a way for you, yourself, to become the piece which completes the jigsaw puzzle of reality? This is the heart of the message which I have tried to convey in this discussion of parts and wholes—which in a sense is the sole theme of this book.

So consider the following table carefully. Try reading each column separately, rather than flipping back and forth between the two. Pause after filling your mind with the attributes of *Parts*, then pause after reading the attributes of the *Whole*. Which column, intuitively, strikes you as being a more attractive alternative? In what direction do you wish your life to move? The choice is up to you, but be warned—the puzzle of Ultimate Reality only has one solution.

Attributes Associated with the Parts and Whole of Existence

PARTS	WHOLE
Mechanical	Conscious
Mind and matter	Spirit
Drop	Ocean
Particle	Wave
Quantum	Continuity
Existence	Non-existence
Analysis	Absorption
Separation	Connection
Symbolic	Non-symbolic
Ego	Humility
Thinking	Contemplating
Activity	Receptivity
Motion	Calm
Division	Addition
Laws	Love
Determinism	Freedom

Outer	Inner
Limitation	Perfection
Subjectivity	Objectivity
Form	Meaning
Complexity	Simplicity
Opposition	Unification
Many	One
Individual	Identical
Material	Spiritual
Physics	Mysticism
Creation	Creator
Probability	Certainty
Chance	Order
Becoming	Being
Illusion	Reality
Partial truth	Absolute truth
I-ness	Thou-ness
Time	Timelessness
Space	Unbroken wholeness

Principle 5.
Laws of cause and effect govern lower levels of creation, including the physical universe.

In reality, He is the Creator of all effects,
but the People of the Skin see nothing but secondary causes.
—Rumi[1]

Having explored four principles of spiritual science, it is possible to see how they are connected. The first principle was *God is One, and present everywhere.* Everything else flows from this penultimate fact of existence: Unity. We saw, in the second principle, that *Spirit is God-in-action, and of the same essence.* Still there is undivided wholeness, with only the form of that divine one changing to bring about acts of creation. The third principle once again echoed the continuity of the cosmos: *God, Spirit, and creation all are consciousness.* This follows logically, since if God's Spirit is conscious and everything is formed from that substance, then all of existence is consciousness.

The preceding principle, *Barriers and veils prevent easy passage between levels of creation, or states of higher consciousness,* explains why unity is difficult to perceive in the physical universe. Most people are able to use only their mind and senses to understand the nature of reality, having lost contact with the faculties of their soul—a purely spiritual entity which alone is able to perceive higher domains of consciousness. Hence the very means used to know this limited reality stand as a barrier to realizing the truth of Ultimate Reality. Since consciousness is conserved, like energy, if used for one purpose it is not available for another. Our mind's usual activities—observing, thinking, and imagining—keep us focused on either the objective physical world or the subjective mental world, which prevents us from perceiving the objective realms of higher consciousness which lie beyond the veils of mind and matter.

God causes all effects

This fifth principle, *Laws of cause and effect govern lower levels of creation, including the physical universe,* also is consistent with the other principles. As the introductory quotation from Rumi indicates, mysticism teaches that there is a single cause lying behind all effects: God. If God projected all of creation through the agency of His Spirit, then this is the first cause of everything in existence. No matter how lengthy and complex the chain of cause and effect has become, a highly evolved consciousness is able to trace any event—from the most minute interaction of subatomic particles, to the formation of gigantic clusters of galaxies—back along a causative chain to that primal act of God.

> *This world and that world are forever giving birth:*
> *every cause is a mother; the effect born is a child.*
> *When the effect was born, it too became a cause,*
> *so that it might give birth to wondrous effects.*
> *These causes follow generation upon generation,*
> *but it takes a very well-illumined eye*
> *to see all the links in the chain.*
> —Rumi[2]

The universality of cause and effect is both entirely scientific and in accord with common sense. Thus this is perhaps the easiest principle of spiritual science to understand from the standpoint of everyday life. Each person sees linkages of cause and effect surrounding them from birth until death. Proverbial sayings reflect this, both popular—"What goes around, comes around"; "You always get what you deserve"—and Biblical: "Whatsoever a man soweth, that shall he also reap"[3]; "By their fruits you shall know them."[4] Some mystical principles appear to be opposed to reason, but this is not one of them.

For if one believes in God's omniscience, omnipresence, and omnipotence, there is no alternative but to accept that every part of creation exists in accord with His will. And that will, say perfect mystics, is not arbitrary. God does not sit on a mountain top and capriciously cast thunderbolts of suffering or joy into the midst of humanity, as in ancient Greek myths. Yet neither is the law of cause and effect the edict of a detached Creator, who observes from afar how the physical universe unfolds. Sawan Singh writes:

A materialist considers this life to be a machine, the parts of which are being run by the blind forces of cause and effect, and he does not admit that there is an ultimate Being who is directing it. But one who knows the reality, while agreeing that the law of cause and effect is working, further knows that this is being done under the orders and directions of a Supreme Being. He, therefore, while keeping an eye on the causes and effects, appeals to the Lord, the Supreme Cause, for help.[5]

"Appeal," in this context, means prayer. Prayer can take many forms: wordless or with words, silent or aloud, spontaneous or ritualized, repetitive or unique. All forms are good, in the view of spiritual science, because they are directed toward a single end—breaking out of the chains of cause and effect which bind us to domains of mind and matter. Prayer is an appeal to the Holy Spirit, God-in-Action, to accomplish something which we are too weak, or too ignorant, or too lazy to do on our own. Thus to some degree, depending upon the purity of that appeal, prayer is a humble recognition of our limited partness. Being so confined by time and space, and the effects of our own actions, we need the aid of the wholeness— Spirit—to loosen our chains.

No freedom in the material universe

This implies the obvious: God is free, and we are not, along with the rest of the physical creation. Rocks, stars, ladybugs, gravity, electrons, ocean currents, bowling balls. Anything and everything we can imagine is subject to laws of cause and effect. As an aside to the reader who remembers the previous discussion of physics' uncertainty principle and probabilistic quantum phenomena, I say, "Yes, everything, even subatomic particles." Uncertainty or chance is a fiction produced by ignorance. It is not a law of nature. But the arguments to support this conclusion must be postponed until the next section. Spiritual science unequivocally states that the law of cause and effect, in the words of Sawan Singh, "is supreme on the material and mind planes, and nothing happens on its own accord, spontaneously, so to say. The law governs the planes; therefore, no haphazard happenings of events takes place anywhere, whether the events are of microscopic or astronomical dimensions."[6]

It is important to note that Sawan Singh's words refer to cause and effect holding sway only on the "material and mind planes," which are levels one to three of our five-story model of creation. This implies that the regions where spirit dominates, especially the highest realm of pure Spirit, are beyond cause and effect. And this is what mysticism teaches. John Davidson says, ". . . within the One Source, there is only one. Without two or more, there can be no causality, for who is to cause what, when all is one?"[7] Perfect freedom only is possible when one rises above the divisions of mind and matter. For the lower spheres of creation are divided into parts, and no part can be absolutely free. Parts inevitably place limitations on other parts. This is so obvious that it largely escapes our attention. We have become so accustomed to our partness that we unthinkingly accept the constraints which have been placed upon our freedom of action.

Imagine that you are driving your car, looking for a parking space on a busy street. You have circled the block several times, growing increasingly fearful that you will be late for an important appointment. Then you see a vehicle pulling away from the curb ahead of you. You speed forward, only to find that another driver got to the space first. He parks while you continue your search. You have just experienced a fundamental law of physics: no two electrons can occupy the same location. This is known as the Pauli Exclusion Principle, and it also prevents two cars from parking in the same space, or two people from sitting in the same seat, or any two bits of matter from merging completely with each other.

If we seek perfect freedom, this example alone should be sufficient to demonstrate that laws of cause and effect prevail in this physical universe. I exercise my apparent free will and head my car toward a parking space, but something prevents me from accomplishing my goal. Another driver had the same desire, and he got to the space first. Where has my freedom gone? Or, rather, was it ever there in the first place? My ego tends to make me believe that I, a single part of the vast cosmos, somehow am the most important part—and that everything else should behave in accord with my wishes. The problem, of course, is that so many other people believe this about *themselves*. We end up with many would-be rulers and very few servants.

However, even if everyone in the world acted in a perfectly loving, humble and gracious manner toward other people, a never-ending chain of causes and effects still would be produced. The actions and reactions would be more pleasant than what is generally experienced by people today, but this simple fact would remain: separate parts exist in the lower reaches of creation. Because these parts interact by bumping into each other, so to speak, rather than merging into a single entity, cause and effect is an inevitable aspect of physical existence. Action produces reaction, which leads to another action, and a reaction. Good or bad, pleasant or unpleasant, constructive or destructive, but still endless cause and effect of one sort or another.

Mysticism teaches that there was a time when souls were not confined by coverings of mind and matter and enjoyed perfect freedom. But once separated from the domains of pure Spirit, it was impossible for the souls to avoid becoming enmeshed in chains of cause and effect. A drop in the ocean moves only when the ocean moves, and has no sense that an external cause is acting upon it. Place the same drop on a tabletop, however, and it is affected by gravity, air currents, evaporative forces, and many other influences. Once it has become a part, rather than the whole, it cannot help but interact with other parts. In a similar sense, Sawan Singh writes that:

> A will is free only so long as it has not acted. Once it acts, then that very act becomes binding on it. The second time it acts, it does not act on a free will, but as a "calculating will," for it carries the experience of the first act

with it. And a calculating will is not a free will, but a limited will. The very creations, or acts of a free will, work as limiting factors upon it, and guide it in its future activity. So the more actions one performs, the more his will is guided and thus limited.[8]

This is most evident, of course, in entities which possess a memory—living beings with at least a rudimentary mind. Inert substances, such as a rack of balls on a billiard table, are affected by only the most recent influences upon them. Say the "2" ball hits the "3" ball in the course of a game, and sends it into a corner pocket. When that game ends and a new one begins, the "3" ball doesn't remember that the "2" ball hit it fifteen minutes ago. Its motion on the table will be determined solely by the forces impacting it moment by moment. If the "3" ball is not touched by any other ball, it remains motionless. Lacking a will, free or otherwise, it moves in a predictable and mechanical fashion.

Other material things, such as the planets in our solar system, *are* in continuous motion. In fact, on the atomic level everything is ceaselessly active, whirling about and vibrating at immense speeds. As John Davidson puts it, "All matter is energy patterns of sub-atomic relationships and interconnections, moving at tremendous speeds, the speed being an integral aspect of matter itself."[9] Action of some kind clearly is unavoidable in material domains of creation. And action entails reaction. At least this much is clear.

Free will is an illusion

Much less certain is the answer to a question which produces endless philosophical and scientific debate: does mind, and the human mind in particular, somehow stand outside these chains of cause and effect? In other words, is *homo sapiens* the only entity in the physical universe which can willfully cause an action that is completely independent of any previous effect? Mysticism answers, "No, not at all. Cause and effect determines human behavior, as well as every other activity in the universe."

Many people find this difficult to believe. To them, unfettered free will certainly *seems* to exist, even though it undoubtedly is constrained by the free will of others. Returning to our earlier example, I desire to park my car in a particular space and so do you. It might be thought that each of us is free to *will* what we want, but only one of us is able to *achieve* that desire. Thus, goes the argument, freedom of action may be limited, but not freedom of will. This is an appealing notion, but mysticism says that it is at odds with a fundamental fact: mind is mechanical and unconscious. As was discussed previously, to spiritual science the mind—human or otherwise—is as much a force of nature as gravity or electromagnetism. Only Spirit is conscious, and thus free.

Hence mysticism teaches that freedom of action is commensurate with the degree to which Spirit, rather than mind, is active within us. Spirit *is* consciousness. Being without parts, it is outside of the sphere of cause and effect. Acting through the intermediaries of mind and matter, Spirit brings about all changes in the physical universe but is not affected by those actions. Our soul is a drop of that all-pervading conscious energy, and consequently enjoys the same qualities. Julian Johnson writes,

> It is in the soul that all consciousness and all power resides. All below the soul, even the mind itself, is unconscious, automatic and mechanical in action. . . . All light, all intelligence, harmony, rhythm, beauty, wisdom, love, morality and power come from the soul. They are all derived from the spirit and are all imparted to the mind by the spirit, just as the electric current gives power to the bulb to make it incandescent.[10]

When the mind is viewed in this light—as automatic and mechanical—it becomes much easier to understand how laws of cause and effect control human behavior, as well as inanimate objects. Consider whether a rock dropped from a high cliff feels itself to be falling because of the force of gravity. No, the rock just falls mechanically. Do you believe that flowers turn toward the sun because they are thinking that energetic photons of light are required for photosynthesis? No, the plant just turns automatically. Yet, if I ask you whether you freely chose to read this book, you probably would say, "Yes, this act was entirely up to me. I had a desire to read it, and I can start or stop reading these pages at will." But what if almost every intention of your mind is controlled by laws as automatic and regular as other forces in the universe? In other words, what you will to do is just as much a function of cause and effect as the motion of a falling rock, or a flower's leaning toward sunlight.

Albert Einstein shared this viewpoint, which is compatible with the principles of both spiritual and material science (even quantum theory, as shall be argued in the next section). Einstein said:

> Honestly I cannot understand what people mean when they talk about the freedom of the human will. I have a feeling, for instance, that I will something or other; but what relation this has with freedom I cannot understand at all. I feel that I will to light my pipe and I do it; but how can I connect this up with the idea of freedom? What is behind the act of *willing* to light the pipe? Another act of willing? Schopenhauer once said, "Man can do what he wills but he cannot will what he wills.". . . when you mention people who speak of such a thing as free will in nature it is difficult for me to find a suitable reply. The idea is of course preposterous.[11]

What makes me will what I will?

The "preposterous" idea of free will is linked to the impossibility of a part of creation being able to exist in complete isolation from other parts. If this were feasible, then perhaps we could speak of free will, for then a person—or whatever independent part exists—justifiably could claim sole credit for his thoughts and actions. Of course, the question would arise as to who would be around to hear that claim. For if a part is able to avoid being influenced by other entities, the price to be paid is total isolation. Earlier I made a tongue-in-cheek comment that those who believe in the possibility of "creating your own reality" apparently would inhabit that realm all by themselves.

After all, if everyone is perfectly free to make of life what they will, then there is no room for anyone else in that existence. For if you and I do not agree completely as to what we want in our reality, either one of us wins out over the other, negating that person's freedom, or somehow we come to be exactly alike, thereby eliminating our differences. But in the second case, there actually is only one being in two forms. This, of course, is how mysticism teaches free will truly can be attained: merge your consciousness with God, thereby eliminating all divisions which limit freedom of action. A whole has no opposition. But so long as we exist as separate parts of creation, the idea of free will indeed is preposterous.

Rumi made that argument in this manner: some people, he said, "say that man is the creator of his own actions, that man 'creates' each and every action that issues from him." But this cannot be so. For human acts come about in only two ways. One is through the medium of some instruments which man possesses: his body, mind, or spirit. However, says Rumi, since these instruments are not within his control, neither are the acts which are produced by them.[12] Can you create your own body and mind? Since the evident answer is "no," how can you believe that what is created by them is the act of a free will. This is a variation of Einstein's perspective: Man can do what he wills, but he cannot will what does the willing.

The second way to act, according to Rumi, is without any intervening medium at all. But he says that "it is out of the question that any act should come from him without such an instrument," because this would entail acting in the absence of a body or mind. Since these are concomitants of physical existence, such an action is impossible. Thus man must act through means that are beyond his control. How can this be called "free will?"

> *Every instant Thou drawest a picture in our brains:*
> *We are the page of Thy script and lettering.*
> *Each image consumes an image, every thought feeds upon thought.*

You cannot flee imagination, nor can you sleep to escape it.
                                              —Rumi[13]

Stop your mind. Right now.

If further evidence of determinism in human behavior is demanded, it is easily produced. Try to stop the workings of the mind. *Your* mind, since this is the only mind with which you are in direct contact. If your will indeed is free, then it should be free to cease willing. This is not a philosophical question, but an experimental one. Close your eyes after you finish reading this paragraph and begin to observe the contents of your consciousness. How many, and what kind of, thoughts and images are produced by your mind over the space of the next five minutes, or even one minute? Do your best to keep them in check. Remember, if you are capable of free will, then you are able to control what enters your mind. Not just for a few minutes, of course, but throughout your entire life—waking and sleeping.

It is a safe bet on my part to predict that after only a few seconds you found yourself thinking about, or visualizing, something. One thought might well have been, "I must not think." But that still counts as an uninvited thought. If you were unable to stop your train of thinking and visualizing, do not feel like a failure. Few people *can*, even those who practice contemplative meditation for years. The mind is a powerful machine. Generally it takes much time and effort to bring it under control. My point simply is this: if you are unable to *stop* your mind, what makes you believe that you are free to *start* it? That is, what evidence could you produce that the desires, thoughts, emotions, volitions, imaginings, dreams, and so on which appear in your consciousness are under the control of your supposedly free will? This would be such evidence: *Stop them from appearing.* Not for an instant, but for—say—three hours. Being able to eliminate all of these contents of consciousness for that length of time would be at least a tentative demonstration of your free will.

I am not suggesting that you actually try to do this, because if you cannot control your mind for three minutes, it clearly will not be possible to do so for three hours. My purpose is only to illustrate how shaky is the foundation of our belief that thoughts and actions are creations of an independent will. The truth, say perfect mystics, is that our mind—and everything else in the physical universe—is under the control of a Universal Mind which mediates the virtually infinite number of cause and effect linkages in the lower realms of creation. This painter, as referred to by Rumi in the quotation below, is not formless Spirit—but a mechanical force which obscures the purity of the canvas of consciousness by drawing upon it endless images of mind and matter.

> *The paintings, whether aware or unaware, are present in the hand of the*
> *Painter.*
> *That Traceless One writes and erases every moment on their thought's page.*
> *He brings anger and takes away contentment.*
> *He brings stinginess and takes away generosity.*
> *Morning and evening this affirmation and obliteration does not leave my*
> *perceptions for half an instant.*
> —Rumi[14]

Puppets controlled by invisible strings

Picture yourself in the studio of a tireless painter. Day and night he is occupied with placing brush strokes on an endless supply of blank canvases. Ceaselessly moving from easel to easel, as soon as he finishes one painting he begins another. Some of those works are beautiful, some ugly; some are uplifting, some degrading. Their varied styles are impressive, and the sheer quantity of the artist's production unbelievable. He *never* stops painting. You wonder how he manages to keep up this frenzied pace. Approaching him, you gently cough to attract his attention, then ask: "Sir, I have been admiring your art, and would like to learn more about your methods and technique. Could you please stop painting for a moment so we can talk without distraction?"

Surprisingly, the artist acknowledges your presence with only a brusque grunt, and continues about his business. Coming to realize that he is some sort of madman, lost in his own world of perpetual painting, you turn your attention to the studio. Perhaps the artist's surroundings contain some clue that explains his unstoppable fanaticism. A curious sight catches your eye. Barely visible against the white walls are a complex network of gossamer filaments, finer than the most delicate spider's web. You trace their course and are amazed to see that each strand is connected to a part of the painter's body. His fingers, arms, shoulders, legs, toes—every limb, large and small, is attached to one of the filaments. It is impossible to tell exactly how the connection takes place, so thin and transparent are those strands.

You are fascinated. What lies behind this mystery? Looking up, you wonder where those filaments lead. The studio's ceiling is high, and shadows obscure your vision. It is impossible to follow the course of the strands beyond a certain point. But there! A tall ladder leans against a corner. You move it to the center of the room and begin to climb, keeping one eye on the filaments rising toward the ceiling, and another eye on the rungs of the ladder. Ascending step by step, the studio's shadowy upper recesses become visible. And what you find at the very top truly is amazing.

Perched on a platform just below the roof, is a puppeteer with unimaginable skills. Holding a complex mechanism that weaves together the myriad gossamer strands, he is able to control those threads with such dexterity that the movements of the painter are as precise and fluid as if that artist had willed them himself. While only a few minutes before you were so impressed with the skill of the painter, now you realize that all of his artistry is the product of the puppeteer. Your perspective changed completely as you moved from ground level—where the artist paints his canvases—up the ladder past the strands which are connected to the painter, to the level of the master puppeteer. In like fashion, the perfect mystic Sawan Singh writes:

> There are two ways of looking at this creation:
>   1. From the top, looking down—the Creator's point of view.
>   2. From the bottom, looking up—man's point of view.
> From the top it looks as though the Creator is all in all. He is the only Doer, and the individual seems like a puppet tossed right and left by the wire puller. There seems to be no free will in the individual, and therefore no responsibility on his shoulder. It is His play. There is no why or wherefore. All the Saints, when They look from the top, describe the creation as His manifestation. They see Him working everywhere. Looking from below, or the individual viewpoint, we come across "variety" as opposed to "Oneness." Everybody appears to be working with a will, and is influenced by and is influencing others with whom he comes in contact.[15]

In the preceding story, the puppeteer is Spirit or God-in-Action, the painter is our mind, and the strands which connect the two are the laws of cause and effect which govern the lower levels of creation. The blank canvas is our pure consciousness, and the many and varied paintings are the never-ending thoughts and actions which fill that canvas every instant of every day of our life. Like all metaphors, this one is imperfect, so do not take it too literally. But it provides us with at least a fuzzy image of what is impossible to visualize clearly. For if we knew what produced the workings of our mind, we would be far beyond this symbolic material reality, having reached the non-symbolic domains of Universal Mind and Spirit.

> *Just so an ant, who saw a pen writing on paper,*
> *Delivered himself to another ant in this way:—*
> *"That pen is making very wonderful figures,*
> *Like hyacinths and lilies and roses."*
> *The other said, "The finger is the real worker,*
> *The pen is only the instrument of its working."*
> *A third ant said, "No; the action proceeds from the arm,*
> *The weak finger writes with the arm's might."*
> *So it went upwards, till at last*

> *A prince of the ants, who had some wit,*
> *said, "Ye regard only the outward form of this marvel,*
> *Which form becomes senseless in sleep and death.*
> *Form is only a dress or a staff in the hand,*
> *It is only from reason and mind these figures proceed."*
> *But he knew not that this reason and mind*
> *Would be but lifeless things without God's impulsion.*
> —*Rumi*[16]

Dreams are real to a dreamer

Each of these ants had a different understanding of what caused the pen to write on paper. Their own "theories," if you like, were in accord with the extent to which they were able to trace the cause and effect linkages that led from God's impulsion to the drawing of flowery images. As Sawan Singh observed, what looks like free will from one perspective appears to be determinism from a higher view of creation. This is the key to making sense of the endless philosophical and scientific debates about the source of human thought and behavior: chance or necessity? free will or determinism? random probabilities or cause and effect? These questions never will be resolved conclusively, because the answer depends upon one's state of consciousness. Just as the events in a dream are reality to the dreamer, but not to a companion who is awake, so does someone enmeshed in physical existence feel that he possesses free will—whereas a more evolved consciousness recognizes this to be an illusion.

This does not mean that each perspective is equally correct, only that questions have different answers on different levels of reality. The higher level always provides the most accurate, objective, inclusive and fundamental answer. However, there is no getting around the fact that we are bound by spiritual geography, so to speak. Whatever level of existence we find ourselves in appears the most real to us—at least until we are able to break out of those confines and reach a higher state of consciousness. Thus spiritual science teaches that it is useless to engage in fruitless debate over questions such as free will versus determinism, because the answer is a shifting target for all but those who have reached the domain of Ultimate Reality—the saints and perfect mystics. In this regard, Lekh Raj Puri says:

> This also settles the question of the free-will of man and the fore-knowledge of God. Both can be true at the same time, for they are truths of different grades of Reality. If the very existence of man is a delusion, then anything about him, his intellect, his free-will, etc. cannot be real. If a thing is unreal, its qualities or attributes are also unreal. Thus in Absolute Truth, man has no existence as man and consequently his free-will is also non-existent, (or rather the question of his free-will does not arise); but for

men and in this world of delusion and appearance, where man has an existence, his free-will also exists. Our existence as men and our free-will are realities of the same order.[17]

Keep in mind, then, that my discussion of this principle occupies a kind of middle ground between the lowest and highest realities: somewhere midway between the illusion of unfettered free will, and the final truth of God's overarching omnipotence and omniscience. In this zone of partial understanding, in one sense we are responsible for our thoughts and actions, and in another sense they are outside of our control.

At first glance it is images, thoughts, and desires—the contents of our mind—which guide our outward behavior. Something enters the mind, then we act upon it. "Some food would taste good," comes the thought. I then put bread in the toaster. Today my mind concludes, "Our car is old and ugly," and tomorrow I find myself standing in an automobile showroom talking to a salesman. People certainly seem to have some control over which thoughts are acted upon, and which remain merely a potential blueprint for action. Yet mysticism says that even our likes and dislikes, what we run toward and what we flee from, are a product of cause and effect.

When we have received positive or pleasurable consequences from some action, we tend to repeat it. Negative or unpleasant consequences cause us to avoid other actions. When a debt is to be paid, we are inexorably drawn to our creditor. Similarly, that which is owed to us is returned naturally and uncontrollably. It is as impossible to avoid these subtle mental laws of cause and effect as it is for material entities to escape the force of gravity. One's belief or non-belief in an objective law of nature has no bearing on how that force affects us. Those who believe that they possess free will are as bound by the mental laws of cause and effect as those who accept the reality of determinism. In either case, destiny—which is nothing other than the net effect of all causes impacting upon us—creates the effects we experience. Charan Singh answered a question about whether someone who is destined to become a doctor still needs to strive for that end:

> When it is in his destiny to become a doctor, he will automatically work to be a doctor; otherwise, he will never become a doctor. We have to do, with our limited intelligence, what we think will be the best; but only that will happen which has to happen. . . . If a thing is in your destiny, then your way of thinking will be to develop it. . . . We may say that this is my decision, this is my free will, but we forget what has led us to make that decision, to think in that direction.[18]. . . we have a conditioned free will—within our own circumstances. If we also include the environment in

which we have been brought up, the way in which we have been molded to think, then we have absolutely no free will.[19]

Once again we see that the issue of determinism and free will only can be examined within a particular context: partness or wholeness, limited or complete truth. One context produces a certain understanding, the other context a different understanding. Perfect mystics advise us to take a practical approach—worry less about this question, and be concerned more about how to reach the level of reality where it can be definitively answered. Sawan Singh says, "Intellect, reason and feeling, being what they have been fashioned to be, now determine our actions and make us choose the predestined course. Thus the acts of one life determine the frame work of the next life. . . . Yet all is not lost if we use the little freedom we have in such a manner as to lead to our ultimate rescue."[20]

Karma is simply cause and effect

The previous quotation refers to actions producing effects that cross several lives, so here I will introduce one of the few non-English terms in this book: *karma*. In general, I am a firm believer that the terms found in any language on Earth can be used to describe truths of spiritual science, which—after all—cannot be accurately reflected by symbols. However, the English language does not contain a term which is as rich and precise in meaning as karma. This is a Sanskrit word which has become recognized worldwide. Still, even though the term is familiar enough, genuine understanding of its meaning is not nearly as common.

That meaning is simple and scientific: Julian Johnson says, "Karma means that law of Nature which requires that every *doer shall receive the exact result, or reward, of his actions.* In its last analysis, it is nothing more nor less than the well-known law of cause and effect."[21] This is clear enough, because we have noted that cause and effect is not only the foundation of material science, but is plainly evident in every sphere of everyday life. Why then does *karma* convey a more accurate understanding of this principle of spiritual science than the words *cause and effect?*

The primary reason is that karma and reincarnation, or rebirth, are closely linked in many mystical teachings. Thus the effects of actions often accumulate over the span of more than one life. The physical body dies, but our mind and soul do not. They return to the material sphere of existence in a new body and carry on the chain of causes and effects experienced in their previous physical incarnation. So karma connotes an expanded time span over which laws of cause and effect operate in living beings.

While I believe in reincarnation, I know that other sincere seekers of God do not, and have valid reasons for their skepticism that the mind and soul of a living entity—plant, insect, animal, or human—usually can return to earth in another form after its physical body dies. However, whether or not one accepts reincarnation as fact, most spiritually inclined people would say that the divine intelligence which governs the universe is greater than our limited intelligence which tries to understand why things are as they are.

Charan Singh gave this advice to someone who had doubts about the reality of reincarnation: "The Saints recommend that instead of engaging in argumentation about the nature of things, we should try to see the Truth with our own Inner Eye. . . . Approach the subject as a scientist, with an open mind, and make research in the laboratory of your own body."[22] This makes good sense. So both believers and non-believers in reincarnation can agree that until we attain a high level of spiritual awareness, many of the causes which produce the effects we observe in both the outside world and our inner consciousness will be unknown to us.

A spiritual uncertainty principle

This means that a spiritual uncertainty principle normally operates in everyday existence which makes it impossible to link every thought or action that occurs in our life with a corresponding cause. Sometimes things just seem to happen to us, and we cannot figure out why. The answer to such questions often lies buried in a chain of causes and effects which extends far beyond one's current awareness. This makes human existence appear to be either almost happenstance, or the result of unpredictable free will.

Neither is true, teaches mysticism. It simply is impossible to identify all of the causes which have combined to produce a particular effect that is manifested in our mind as a thought, intention, emotion, perception, or other content of consciousness. People often put considerable effort into trying to understand themselves by associating their present personality with childhood influences. "Perhaps I am the way I am because of the way I was brought up by my parents," goes the reasoning. Partly true, according to spiritual science. But at the moment of your birth you were not a blank slate, and neither were your parents. Everyone brings with them into the world a myriad of karmic influences, some of which become evident in one's personality as a young child, while others appear at various times during the span of life.

If it seems that how your parents treated you as a child is the cause of some present trait or behavior of yours, then what made your mother and father act the way *they* did? Their own parents? If so, what were the causative influences on your grandparents? There is no end to the questions that arise when we try to understand the "whys" behind the "whats" of existence. Whether reincarnation is considered to be fact or fancy, the impact of previous generations on present lives is obvious and inescapable.

Thus it appears that from the standpoint of either karmic law or developmental psychology, it is most difficult to break out of personality and behavior patterns which have their roots far in the past. "Difficult, but not impossible," teaches spiritual science. In large part freeing oneself from these chains of cause and effect is a matter of grace, of being receptive to the touch of Spirit—which is beyond mind and matter, and so is able to free us from their binding influences. A perfect mystic, being one with Spirit, is needed to connect us with that all-pervading conscious energy.

Breaking karmic bonds

Mysticism teaches that it is impossible for a person to scale the high prison wall of material existence with a heavy ball and chain—extensive karma—attached to his ankle. He still needs a ladder—Spirit—to get over the top, but a precondition for escape is lightening the burden of karmic cause and effect. How is this done? By eliminating most, or all, of the connection between a cause and its effect. Life then becomes a series of causes without effects, of actions without karma.

At first glance this appears to be nonsense. For if laws of cause and effect govern the physical universe, then how it is possible to toss aside half of the equation and be left with a well-regulated world? The key lies in understanding where karma is "stored" in living beings: not in any material form, but in the mind. Thus by itself a physical action does not produce karma, while the mental desire or intention that precedes and accompanies such action does. Sawan Singh says:

> The individual thinks he is the doer and thereby becomes responsible for his actions and their consequences. All the actions are recorded in his mind and memory, and cause likes and dislikes which keep him pinned down to the material, astral or mental spheres [levels one to three of our model of creation]. . . . The individual in these regions cannot help doing actions and, having done them, cannot escape their influences. The individual acts as the doer and therefore bears the consequences of his actions.[23]

A billiard ball does not accumulate karma. It is bound by the physical laws of cause and effect when struck by another ball—or the cue stick—but the effects of such an action last only as long as the force being applied. When the physical force ends, the effect ends. In other words, a primary impact is transformed into secondary effects—rotation and movement across the table—which are potential causes that can produce effects of their own. This is how a combination shot is made. But at some point all motion on the table stops. The chains of cause and effect are ended until the next shot is made by a player, because the billiard balls have no mind. If they *did* have a memory, as do humans, we can imagine them thinking after each shot: "That six-ball hit me too hard. When I get the chance, I'm going to pay him back;" or, "The nine-ball who just brushed by me is cute. I hope she rolls by this way again soon."

Such thoughts would entail karma. And this is why the behavior of people is so much more complex and difficult to predict than the motion of billiard balls on a table. When *we* are impacted by a cause, generally we not only experience that "force," but also various mental reactions. Walking along the street, someone runs into me from behind. I fall down onto the sidewalk. This is pure physics. Action and reaction. The physical effect of that collision ends soon after our bodies make contact, leaving aside some bruised knees or other minor injuries. However, the mental effect is another matter entirely. If I immediately start cursing at the other person as soon as I hit the ground, a host of other effects may arise from what began as a simple interaction between two material entities.

If a mean-spirited professional boxer was the one who ran into me, I soon may regret my hasty words—and my broken nose. Or if the other person involved in the collision was a nun rushing so she wouldn't be late to serve lunch to the homeless at her parish church, I also am going to experience some unpleasant effects: guilt, remorse, embarrassment. The effects of a simple interaction soon disappear in the case of billiard balls, but can persist for a lifetime in the case of people. Everyone has vivid memories—good or bad—of significant events in their life which took place many years, or decades, ago. Such memories are the visible traces of karma. Yet even forgotten actions have produced karma if they were performed with desire, or intent.

The true cause of any action performed by human beings is the mental *idea or motive* which precedes that action. The physical action itself does not produce karma, only effects consistent with the laws of physics and the other material sciences. This is why the dynamics of mindless objects— planets, billiard balls, gas molecules—can be described so accurately by purely physical laws. The actions of any being with a mind, on the other hand, are much more difficult to fathom. For most of the motivating force behind any action stems from karmic influences which cannot be known by anyone lacking access to higher planes of consciousness, where the operation of the law of karma can be observed directly.

From the standpoint of knowing Ultimate Reality, the problem is this: we have already seen that it is mind and matter which obscure the unbroken wholeness of Spirit. The unending cause and effect linkages created by the law of karma keep us bound to mind and matter, for every action creates a reaction, which causes an action, which creates a reaction— and so on without end. Further, we are unable to tell how much, or what kinds of, karma is being produced by a particular action. All we are able to observe are the immediate mental and physical effects, which often are only a fraction of the total karmic result of that action. Someone robs millions of dollars from a bank and is never caught. The apparent effect of that crime is a life of riches and leisure. Has he escaped punishment completely? Perhaps in this life, but the law of karma assures that divine justice eventually will be meted out.

> *That which God has decreed from all eternity, ill for ill and good for good, can never change because God is the decreer. Who would say to do evil in order to have good? Does anyone ever plant wheat and reap barley, or plant barley and reap wheat? It is not possible. All the saints and prophets have said that the recompense for good is good and the retribution for evil, evil.*
>
> —Rumi[24]

Some form of learning—or reformation of character—occurs in line with karmic laws, but generally this happens unconsciously. That is, on a deep level our mind understands that a positive or negative effect is being experienced because of a like cause resulting from some action in the past. Thus automatically we are drawn toward, or away from, certain ways of acting as a result of these unconscious karmic influences. In this way we may become more virtuous, but not more godly. For now we are drawn toward good actions, and accumulate good karma, which produces more good actions. Yet all of this activity still is taking place mechanically within the spheres of mind and matter. Perfect mystics point out that we have exchanged iron chains for golden ones, but are just as firmly bound.

Desire nothing, get everything

What is the solution? In part, to avoid incurring the effects of actions. Then we experience the results of previous causes without adding new links to the karmic chain of cause and effect. Remember that I am referring here to mental causes and effects. That is, while continuing to perform physical actions in the world—being motivated to do so by pre-existing karmic causes—it is possible to act with an inward attitude of indifference to the effects of our actions. "Indifferent" does not mean "uncaring." We continue to strive to do the best we can in any situation, but stop dividing outcomes into dualistic categories: good/bad, pleasant/unpleasant, deserved/undeserved, welcome/unwelcome. It is our ego which makes us believe that how *we* perceive the world is the whole truth of the matter. I accidentally drop a twenty-dollar bill on the sidewalk and feel miserable about having lost that much money. A poor woman finds the bill and is overjoyed that now she can buy food for her hungry children. Was this episode good luck, or bad luck? It depends upon one's perspective, and from God's point of view it was neither—or both.

Sawan Singh tells a story about a female saint, Rabia Basri, who was sitting with some devotees and discussing the attributes of a true lover of God. One person said that "Only he is truly in love with God who, if he were subjected to continued pain and suffering, would bear it patiently, never uttering a single word of complaint." Rabia said that this definition contains too much of egotism. Another suggested that "He is a true lover who, if pain and trouble come from the Friend, will accept it with joy as a boon." Rabia said that this also fell far short of the truth. "A true lover is of a far higher order than any of your definitions, friends," she told them. "He is a lover who is so steeped in His love that he makes no distinction between pain and pleasure. Everything, whatever it may be, he welcomes as if coming from the Friend."[25]

> *He who seeks felicity is one thing,*
> *the lover something else—*
> *he who loves his head has not the feet for Love. . . .*
> *The world has two nests: good fortune and affliction—*
> *by God's Holy Essence, the lover is outside them both!*
> —Rumi[26]

Strange as it may seem, desiring nothing is a key to obtaining everything: God. Because karma—like all other laws of cause and effect—operates mechanically, the only way to become truly free is to break the machine of the mind. Throw a wrench into its gears. Stop its habitual patterns. Steer clear of ruts. Rather than blindly treading along the paths traveled in life after life, it is possible to head for open ground, to act more spontaneously and naturally, to cease being driven by egotistical desires. Instead of playing the part of, well, a *part*, our movements on this stage of life begin to be a conscious reflection of the *whole* that is God. Of course, in truth we never have been separated from that Oneness, but our sense of ego—partness—leads us to believe that we possess free will and the capacity for independent action. This is simple ignorance. Illusion.

> *We are like bowls floating on the surface of water. How the bowls go is not*
> *determined by the bowls but by the water. . . . There is no doubt that all bowls are*
> *floating on the water of divine might and will. . . .*
> —Rumi[27]

To realize this, stop paddling! Become an experimental spiritual scientist. Cease trying to reason out the whys and wherefores of human destiny and free will, determinism and chance. You will never succeed in knowing final truth in this manner. If you doubt that there is a current which unerringly guides your actions, take your hands off the rudder of your ego. Release the throttle of your desires. Sit calmly in the seat of your consciousness and observe what happens to the boat of your mind and body. Does it stop moving entirely? No. You will find yourself still engaging in actions, but without the anxiety that comes with wrongly believing that you are the navigator and steersman of your life.

Carnivals often have a ride for little children that features boats circling in a small pool of water. When my daughter was three years old, she used to enjoy this attraction immensely. She would sit happily in the moving boat, twirling the steering wheel this way and that, ringing a bell to warn other boats that she was coming. Of course, the wheel was not attached to anything, and any sort of warning was unnecessary, because all of the boats were connected to rods that kept them firmly in line and attached to the central machinery of the ride. But this did not matter to her. In her child's mind, *she was steering a boat!* And in our adult minds, a similar feeling is present: *we are controlling our lives!*

> *Those who use the expression "if God wills" are the true lovers of God. For the lover does not consider himself in charge of things and a free agent; he recognises that the Beloved is in charge.*
>
> —Rumi[28]

Actionless action

Perfect mystics teach that so long as our store of karmas, or accumulated causes, has not been exhausted, it is impossible to stop performing physical actions. And because thinking almost always precedes acting, this makes it equally impossible to cease all mental activity. Recognizing these facts, a person still can take steps toward his eventual release from these chains of cause and effect. The key is what Shanti Sethi calls actionless action:

> It is the nature of the mind and senses to indulge in actions. So long as the mind is not detached from this world, it is impossible for it to free itself from actions. Mind is restive by nature. It is difficult to still it even for a moment. . . . Consequently, it is of the utmost importance to remain desireless while performing actions. In other words, one must learn the art of performing "Actionless Actions." So long as man considers himself the doer of his actions he has to bear the weight of his karmas. When he gets rid of this feeling, he also gets rid of the burden.[29]

Action without desire is a subject of great subtlety. It requires a lifetime of dedication to the study of spiritual science to perfect. So only a hint of the inner meaning of actionless action can be provided here. As we have been discussing, this entails breaking the chains of cause and effect by ceasing to desire the results of actions. This most certainly does not mean stopping all mental and physical activity, for such is impossible as long as we exist on earth. Rather, one still has goals in mind—for these are determined by uncontrollable karmic influences—but tries diligently to stop having goals for those goals.

"I want to become president of my company." Fine, professional ambition is a worthy goal. "I want to be president of my company so people will be impressed by how rich and powerful I am." Not so fine. These desires for money and influence are permeated by the scent of ego. It is better to stick with as simple a motive for action as possible: the Zen saying goes, "Chop wood and carry water." There undoubtedly is a purpose behind these actions, but it is not necessary to keep that purpose in mind. Just chop wood and carry water. Then, light a fire and boil vegetables. One step at a time. You may die before being able to take the next step, and mysticism teaches that any unfulfilled desires at the moment of death attach us that much more firmly to the illusory reality of this physical universe.

In addition to acting in the present moment, our motive for taking action should be as selfless as possible. Perfect mystics advise us to try to act for the good of the whole, rather than for the benefit of ourselves—an insignificant part. When actions are performed from an attitude of doing one's duty toward God, they are not binding in a karmic sense. That is, because there is no ego-centered desire to personally derive some benefit from those actions, it is immaterial what kinds of effects are produced by the causes. The mind of the "doer" has not become attached to any particular result. He only is concerned that his actions are performed to the best of his ability. Whatever results accrue is in the hand of God.

The motive lying behind any action, as we have seen, is all important from the standpoint of spiritual science. This fact is reflected in secular laws. Imagine yourself waiting for a train to arrive. Some distance away two men are standing on the edge of the railroad platform. You cannot hear what they are saying, but they are talking and gesturing animatedly. Suddenly one man appears to reach out and push the other man backward, who falls off the platform onto the tracks. An approaching train crushes him under its wheels. He dies instantly. Now, the physical action which preceded that death is indisputable: the arms of one man make contact with the body of another. But the motive for this action makes all the difference in determining what effects are attached to it.

If the push was playful, a result of two friends jostling each other in fun without being aware of how close they were to the edge of the platform, then the death was tragic but accidental. No legal action would be taken against the survivor. If they were arguing and the dead man was pushed in anger, but not with a desire to do serious harm, then this might well be considered manslaughter. However, if the survivor *intended* to kill his companion by throwing him into the path of an oncoming train, the police will charge him with murder. If that intention was long-standing, then the homicide was premeditated. This would result in a more serious sentence than if it was a crime of passion—an urge to kill arising in the heat of the moment.

And there is yet another possibility. Contrary to how the scene appeared to witnesses, the survivor actually was reaching out *to save* his friend just as he was stepping backward in a suicide attempt. The reach and the step occurred so close together that an attempt to grasp could be perceived from a distance as a push. In this case, the survivor is a hero of sorts, for he easily could have been pulled along with the dead man onto the tracks. So we see that our understanding of the situation, and the consequences which should justly accrue to the man left standing on the platform, entirely depends upon the reason he reached out with his hands. By itself, that physical action has no meaning at all. This is a reflection of the law of karma. As Sawan Singh says, "All actions are performed with a motive, and it is the motive that is binding."[30]

In conclusion, then, the message of mysticism is clear: Think rightly, and you will come to act rightly. Act rightly, and eventually you will experience the grace of God. Then you will not consider yourself to be acting at all. Some other force—Spirit—will be guiding your thoughts and actions. Further, the effects of whatever causes impact upon your life will be felt imperceptibly, if at all. It is not a law of nature that we are bound to feel pleasure or pain, happiness or sadness, hope or despair as a result of the myriad events of life. These subjective emotions are within our control, and by eliminating them from our mind we come that much closer to a state of objective higher consciousness.

As Charan Singh explains, karma determines only what one experiences in life, and not our reaction to those events:

> Some people go through a tragedy without being much affected by it; other people pass through the same tragedy and start howling and crying. For both of them the tragedy they have to face is the same, but not its effect; that is dependent on their own spiritual and mental development. Certain people are so spiritually developed that they are not bothered by what is happening, though they may be facing a tragedy day and night. . . . And meditation helps us in that we become spiritually strong within ourselves and do not lose our balance in undergoing such karmas, we do not feel their effect.[31]

Freedom begins with longing to be free

I have devoted quite a few pages to describing the workings of karma, the law of cause and effect in spiritual science. Given this book's focus on Ultimate Reality, one might question the need to place so much emphasis on what admittedly is an attribute of limited reality—regions of mind and matter. Well, the plain fact is that God is there, and we are here. So we need to understand what keeps us bound to physical existence in order to be free. This is not easy to do. The strings of the puppeteer are so subtle and fine that we do not feel them controlling our actions. Those strands of karma which bind us must be recognized before they can be cut.

Tragically, most people do not even realize how enslaved they are. This prison house of materiality has been cleverly designed: it is exceedingly difficult to escape from a jail with no apparent walls. Prisoners do not attempt to gain their freedom if they are not aware of being confined. This is why it is almost impossible to free oneself from the law of karma without the guidance of a perfect mystic who knows the reality of liberation. As limited human beings, we have no basis of comparison, so the pitifully small amount of freedom we possess satisfies us.

Mysticism teaches: Raise your sights. Expand your horizons. Consider greater possibilities. Do not be content with playing endless games of cause and effect: "I do this to that, then that does this to me, and I do this to that again." How tiresome. *Become both this and that. Stop acting, and become action.* These words may sound mysterious, even nonsensical, but they point toward a way of being—a state of consciousness—which is eminently possible to attain. And once attained, the activities of life which once seemed so important appear inconsequential.

Once the part becomes a whole, partness loses its allure. After a taste of spiritual freedom, material bondage is insipid. Become more aware of the linkages of cause and effect which constrain your thoughts and actions. Have faith that these are but temporary chains. God has imprisoned you for a day so you may be free for eternity. He is waiting for you to chafe at your bonds and yearn for His freedom. That very yearning is the sound of Spirit's key beginning to open the padlock of mind and matter. It takes some time for the rusty tumblers of karma to turn and allow the door of our consciousness to open into Ultimate Reality. But this will happen if you want it badly enough. It will.

> God said, "It is not because he is despicable that I delay My gift to him:
> That very delay is an aid.
> His need brought him from heedlessness to Me, pulling him by the hair to My lane.
> Were I to satisfy his need, he would go back and immerse himself in that game.

*Although he laments to the bottom of his soul: "Oh Thou whose protection is
sought!"—let him weep with broken heart and wounded breast.
For I am pleased by his voice, his saying, "Oh God!" and his secret prayers. . . .
People cage parrots and nightingales to hear the sound of their sweet songs.
But how should they put crows and owls into cages?
Who indeed has heard tale of that? . . .
Know for certain that this is the reason
the believers suffer disappointment in good and evil.*

—Rumi[32]

Echoes in the new physics.

To understand how cause and effect is perceived by the new physics,
we need to begin with a brief look at the old physics. In deference to the
pioneering seventeenth-century scientist, Issac Newton, this often is
referred to as Newtonian science. Robert March says:

> In Newton's physics, motion is governed by perfectly deterministic laws.
> Early in the nineteenth century, the mathematical physicist Pierre-Simon
> de Laplace speculated that if one could only observe at some instant all of
> the atoms in the universe and record their motions, both the future and the
> past would hold no secrets. Put another way, all of history was determined,
> down to the last detail, when the universe was set in motion. The rise and
> fall of empires, the passion of every forgotten love affair, represent no more
> than the inevitable workings of the laws of physics; the universe marches to
> its unalterable destiny like one gigantic clockwork.[33]

A clockwork universe

For Newton, this clockwork universe was by no means Godless. He
was a profoundly religious man who saw divine omnipotence and
omniscience reflected in the perfectly designed laws of nature. However,
eventually this deterministic perception of reality came to be an argument
for leaving God out of material science: if He "winds up" the cosmos like a
clock, and then watches events mechanically unfold, the workings of the
universe can be understood without taking God into consideration. That is,
God certainly is needed to explain how the world came into existence. But
if one knows the laws of nature He set into motion—and the conditions
that prevailed at the moment of creation—then all events, past and future,
are an open book.

From this perspective, each instant of time can be thought of as a page
in the vast volume that contains the history of creation. Even though we
only can observe directly one page at a time—the present moment—
Newtonian physics implies that in theory one could reproduce the content
of all the pages by knowing what was on just one page. (As I write these

words, the world is on page 3:17 pm on March 11, 1994. Given that I know precisely and completely the conditions prevailing in the universe at this instant, and how the laws of nature operate to bring about changes in those conditions [admittedly two big "givens"], I could move either backward or forward through history and tell you—for example—what is on page 2:29 am on October 7, 1066, or page 9:23 pm on June 21, 4025.) In the eyes of the old physics, the fact that no scientist *actually* could do this was due solely to incomplete knowledge, and not because of any theoretical barrier. The book of existence was there to be read from cover to cover, beginning to end, and all material science needed to do was learn how to turn the pages.

The clock turns into a pinball machine

With the ascendance of quantum theory, however, the new physics began to come to different conclusions. Remember that the behavior of individual quantum phenomena—such as the emission of a radioactive particle—cannot be predicted exactly. The equations of quantum physics assign *probabilities* to subatomic events: "There is a high probability that the particle will be emitted at this moment, and a low probability that it will be emitted at that moment" (the actual mathematical language, of course, is much more precise). Given this perspective, Heinz Pagels says that "The world changed from having the determinism of a clock to having the contingency of a pinball machine."[34] This is what makes playing pinball so much more interesting than watching the hands of a clock go around.

Mostly—in my experience at least—a pinball drops out of play in a few seconds. But once in a while it hits just the right place, and a burst of light and sound occurs as points are accumulated rapidly. Then the ball falls out of play, as always, but those rare and unpredictable high scores are what seem to keep people playing the game. A clock, on the other hand, never does anything very interesting. Second by second, hour by hour, it moves its hands at the same measured pace. Even when a clock breaks, it generally just stops working rather than exploding dramatically. So it is not difficult to understand why the unpredictable probabilities of quantum physics produce so much discussion and interest among both professional scientists and interested laypersons. To some, they even represent a key to understanding the universe.

Pagels for example, goes on to say that with this new view of a pinball world, "Physicists realized that the concept of the perfect all-knowing mind of God has no support in nature. . . . Like us, God plays dice—He, too, knows only the odds."[35] Now, this is a strange bit of metaphysics which not infrequently is echoed by other material scientists: since *we* perceive reality in a certain fashion, this must be the way *God* sees things as well. Everyone certainly has a right to his own opinion, but it is difficult to fathom how a physicist could use quantum theory to draw the conclusion that God is not omniscient and that, like us, He knows only the probability that a particle will be emitted at a certain instant.

Randomness or ignorance?

Using more subtle philosophical reasoning, B. Alan Wallace notes that when physicists encounter natural events for which no definite causes could be identified, "They are then faced with two options: (1) to admit that, at least for the time being, the causes of those events remain hidden from our understanding; or (2) to claim that those events occur at random, without cause. . . . The choice is a metaphysical one, and the prevailing interpretation of quantum mechanics has projected our human uncertainty upon the physical world by claiming that it is inherent in nature."[36] This is the view of spiritual science. Even though the human mind sees uncertainty in the subatomic realm, this does not mean that God—or Spirit—fails to guide every quantum particle, as well as all other entities in creation.

As Wallace implied, logic and reason are of little help in deciding whether cause and effect, or causelessness, applies in an ambiguous situation. For it turns out that there is no way to prove that an event is random. In fact, conclusive support for the very existence of what we call randomness does not exist in either mathematics or physics. The reasons for this are somewhat arcane, but Heinz Pagels says that the view of material science is that "a precise mathematical definition of randomness for finite sequences [a list of numbers] simply does not exist. . . . The mathematical theory of probability begins *after* probabilities have been assigned to elementary events. How probability is assigned to elementary events is not discussed [by mathematicians], because that requires an intrinsic definition of the randomness of events—which is not known."[37]

So while some physicists argue that God lacks perfect knowledge because He "only knows the odds," these scientists do not know what produces those seeming probabilities. One would think that a force such as the uncertainty principle that supposedly prevents even the Creator from knowing about what He has created would have a more solid foundation. When mathematics and physics do not know what randomness *is*, how can material scientists be so confident that causelessness is a fundamental principle of existence, rather than simple ignorance of the nature of Ultimate Reality? In other words, does randomness even exist? If chance is an illusion, then clearly cause and effect rules the physical universe, whether in the guise of certainties or of probabilities.

A free lunch really isn't free

Recall that adherents of quantum cosmology believe there is a certain probability that a physical universe such as ours can be created by chance from a random fluctuation of vacuum energy. One of the criticisms of this purported creation out of nothing, the so-called "free lunch" hypothesis, is that there is nothing free about it. If you go to a restaurant which serves a complimentary lunch to one out of every thousand patrons, your ability to eat without paying is controlled by statistical probability. Those probabilities are what produce the occasional free meal, not "nothing." Take away the concept of *one in a thousand* and you take away the possibility of *free lunch*. The one determines the other. This is why Heinz Pagels says that probability distributions, which control the outcome of games of dice as well as the number of dog bites each year in a large city, constrain human liberty:

> You may think you exercise your freedom by promoting a specific political opinion or deciding to wear blue shoes, but in fact, your actions are just part of a probability distribution. . . . What is perceived as freedom by the individual is thus necessity from a collective standpoint. The die when it is thrown may 'think' it has freedom, but whatever it does it is part of a probability distribution; it is being influenced by the invisible hand. We cannot act without being part of a distribution—it is like being in an invisible prison held by invisible hands. Even the very act of trying to escape is again part of a new distribution, a new prison.[38]

This sounds very much like the mystic conception of karma: every action produces a reaction, which limits one's range of possible future actions. Further, our physical and mental attributes bind us with chains of statistical probability—which may be thought of as a sort of summary of individual karma. My life insurance company knows how likely it is that I will die this year. This is how they calculate my premium. They do not know *which* of their forty-five year-old male policy-holders will die in 1994, but with surprising precision actuaries can predict the total number. If I am killed tomorrow in an automobile accident, friends and family will say, "how untimely was his death." But when that fact appears in my insurance company's year-end statistics, likely no irregularities will be apparent. Perhaps 429 deaths in my age and sex group were expected, and I was one of 435. If man proposes, and God disposes, probabilities are a summation of divine will.

Mathematician Ivar Ekeland says that while philosophers and scientists argue about whether the nature of the world is random or predictable, determinism is unavoidable: "If the final reality is described by probability theory, the world will be subject to the laws of statistics. With these laws, we can accumulate independent events that are highly uncertain on the microscopic scale and obtain almost certain facts on the macroscopic scale. Determinism here is a matter of experience. . . . We can't get away from determinism. Chase it out the door, by postulating total incoherence, and it comes back through the window in the guise of statistical laws."[39]

Determinism rules (but slyly)

Thus many physicists and mathematicians are coming to the conclusion that certainty and probability, predestination and free will, causation and randomness, order and chaos—whatever words we wish to use—actually are two ends of the same pole, varying manifestations of a single deterministic reality. Referring to Albert Einstein's objection to the belief that "God plays dice with the universe," mathematician Ian Stewart writes:

> The very distinction he [Einstein] was trying to emphasize, between the randomness of chance and the determinism of law, is called into question. Perhaps God can play dice, and create a universe of complete law and order, in the same breath. . . . For we are beginning to discover that systems obeying immutable and precise laws do not always act in predictable and regular ways. Simple laws may not produce simple behavior. Deterministic laws can produce behaviour that appears random. Order can breed its own kind of chaos. The question is not so much *whether* God plays dice, but *how* God plays dice.[40]

Consider this example, provided by Heinz Pagels, of a seemingly random sequence of numbers:[41]

31415926535897932384622643383279502884197l. . . .

This sequence could go on for a million digits, or two million, or a billion. Regardless of how many numbers are in the list, various mathematical tests could be carried out to try to determine whether the sequence is "random." Pagels notes that one test for randomness would be that each of the ten integers (0,1,2,3,4,5,6,7,8,9) appears in the list on average one-tenth of the time. Suppose we assume that this list passed every possible test for randomness. Pagels asks, "Can we then conclude that the sequence of integers is random? Unfortunately we can never reach such a conclusion, even if all the tests are passed, because the sequence may be specified by a rule and then it does not correspond to what we think of as random—not governed by any rule."[42]

As the reader may have surmised, the list of numbers above is the decimal expansion of $\pi = 3.14159$, the ratio of the circumference to the diameter of a circle. Obviously this is not a random number at all, but that cannot be known by studying the sequence. If you know the *rule* that produces the list of integers that constitutes π (pi), it would be possible to amaze a friend who only is aware of the number sequence itself. Your friend calls on the telephone and says, "Try to guess the five integers that come after the 4278th number in this list I have before me." "Wait a minute," you reply, turning to a computer programmed with the equations needed to calculate pi to a particular accuracy. And almost instantly you have the correct answer. Your ability to predict the exact sequence of numbers is not miraculous—you simply know the rule that generates the list, and your friend does not.

What if that friend fails to phone you and continues to believe that the list of numbers before him is random? As was already noted, he has two options: (1) assume that he lacks complete knowledge about that sequence, so cannot determine whether the list was produced by a rule; or (2) erroneously jump to the conclusion that the sequence is random, and no one is able to fathom it. It would seem irrational for a scientist to choose the second option, but this is what many physicists do when confronted with the seeming uncertainty of the quantum world.

B. Alan Wallace writes that those scientists also could "assert that we simply do not know whether atomic processes are subject to strict determinism. We do not even know whether the only possible causes are necessarily of a physical nature. The enormous success of quantum mechanics in accounting for a wide range of phenomena does not rest on the metaphysical denial of strict causality. It would work just as well if physicists acknowledged certain limits to their domain of knowledge; and this might be more responsible than informing society that the fundamental nature of physical reality is irrational and chaotic."[43]

The pi example contains an important lesson for both everyday life and quantum physics: it cannot be assumed that something happens by chance just because it appears to be random. This, of course, is a central teaching of spiritual science. The events in our lives often *seem* to take place in a happenstance manner, just as our "free will" apparently can choose thoughts or actions which are unrelated to previous mental or physical activities. However, mysticism says that this freedom of choice is an illusion, similar to the seeming randomness of the integers above. One is produced by karma, the other by the decimal expansion of pi. In each case a deterministic rule—or law—generates unpredictable effects.

Randomness is hidden order

Unpredictable, that is, given the limited conceptual and technological tools available to man. Not unpredictable to God, or one who has reached a highly evolved state of consciousness. For the essence of creation is order. This is not only a tenet of mysticism, where the conscious energy of Spirit is recognized as the all-pervading manifestation of God's law, but a widely-accepted conception in material science. Randomness can be viewed as a special case of extremely complex order. Physicists David Bohm and F. David Peat liken this to a flowing river, where in some places the water moves smoothly and predictably. But if rocks obstruct the flow, or the channel twists and turns abruptly, the river forms eddies, whirlpools, bubbles, spray. Each of these elements is produced by a definite cause which could be well understood in isolation.

However, when all of those elements are combined in the reality of a fast-flowing river, the resulting motion of the water is so complex as to be unpredictable, "random." Try this for yourself: place two corks side-by-side into the water just above some rapids, and watch them float downstream. After traveling even a short distance, they probably will be far apart, even though their journey started at exactly the same point and the same time, and each was acted upon by totally deterministic forces. A tiny eddy may catch one cork and move it into a backwater, while the other cork is pushed further into the main current. Bohm and Peat write, "Randomness is thus understood as the result of the action of the very small elements on each other, according to definite orders or laws in an overall context that is set by the boundaries and the initial agitation of the water. By treating randomness as a limiting case of order, it is possible to bring together the notions of strict determinism and chance (i.e., randomness) as processes that are opposite ends of the general spectrum of order."[44]

In other words, conclude Bohm and Peat, "order pervades all aspects of life."[45] Strict determinism—one billiard ball knocking another into a corner pocket—is order of a low degree. Chance—the random motion of a gas caused by billions upon billions of molecules colliding with each other—is order of an infinite degree, because the details of all of those myriad interactions cannot be taken into account. Yet each is a manifestation of the universal order of cause and effect. Physics thus finds in nature the same dynamics as mysticism recognizes in the human mind: determinism is to chance, as fate is to free will. Seeming opposites are reconciled through the overarching concept of order.

When I hit my thumb with a hammer, the cause and effect relationship is obvious. Simple order prevails. When I am diagnosed with a rare disease that afflicts one in a million people, the causal chain is hidden, and I may say, "What terrible luck has befallen me." Complex order prevails. Anatole France said, "Chance is the pseudonym of God when he did not want to sign." Like corks floating down a river, we are aware only of the most immediate forces acting upon us. When a current of life throws us against a hard rock of suffering, we often consider this to be a random and uncontrollable event.

Uncontrollable it may be, but not at all random. For our limited minds are not aware of the uncountable influences which have acted upon us during our journey down the river of existence. Each influence operated completely in accord with laws of cause and effect. Taken as a whole, those actions and reactions combine to produce behavior of virtually infinite complexity, and hence seeming randomness. Just as human beings are unable to tell what has made them think and act in a certain way, and term this ignorance "free will," so has the new physics recognized that the dynamic behavior of physical entities may be so unpredictable as to appear "chaotic." Determinism—cause and effect—always rules, but usually from behind the scenes. Physicist Doyne Farmer says of these complicated dynamics: "On a philosophical level, it struck me as an operational way to define free will, in a way that allowed you to reconcile free will with determinism. The system is deterministic, but you can't say what it is going to do next."[46]

Mathematician Roger Penrose emphasizes that not every cause can be known, nor every effect predicted. He writes:

> The issue of *determinism* in physical theory is important, but I believe that it is only part of the story. The world, might, for example, be deterministic but *non-computable*. Thus, the future might be determined by the present in a way that is *in principle* non-calculable. . . . Computability is not at all the same thing as being mathematically precise. There is as much mystery and beauty as one might wish in the precise Platonic mathematical world, and most of this mystery resides with concepts that lie outside the comparatively limited part of it where algorithms and computation reside.[47]

For want of a nail, a kingdom was lost

In general, these mysterious mathematical concepts to which Penrose is referring are *non-linear*. This term has a simple meaning—not-straight. James Gleick says, "Linear relationships can be captured with a straight line on a graph. Linear relationships are easy to think about: *the more the merrier*."[48] That is, what goes in basically is what comes out. If you have a linear cake recipe, and add two cups of sugar instead of one, you can expect that the cake will be twice as sweet as usual. By the same token, if you fail to measure the sugar precisely and put in 1 1/16 cups of sugar, that cake is going to taste just about the same as if you had followed the recipe exactly. With a linear equation, as with a recipe, small changes in a variable—or ingredient—produce correspondingly small effects. Big changes, of course, result in big effects.

But non-linear relationships are much less predictable, for they are not strictly proportional. Put a tiny bit of extra sugar into a non-linear cake recipe, and when that dessert comes out of the oven you may get a big surprise. Perhaps the cake shrunk to the size of a marble, or exploded, or even turned into a pie. As Ivars Peterson and Carol Ezzell put it, "a nonlinear system offers the possibility that a small change can cause a considerable difference in the final result."[49] This is reflected in the saying,

"For want of a nail, a kingdom was lost," based upon the story (as I recall it) of a king being fatally thrown from his horse as the result of a loose horseshoe, which in turn was caused by a single missing nail.

Mathematician Ivar Ekeland offers a Norse version of that tale: During the battle of Svolder, expert archers on two warring ships were trading shots. First, Einar—one of the archers—shot an arrow which passed only a few centimeters from the head of Earl Eirik, the commander of the warship *The Long Serpent*. Then, Eirik's master archer shot, and *his* arrow struck Einar's bow in the middle, bursting it in two. The result of this was known immediately to Einar: "Then said King Olaf, 'What cracked there with such a loud report?' Einar answered, 'Norway, out of your hands, sir king.'"[50] Indeed, notes Ekeland, those two arrow shots determined the course of the battle, and with it the history of Norway:

> We marvel at the fact that a few centimeters' deviation in the course of an arrow could change human destinies and decide the fate of a kingdom—in the final analysis, this translates into a few tenths of a millimeter to the left or to the right in the position of the fingers on the bow, and a few tenths of a second earlier or later in releasing the arrow. . . . What this story shows . . . are the major consequences that slight modifications can produce in the normal unraveling of a temporal process. The phenomenon is well known in mathematics under the name of *exponential instability*.[51]

Ekeland explains this term with a meteorological example: the magnitude of a weather disturbance doubles every three days if nothing interferes with its development. Thus over brief periods of time, small changes—such as a butterfly batting its wing or someone lighting a candle—will have no appreciable effect on large-scale weather patterns. However, says Ekeland, if an effect doubles every three days, it will be multiplied by 1,000 every month, 1,000,000,000 every two months, and 10^{36} every year: "this is an enormous number, which means that the flapping of a butterfly's wing or the flame of a candle can cause a cyclone at the end of a year, in the sense that in a test atmosphere, in which everything else remains the same, if this butterfly or this candle hadn't existed there wouldn't have been a cyclone at that moment."[52]

Chaos is everywhere—and that's OK

This is a product of *chaos*. Not in the sense of total confusion, but the special mathematical meaning of this term: random behavior occurring in a deterministic system.[53] More precisely, behavior that is so complex, so sensitive to small influences, that it appears random. Is chaos limited to the weather, or stock market prices, or other systems composed of vast numbers of interacting entities? No, material scientists are finding that chaotic behavior, rather than being an exception to predominant linear laws of nature, basically is the way the world works. This should come as a relief to anyone—which includes almost everyone—who is unable to make perfect sense of his life. Relax, this is normal.

Mathematician Ian Stewart writes, "Today's science shows that nature is relentlessly *non*linear. So whatever it is that God deals in, it's not explicit formulas. God's got an analogue computer [which do not use digits, 0s and 1s] as versatile as the entire universe to play with—in fact it *is* the entire universe—and He finds little satisfaction in formulas designed for pencil and paper. Less blasphemously: it's no surprise that nature is nonlinear. If you draw a curve 'at random' you won't get a straight line."[54] Look at nature for proof of this. How many tree branches are unbent, how many rivers take a direct course to the sea? Only human beings, it seems, expect that life will be linear and predictable. Such is impossible, for both spiritual and material science view the world as seamlessly interconnected. When one piece of the cloth of existence moves, the *whole* garment moves. And vice versa.

As James Gleick puts it, "nonlinearity means that the act of playing the game has a way of changing the rules."[55] This is a central feature of both the law of karma and physical laws of nature, such as gravity. Take three gravitating bodies—say the sun, earth, and Jupiter. Each body attracts the other two. The sun tugs on the earth and Jupiter, while the planets pull on each other and the sun. Picture one instant in this ceaseless tug-of-war: the sun pulls the earth toward itself, which changes the gravitational force involved. At the same time, Jupiter is moving either toward or away from the earth, which also alters the interplay of gravity.

To be more precise, Einstein's equations of general relativity demonstrated that gravity does not so much "tug," as "curve." That is, no gravitational force passes between two bodies—such as the sun and earth. Rather, says physicist David Lindley, Einstein's theory states that "the presence of mass makes space curved, and that the curvature of space makes massive bodies roll toward each other."[56] This is relatively easy to visualize, because two people sleeping on a soft mattress do the same thing, but with the cosmos there is a complication: the cause of gravitational movement creates effects which alter that cause. Lindley writes: "Curvature [of space] makes bodies move, and moving bodies make a changing curvature, and thus the way a system changes with the passage of time results from an intricate feedback."[57]

Such is the nature of chaos. Non-linear systems constantly change, and those changes change what is producing the changing. This is a reflection of the mystic conception of karma. Action is inescapable in regions of mind and matter. Once an action is performed, a reaction is inevitable, which leads to another inescapable action. Interestingly enough, a mathematical proof of this tenet of karma theory can be found in the aforementioned equations of Einstein. In Lindley's words:

> A universe constant both in time and space was not a solution of the equations of general relativity—unless the density was exactly zero. In other words, you could have a perfect unchanging universe, but only if there was nothing in it. As soon as some matter was introduced, the influence of that matter on the curvature of space, and the feedback of the curvature on the distribution of matter, forced the universe to change somehow with time, to be inconstant.[58]

All of this activity ended up producing the present conditions in our solar system. And those actions and reactions will continue as long as our corner of the universe exists, yet in an unpredictable manner. Even the slightest "random" perturbation—such as a small asteroid hitting earth—can throw long-range astronomical predictions way off base. Similarly, any inaccuracy in a calculation, no matter how small, will become vastly magnified over cosmic lengths of time. Ivar Ekeland says that studies of the entire solar system have found that disturbances are multiplied by 10 billion in 100 million years. "That is," he writes regarding computer simulations, "a fluctuation of one-tenth of a meter in the initial position can ultimately translate into a displacement of one million kilometers. . . . "[59]

Implications of chaos for everyday life

"Well, so what?," you might be asking yourself, "I'm not worried about what happens to the universe in a few million years, so all this does not mean much to me." Not true. Chaos theory has far-reaching implications for everyday life. First, *it is senseless to try to isolate the ultimate cause of something.* Without being God, and knowing the complete truth about the totality of existence from the beginning to the end of creation, reasons for why things are as they are cannot be found. Yes, it is possible to put forward causative theories, but these invariably will be simplistic approximations of the complex dynamics which actually produced some event, or behavior. This applies both to living and non-living entities, but is most clearly visualized in the case of physical systems. In the words of Ekeland:

> If a demon displaces the earth a few centimeters from its orbit today, over a long enough period of time he will affect all the planetary orbits, and this effect can only be calculated, or even envisioned, by considering the solar system in its entirety. . . . We cannot calculate the effect of a single terrestrial palpitation if we limit ourselves to simply considering the earth's orbit; we do, of course, have to take into account the primary effect on

this orbit, but also the secondary effects that changes in the earth's orbit will have on other planetary courses, because of Newton's law of gravitation, and the inevitable tertiary effect that the secondary effect will have once the changes in planetary paths have reverberated back onto the earth's motion, and so on.[60]

This is the material manifestation of the law of karma, never-ending chains of cause and effect. The earth moves. This changes the course of the planets. The altered planetary orbits affect the motion of the earth. That in turn affects the planets. Interactions without end. After several hundred million years, when the positions of the planets are completely jumbled around, no one could trace the cause back to an imperceptible twitch of a fraction of an inch in the earth's orbit. Too much has happened since, and it could easily be argued that the cause is the dynamics of the entire solar system. Ekeland says, "As a general rule, we cannot isolate a subsystem in a deterministic system; hence we cannot attribute a certain effect to a certain cause."[61]

This brings us to the second implication of chaos theory: *everything in the universe is connected.* From the broadest perspective there are no independent parts in the cosmos, only the whole. Once again we come face-to-face with the unbroken wholeness of existence. No matter which way we turn in our explorations of either spiritual or material science, unity is evident. Ekeland observes that chance often is viewed as the intersection of independent causal sequences: "A person is walking down the street just as a shingle comes loose from a roof; it hits him on the head and he dies on the impact. . . . Two sequences of events, each with its own logic; they are so clearly separate, and their common result is so out of proportion, that we immediately bemoan his bad luck—a matter of chance."[62]

This, says Ekeland, is an illusion. There is no such thing as independent causal sequences in the universe. Put more simply, nothing acts on its own. "To talk about independence is only a convenient approximation, a myopic view of events which we are forced to abandon if we are looking for more refined analyses or more distant horizons."[63] Ekeland observes that the movement of a single electron on a distant star changes the forces of attraction exerted on all the other particles in the universe, including the gaseous molecules of the Earth's atmosphere. Amplified by meteorological instability, a slight breeze in the Caribbean becomes a cyclone in the eastern United States. Then a shingle blows off a roof. And a man dies. He concludes:

> To try to isolate the causes of an event that has affected us is a necessarily limited venture; if pushed to an extreme, we might find ourselves investigating the movement of electrons on Sirius. We can only apprehend

a small piece of the vast universe at a time, and we don't know when what we've forgotten is more important than what we see. We are like travelers lost in the fog; our gaze defines a small area that is reassuringly familiar, but beyond the gray walls that surround us the realm of the spirits begins.[64]

From the standpoint of either spiritual or material science, then, this conclusion is inescapable: laws of cause and effect rule the physical universe, but not in an obvious manner. As the perfect mystic Jalaluddin Rumi said in a previous quotation, "it takes a very well-illumined eye to see all the links in the chain" that connect an action and its reverberating reactions. This results in behavior which appears chaotic, random, unpredictable, capricious, and yet actually is lawful, determined, just, orderly. As determinism lies under seeming randomness, so is Ultimate Reality the foundation of—and something far different from—either necessity or chance. When pure consciousness is liberated from mind and matter, cause and effect becomes a meaningless concept. After the drop of soul merges in the ocean of Spirit, there is no part to act, and no part to react.

Let's see some proof of free will

I realize that these arguments are not sufficient to convince the reader who firmly believes in the evident fact of his own free will. Fine. If we were conversing in person, at this point we might agree to disagree. But allow me to put forward a few more arguments for your consideration: virtually all of the scientific evidence, all of it—both spiritual and material— is on the side of the determinists. Every tenet of mysticism and physics points toward a universe that is lawful, seamlessly interconnected, supported by unbroken wholeness. Believers in free will have only an inner *conviction* that their thoughts and actions are caused by an unfettered individual consciousness that is isolated from outside influences. No facts exist to support this blind faith.

I may appear to be belaboring this point. However, it is important for the seeker of Ultimate Reality to come to grips with this issue of determinism. As long as we consider ourselves to be as free as we can be, we will not try to loosen the snare of cause and effect. Recognizing one's bondage is a prerequisite for liberation. So carefully consider your present condition. Moment by moment, are you not enmeshed in chains of action and reaction? Are you in perfect control of your mind and body? Can you stop all mental and physical activity whenever you wish, and transport your consciousness to domains of pure Spirit? If you cannot do this, speak no more about your free will. You are deluding yourself.

Nick Herbert observes that the psychological theories of behaviorism assume that people act much like robots, responding to stimuli without any sort of conscious experience. What we call consciousness, in other words, is an illusion. Herbert says that:

> Antibehaviorists sought a kind of human action that would convincingly show that a human being is more than a robot. . . . Although psychology has returned to a more "humanistic" orientation, it is important to realize that behaviorism was never refuted. In particular, no enterprising antibehaviorist was able to come up with a type of behavior for whose explanation consciousness was essential. . . . The behaviorists in effect issued a challenge to modern experimental science to come up with an objective way to measure the presence of subjective experience. So far science has utterly failed to meet this challenge.[65]

Could this be because humans indeed are not conscious, insofar as our thoughts and actions are controlled by the unconscious workings of karma? Michael Talbot and Roger Penrose each cite the results of an intriguing study conducted by researchers at the Mount Zion Neurological Institute in San Francisco. Benjamin Libet and Bertram Feinstein measured the time it took for a touch stimulus on a patient's skin to reach the brain as an electrical signal. The patient also pushed a button when he or she was aware of being touched. It was found that on average the brain registered the stimulus within a ten-thousandth of a second, and the patient pushed the button a tenth of a second after being touched.

But it took almost half a second for the patient to be consciously aware of the stimulus. "This meant," says Talbot, "that the decision to respond was being made by the patient's unconscious mind. The patient's awareness of the action was the slow man in the race. Even more disturbing, none of the patients Libet and Feinstein tested were aware that their unconscious minds had already caused them to push the button before they had consciously decided to do so. Somehow their brains were creating the comforting delusion that they had consciously controlled the action even though they had not. This has caused some researchers to wonder if free will is an illusion."[66] Penrose is not sure what to make of this finding, but suggests one possibility: "Perhaps consciousness is, after all, merely a spectator who experiences nothing but an 'action replay' of the whole drama."[67]

Spiritual science agrees that our present state of awareness functions much more as an observer, rather than a creator, of both external and internal events. Mysticism says that reaching a state of pure consciousness, consciousness without mental content, is the only way to prove conclusively that anything other than programmed sequences of cause and effect hold sway both in the worlds of "out there" and "in here"—the realms of matter and mind. Since material science has provided no evidence that refutes the notion that human behavior is robot-like, and spiritual science teaches that we are bound by the law of karma, every sign points toward contemplative meditation as being the only way to achieve genuine freedom. It comes down to this: we only are free to *be*, not to *do*. Being is a liberating quality of Spirit, action a confining byproduct of mind and matter.

Who creates reality? One mind, or many minds?

Still, some people believe that there is a thread of support for free will in the probabilistic uncertainties of quantum theory. That is, since individual quantum jumps cannot be predicted in a deterministic manner, Roger Penrose says that "Early on, some people leapt at the possibility that here might be a role for free will, the action of consciousness perhaps having some direct effect on the way that an individual quantum system might leap. . . . But if [this leap] is *really* random, then it is not a great deal of help either, if we wish to do something positive with our free wills."[68] Penrose cogently pinpoints the problem with using quantum uncertainty as a philosophical prop for free will: what kind of freedom is it to be uncertain about what will happen? Being free implies the ability to do exactly what one wants, rather than merely not knowing what will happen next.

Yes, it might appear that an observer of quantum phenomena is free to produce certain effects: given the wave/particle duality of subatomic entities, when one looks for a particle of light, a particle is found. When the wave like nature of light is sought, then waves are observed. However, there is no evidence that any force other than the law of karma is operating here. That is, a chain of causes and effects leads an experimenter to desire to observe a particle of light, and voilá, that effect is manifested. While it appears that an individual conscious choice is responsible for bringing some potential aspect of the quantum realm into actuality, spiritual science says that in reality this choice emanates from the non-physical Universal Mind—which encompasses and controls the experimenter's personal mind.

Physicist Amit Goswami has arrived at a similar conclusion. He believes that a nonlocal unitive consciousness determines the result of a quantum measurement—not the local divided consciousness of a single person. Strong, albeit unscientific, support for this theory can be found by spending an evening in any pizza establishment. Find a seat near the counter where customers order, and listen to some discussions: "Kids, what do you want on your pizza?," a parent will say to a group of children. "Canadian bacon and sausage." "Anchovies and onions." "I don't like anchovies!" "I'm a vegetarian. No meat." "Everything, put everything on it." Eventually some agreement is reached, but to please everyone several different types of pizzas often end up being ordered.

There is, then, what might be termed the Pizza Choice Objection to the traditional view of quantum physics that a freely chosen individual measurement determines what quantum phenomena will be observed. Imagine what a mess it would be if every person was able to create their own little slice of reality on earth. How would those slices ever fit together into a coherent whole? Goswami writes, "The world would be pandemonium if individual people were to decide the behavior of the objective world, because we know subjective impressions are often contradictory. The situation in such a case would be like that of people coming from different directions and choosing the color (red or green) of a traffic light at will."[69]

Rather, he believes, outcomes in the quantum realm are the result of a *single* unitary consciousness. This is why the laws of nature manifest such order and harmony instead of confusion and dissension. A whole cannot be divided against itself. It is the individual ego, our sense of partness, which creates the illusion of free will. As Goswami puts it, ". . . we choose our conscious experiences—yet remain unconscious of the underlying process. It is this unconsciousness that leads to the illusory separateness—the identity with the separate 'I' of self-reference (rather than the 'we' of unitive consciousness."[70] So far from being a demonstration of free will, bringing potential quantum phenomena into actuality occurs within the bounds of cause and effect. For these actions and reactions are mediated by universal laws which are beyond the control of the individual self. The same holds true for any sort of positive thinking or affirmation.

I'm free . . . to choose my prison

From the standpoint of mysticism, it is strange that so many people believe that desiring something, and then attaining it, is a sign that we are free to "create our own reality." Actually, this means that we are free to decide what kind of bars keep us in prison. Desire and attainment, wish and fulfillment—all this is considered by spiritual science to be part of the law of karma. Sow and you shall reap. Certainly it is possible to have more money, a bigger house, an adoring spouse, better health. Just want any of these enough, and the universal law of cause and effect assures that at some point your desire will be fulfilled. And once that goal is attained, your sights will be set on another point over the horizon. If this is your vision of perfect freedom, then you already have it.

To reiterate, mysticism teaches that when we think and act, we are not aware of what lies behind these thoughts and behaviors. In order to confirm that we are free to act, we need to be able to stop acting. And this includes mental actions, thinking. When you can start and stop your mind at will, then only can it be said that you possess a modicum of free will. Otherwise that mind is simply a machine which carries out the activities determined by the mental law of cause and effect—karma. John Bell, the developer of Bell's Theorem—which has been termed the "most profound discovery of science"—had this to say about determinism and free will in the course of an interview concerning the meaning of quantum theory:

> *Interviewer:* Yes, I was going to ask whether it still is possible to maintain, in the light of experimental evidence, the idea of a deterministic universe?
>
> *Bell:* You know, one of the ways of understanding this business is to say that the world is super-deterministic. That not only is inanimate nature deterministic, but we, the experimenters who imagine we can choose to do one experiment rather than another, are also determined. If so, the difficulty which this experimental result creates disappears.
>
> *Interviewer:* Free will is an illusion—that gets us out of the crisis, does it?
>
> *Bell:* That's correct. In the analysis it is assumed that free will is genuine, and as a result of that one finds that the intervention of the experimenter at one point has to have consequences at a remote point, in a way that influences restricted by the finite velocity of light would not permit. If the experimenter is not free to make this intervention, if that also is determined in advance, the difficulty disappears.[71]

Better to be a part of the whole, than just a part

It is not necessary to know the details of what Bell is talking about to understand the essence of his point: both physical and mental actions in this universe are determined, not random, or the result of free will. If his words affront your sense of human dignity and responsibility, consider this way of looking at the issue: would you rather be isolated from everything in existence, or a part of the whole? To avoid being caught up in chains of cause and effect, you would have to be unaffected by any and all influences. People, pets, sunrises and sunsets, births and deaths, music, art, literature, conversation, the opposite sex. None of these things could move you in the slightest if your aim was to be completely self-willed. Is this the sort of life you want to lead? Or, failing to be one with God, is it better to be an integral part of Creation?

John Davidson says, ". . .the One creates the many and yet remains both undivided within Himself and present within the myriad energy patterns and forms of His creation. The presence of the One within the many thus gives rise to connection and relationship between the many 'parts'—that is, to causality. . . . It is an expression of the fact that the One is still present within the many. Connection and relationship is an expression of His presence within the creation."[72] So just as mathematicians have come to realize that necessity and chance are both part of an underlying deterministic order, cause and effect can be seen as a reflection of God's immanence in His creation.

Only God or a highly-evolved mystic truthfully can say that an event was predetermined. As limited human beings, this remains a belief or hypothesis until we are able to reach a state of consciousness beyond time and space as we know them. For the *pre* before *determined* implies foreknowledge. While many people claim to be able to foretell the future, almost all of them are hucksters. Those mystics who have this ability do not use it for public entertainment, so anyone who advertises himself as a seer almost certainly is not. Hence, it is more appropriate for us to say that events are determined, rather than fated.

In discussing karma with a skeptic over lunch, I know that I have struck a raw nerve if he says: "Oh, so you believe that everything is predestined. Well, maybe I am destined to stick this fork into the back of your hand." My companion, you see, wants to show me that he possesses sufficient free will to spontaneously and painfully alter the condition of one of my body parts. Unfazed, while prudently shifting my chair a bit further away from him, I reply: "All right, forget the idea of things being predetermined. Let's just say they are determined. After all, until we had this discussion you had never thought of sticking a fork into my hand. Where did that notion come from? It was caused by my comments about karma, and your indignant reaction to those words. Thus this very intention to demonstrate your free will is the result of a deterministic chain of events." This either convinces the other person that my argument is valid, or makes them yearn for a longer fork.

As the mathematician Roger Penrose pointed out, events can be determined yet completely unpredictable. So we lose nothing by realizing that laws of cause and effect control all mental and physical activity in the material universe. Life still will be mysterious and rich with surprises. Knowing that something is bound to be coming around the corner is not the same as seeing it before it arrives. What we gain by surrendering a belief in our illusory free will is infinitely more valuable than what we lose. Gained are peace of mind, a sense of connection with the cosmos, and an understanding that everything in life happens as it should. Lost are anxiety, a feeling that we are an isolated part of a fractured universe, and the self-pity that comes from believing that we deserve better than we are getting.

Raising the veils over Ultimate Reality

Something else may be gained by realizing that unseen laws of cause and effect are operating within and without us: we want to perceive those laws, to know from where they emanate, to experience the force which guides all in existence. I hope that this overview of karma and chaos theory has demonstrated why the study of creation never will reveal the Creator. The operator is too well hidden behind the workings of the machinery. It is impossible to penetrate those cogs and gearwheels which drive the laws of nature. Our analytical mind is chewed to bits when it tries. Science writer John Briggs has described what chaos theory means for material science (he mentions "fractals"—these are the visible patterns of mathematical chaos reflected in the appearance of trees, mountains, the scattering of leaves).

> Chaos theory and fractal geometry extend science's ability to do what it has always done: find order beneath confusion. However, the order of chaos imposes a definite limit on our ability. With the use of computers, scientists can see chaos, can understand its laws, but ultimately can't predict or exert control over it. The uncertainty built into chaos theory and fractal geometry echoes two earlier scientific discoveries of this century: the fundamental uncertainty that Gödel's theorem found skulking inside mathematics and the array of essential atomic uncertainties and paradoxes unearthed by quantum mechanics. Science, in this century, seems destined to learn about nature's intention to remain behind a veil, always slipping just beyond our understanding, imposing a subtle order.[73]

The complex determinism of chaos thus is another barrier between the new physics and Ultimate Reality. Material science appears destined to run into blind alleys. Mathematics promises to mirror precisely the laws of the universe until Gödel's Theorem—like a strict judge—decrees that no mathematical system is able to provide sufficient logical evidence that it contains the complete truth. Quantum physics delves ever deeper into the subatomic world, almost reaching the boundary between materiality and whatever mysteries lie beyond. Then the uncertainty principle—a sly jester—stops the progress of quantum mechanics dead in its tracks. "Halt! Stay away from the stage of quantum phenomena," it says. "You can sit in the audience and watch the magic being performed, but keep your distance from the trick by which the magician makes matter appear."

And now we find that chaos theory places an amazingly skillful juggler alongside the strict judge and sly jester. Thus a crowd is blocking the march of material science toward Ultimate Reality. "Watch me closely," teases chaos. "My myriad hands of cause and effect are much quicker than your eye of reason. I'm constantly tossing and turning the countless parts of existence, but you never will be able to keep track of my movements. In fact, I am juggling you along with everything else." John Briggs says, "Nature's chaos is profound—because the only way we can ever gain enough information to understand it will be to include the influence of even our attempts to gather the information itself."[74] Once again we find the dog of material science chasing its own analytical tail.

At this point in the book we can move to cut off that tail of mind and matter with the incisive blade of Spirit. This appendage is too distracting, and leads us in circles. The final principles of spiritual science will take us in the direction of the open reaches of Spirit. Detours are over. Have you noticed that my discussion of those principles has borne testimony to Marshall McLuhan's adage, "the medium is the message"? Moving from the simple truth of *God is one, and present everywhere,* to the complexities of *Laws of cause and effect govern lower levels of creation, including the physical universe,* it has taken more pages to describe the meaning of each principle. This, I believe, reflects cosmic geography—not the author's accelerating verbosity.

The further one descends into realms of mind and matter—and our universe is as low as you can go—the more veils are placed over Ultimate Reality. It takes considerable energy and effort to penetrate those barriers, intellectually or actually. Mysticism teaches that this striving serves certain purposes, such as exhausting our cleverness. We grow weary of journeying down blind alleys, exploring mazes with no exit, tackling questions that cannot be answered. Out of this spiritual frustration comes a wonderfully liberating realization: *I cannot light a fire with water, get myself clean with soap made of mud, or become sober by drinking whiskey. Nor can I think the Truth that lies beyond thought; my mind is the problem I am trying to solve, not the solution.*

With this insight new vistas open up. The mind finally has begun to realize its own limitations. It possesses a fresh humility, a willingness to stand aside and let the soul take center stage in the play of existence. The realm of Spirit beckons. Gradually roadblocks crumble. The path that leads to the Ultimate Reality of God stands revealed, if not yet traveled.

> *With every breath the sound*
> *of love surrounds us,*
> *and we are bound for the depths*
> *of space, without distraction. . . .*
> *Out beyond duality,*
> *we have a home, and it is Majesty.*
> *That pure substance is*
> *different from this dusty world.*
> *We once came down; soon we'll return.*
> —Rumi[75]

Principle 6.
Spirit creates and sustains
all levels of creation.

This supreme joy has no resting place—
It enters one form then another,
from box to box—an eternal movement
between heaven and earth.
Here it comes, pouring down from the sky,
seeping into the earth,
and rising again as a bed of roses.
Now it is water, now a plate of rice,
Now the swaying trees, now a horse and rider.
It lies within these forms for awhile
then bursts forth to become something new. . . .

—Rumi[1]

This principle is an elaboration of previously-discussed tenets of spiritual science: *God is one and present everywhere*, and *Spirit is God-in-action, and of His same essence.* Here we will examine in more detail some key questions: By what means does Spirit carry out and sustain acts of creation? What has Spirit created? Clearly these are large questions which cannot be addressed adequately in a few pages. In fact, they could not be answered in an infinite number of pages. For it must be remembered that Ultimate Reality—which encompasses the realm of Spirit—is non-symbolic. No word, number, concept, image, or thought is able to mirror that which has no opposite.

By definition, a symbol stands for something else. What can stand in the place of unity? Only something that is not part of that one. Then we lose the essence of the very unbroken wholeness which we are attempting to describe. So perfect mystics say that God is beyond speech, and write that He is beyond words. Whatever descriptions they give us of higher spheres of consciousness are intended to spur us to enter those domains ourselves. Using the subtle faculties of our soul, we are encouraged to prove the truths of mysticism by seeing and hearing the sights and sounds of spiritual regions. Blind faith is condemned as severely in spiritual science as in material science. Seth Shiv Dayal Singh tells us, ". . . a man can behold Reality with his own eyes, when his inner eye is opened. . . . O know thou thyself by thyself; believe not at all what others say."[2]

Spirit is all there is

Mysticism takes an operational approach to Ultimate Reality. Much more of an emphasis is placed on what actions must be undertaken to realize that final truth for oneself, rather than on static descriptions of what cannot be described. Still, a broad understanding of the levels of creation which have been created by Spirit is useful for this reason: the soul returns to God by the same means it was brought to this physical universe. What comes down, goes up. And in the same manner. As was noted previously, the elevator which takes us from the roof of a building to the basement still is available for the return trip—assuming, of course, that we are able to find and enter that means of transport.

The elevator of spiritual science is Spirit, the all-pervading conscious energy which both creates and sustains all levels of creation. This force, if we may even call it that, actually is much more than an "elevator." It also is the floors that make up the structure of creation, the laws which govern the activities at each level, and the root of the consciousness which attempts to understand the workings of the elevator, the floors, and the activities. Spirit is everything in creation. This one sentence uttered by Seth Shiv Dayal Singh encompasses the entire truth of existence: "Save Spirit is no other Reality; this do I tell thee over and over again."[3]

Our mind finds it difficult to grasp such a simple truth. We are accustomed to division, complexity, specialization. How could Spirit simultaneously be all-pervading *and* conscious *and* energy? We are used to having things fit neatly into mutually exclusive categories: "Electricity is energetic, but not conscious. Humans are conscious, yet possess little energy." "A hydrogen bomb is mindless and physically powerful; a genius is full of mental power and virtually devoid of material power." "Empty space is everywhere. Something with energy or consciousness has to be *somewhere.*"

When you see a large muscular man, brimming with physical power, are you more likely to suspect that he is a lumberjack or a theoretical physicist, strong and dumb or strong and brilliant? We say of someone, "He's a lover, not a fighter," as if it were impossible to be both at once. Many more examples could be given, but I am sure you get the point. Living in a domain of opposites—such as male and female, hot and cold, empty and full—we have no other way of viewing things. Lacking any direct experience of unity, our minds tend to assume that the separateness which distinguishes the physical universe must be a hallmark of *all* planes of existence. Such is not the case.

> God created suffering and heartache so that joyful-heartedness might appear through its opposite.

> *Hence hidden things become manifest through opposites. But since God has no opposite, He remains hidden.*
>
> *For the sight falls first upon light, then upon color: Opposites are made manifest through opposites, like white and black.*
>
> *So you have come to know light through light's opposite: Opposites display opposites within the breast.*
>
> *God's light has no opposite within existence, that through its opposite it might be made manifest.*
>
> —Rumi[4]

God's laws are conscious

Along these lines, both everyday experience and material science lead us to believe that there is a clear distinction between an unconscious *law* and conscious *beings*. In biology, there are laws which are believed to regulate the workings of cellular DNA and RNA. These, in turn, are held to be largely responsible for why beings—humans, cows, mice, spiders— look and behave as they do. While the cells containing DNA and RNA are part of a physical body, it is generally assumed that the laws governing heredity and development are independent of any particular being. Similarly, in jurisprudence homicide laws determine whether a person who kills another is convicted of a crime, or absolved by reason of self-defense. Again, it is evident that societal laws are separate from people, even though people make those laws. The saying goes, "No one is above the law"— which implies not only that laws and beings are individual entities, but also that no one *is* the law.

Mysticism teaches that viewed from the standpoint of Ultimate Reality, the situation is quite different. The laws of existence cannot be distinguished from the consciousness which created them, and in truth that being *is* the very essence of those laws. God, in other words, is inseparable from His edicts. As was discussed under another principle, God-as-God, God-in-Action, and God-in-Creation appear to be three entities only from our limited perspective. In reality, God and His will are one. Sawan Singh says, "The Supreme Lord, the soul and Spirit are a holy Trinity. The One Lord exists in all the three forms. . . . From the Will of the Lord there originated the Lord's Law, as the Name (Spirit). This became the Creator of all the universe. In this way, the Law is the connecting link between the Lord's Will and the putting of the Lord's Will into creative action."[5]

Spirit is God's primal law, and the source of all secondary laws— including the physical laws of nature. Spirit is like a soldier obeying his general. When the general orders an advance, his troops carry out that command. When God orders a drop of rain to fall from the sky, or an earthquake to swallow a city, the laws of existence cause such to happen. Certainly from one point of view there is a long and involved chain of causes which leads from a raindrop falling on a single insignificant planet of the physical universe, all the way back to the ineffable heights of divine consciousness. However, from another perspective there is no distance at all between materiality and spirituality, the creation and its creator. Higher consciousness sees the divine Will operating in every aspect of existence. The story is told of an evolved mystic who was traveling with a group of devotees, and one complained of the heat. She asked him to leave the group because he had found fault with her Lord.

> *This at least is notorious to all men,*
> *That the world obeys the command of God.*
> *Not a leaf falls from a tree*
> *Without the decree and command of that Lord of lords;*
> *Not a morsel goes down from the mouth down the throat*
> *Till God says to it, "Go down."*
> *Desire and appetite, which are the reins of mankind,*
> *Are themselves subservient to the rule of God.*
> —Rumi[6]

So material science—and a lower form of spiritual science—is the study of secondary causes, while mysticism is the study of the First Cause, Spirit. Pure spirituality makes no distinction between laws of existence and the Lawgiver. In part, this is why "miracles" are performed so rarely by perfect mystics. Though possessing all of the powers of God, they live in accord with the laws of this physical universe, for such are the manifestation of divine will. Whenever we say, "this should have happened otherwise," we are telling God that He has made a mistake. What fools we are. If the creator of this universe had wanted things to be different here, then the laws which govern materiality would not be as they are. What we experience is what He wanted. Sickness and health, cold and heat, happiness and sadness. These all are produced by Spirit, and no amount of weeping and wailing—or smiling and laughing—will make them otherwise.

Fate does not imply fatalism

Yet when mysticism teaches that everyone and everything is subject to the will of God, or Spirit, this does not imply that we should become weak and fatalistic. There is a subtle yet simple distinction here—fate is not the same as fatalism; accepting that one's actions may fail is different from failing to act. Playing God when we have not yet reached His level of consciousness is inappropriate. In Ultimate Reality, all is destined. In limited reality, all is action and reaction. Being part of physical existence, we must act our role, not God's. So long as you believe you possess free will, exercise it—even if this is an illusion from the standpoint of final truth.

> *When God assigns a particular lot to a person, this does not preclude his consent, desire, and free will. God makes an unfortunate man suffer and he flees from Him in ingratitude. But when He sends suffering to a fortunate man, he moves closer to Him. In battle, the cowards fear for their lives and freely choose retreat. The brave also fear for their lives, but they charge the ranks of the enemy.*
> —Rumi[7]

While conducting an experiment concerning gravity, no material scientist worries about whether he is acting in accord with the "destiny of gravitation." He simply goes ahead and carries out the research, knowing that his own body—along with whatever other materials are being utilized in the experiment—is subject to gravity and also the means of learning about gravity. In other words, even though the physicist is in the grip of the very force he is studying, he correctly does not let this keep him away from his research. Gravity constrains his movements, but still allows him to move around his laboratory. Unable (on earth at least) to study gravity while floating weightlessly, he learns about gravitation while unavoidably being subject to it.

Students of spiritual science sometimes worry about whether they are acting in line with their destiny. How senseless. If a physicist gives no thought about acceding to the law of gravity, why should mystics be concerned about accepting God's edicts? There is no choice about the matter in either case. To be unaffected by gravity, one would have to be outside of the physical universe. To be unaffected by the will of God, one would have to be outside of Him. And since everything in existence *is* His conscious projection, such is impossible. Consider carefully these words:

> *Whosoever is bewildered by wavering will*
> *In his ear hath God whispered His riddle,*
> *That He may bind him on the horns of a dilemma;*
> *For he says, "Shall I do this or its reverse?"*
> *Also from God comes the preference of one alternative;*
> *'Tis from God's impulsion that man chooses one of the two.*
> —Rumi[8]

Yes, Spirit is the force that activates, energizes, and guides all of creation—including you and me. At this moment are we completely merged with this ethereal essence of God? No. So we must continue to act as parts, rather than the whole. Certainly it is good to aspire to being the ocean and not a drop. But this aspiration must become an actuality before our lives are fully and completely a reflection of the divine Will. Until then, as the saying goes, "Tie the legs of the camel, and then rely on God" (updated version: "Lock the car doors, and. . . . "). Even though that camel might wander off anyway—or your car be stolen—use your God-given mind and body to best advantage.

I was asked once, "Does believing in karma mean that a woman who has been kidnapped must submit to being raped and murdered by her attacker?" No! It means that if she desires to remain alive, she must do her best to *escape*. If this means killing her assailant in self-defense, then so be it. That desire—as Rumi says—is an impulsion from God, and she will act in accord with it. Rumi also writes that "God wills both good and evil, but He only approves of the good."[9] Hence, mysticism teaches that one must not use impulsion as an excuse for wallowing in sin. It may be that the actions of both the kidnapper and his victim were motivated by forces beyond their understanding, but this does not absolve a wrongdoer from the consequences of his actions, nor an innocent person from their duty to resist evil.

> *When you place your goods upon a ship, you do so while trusting in God.*
> *You do not know which of the two you will be: drowned on the voyage, or one of the saved.*
> *You cannot say, "Until I know who I am, I will not hurry to the ship and the sea. Am I saved in this way or drowned. Reveal to me to which group I belong! I will not go on the journey in doubt and with empty hopes like the others."*
> *If you say this, you will not accomplish any trading, since the mystery of these two possibilities lies hidden in the Unseen.*
>
> —Rumi[10]

Spirit—the law behind all other laws

What, then, is the nature of that unseen Law of Spirit which decides the outcomes of ocean voyages, and every other event in the physical universe? To reiterate, we are accustomed to viewing laws as unconscious, and only the lawmaker as conscious. But upon closer examination this assumption is found to be untrue. Consider cars traveling on a road clearly posted with a speed limit: 65 mph. That is the law: drive no faster than this. It is obvious, however, that this law does not control how fast the cars are going—many go faster than the limit, some much faster. The rate at which each vehicle travels is dependent mostly upon the intention of the driver, and secondarily upon the top speed of his car. So we can see that in the case of human beings, awareness of a societal law has to be present in consciousness for that law to control behavior. (When a highway patrol vehicle—a symbolic reminder of the law—is within view, drivers are much more likely to stay within the speed limit.)

Similarly, when a revolution takes place, what happens to the laws in that country? Often they are ignored or replaced. Perhaps a law proclaims that no one can be put in jail without just cause. If a rebel rises to power, this statute—though it still can be found in legal libraries—will be of little comfort to those who find themselves in prison for no reason other than their political beliefs. The conscious will of a dictator supersedes unconscious words in a lawbook. After such a revolution, functionaries at every level—from the lowliest customs inspector to the highest-ranking cabinet members—soon realize that their authority stems from the government in control of the nation, and does not stand on its own. Rebel tanks about to overwhelm the army defending a capital do not stop when a white-gloved policeman raises his hand.

Societal laws thus are rooted in consciousness. If I am not aware that a law exists, or am not willing to obey it, then that edict has no effect on my actions. But what about "laws of nature?" Mysticism teaches that whether material or spiritual, these laws also are founded in consciousness—the all-pervading conscious energy of Spirit. Mind and matter conform to these laws because they are obedient subjects of God, their ruler. No revolutions, no coup d'etats take place in the Lord's domain. For the ruled are not separate from the ruler. Everything in existence obeys His command because in truth all *are* that Law—all are Spirit. Julian Johnson writes:

> Without Spirit, nothing could live for a single moment or even exist. All life and all power come from it. From the crawling ant to the thunderbolt, from the tidal wave to the solar cycle, every manifestation of dynamic energy comes from Spirit. . . . The pull of gravity, the flash of lightning, the building of thought-forms, and the love of the individual soul all come from this Current primordial. . . . It is God himself in expression. It is the method of God in making himself known. It is his Word. . . . It includes all of his qualities.[11]

Just as a minor official remains in office by virtue of his allegiance to the leadership of his country, every law of nature—material or spiritual—is subservient to Spirit. Sawan Singh says, "There is no power on earth or heaven greater than the power of Spirit. It is the Primary Power. All other powers are derived or secondary. . . . Spirit is the source of all else. It is the Power at the back of all other powers, and from which all else has been derived. . . . "[12] This simple truth answers questions which have perplexed philosophers, theologians, and material scientists for many centuries: How is God related to His creation? Through what means does He communicate with and guide sub-atomic particles, people, animals, plants, angels, and all the other entities in existence? What links mind and matter?

Linking mind and matter

This last question has been, and is, particularly vexing. I think, "raise my right index finger," and it does! We take this for granted, but that movement is a miracle of sorts. Daniel Dennett notes that since the mind and its thoughts certainly seem to be something other than the observable forces of nature (mysticism assures us that this is indeed true), no physical energy or mass is associated with them. "How, then," he asks, "do they get to make a difference as to what happens in the brain cells they affect, if the mind is to have any influence over the body?" Since this seemingly would require an expenditure of energy, says Dennett, "where is this energy to come from?" [13]

Like most material scientists who study the mind and brain, Dennett concludes that if there is to be a link between the two, that link must be material. Hence mind is reduced to matter. Otherwise, how could something non-material (mind) affect something material (brain cells)? With a touch of humor he likens this to a ghost who can effortlessly pass through walls, but also can tip over a lamp or slam a door. How could that spirit be ethereal and unaffected by matter at one moment, and capable of pushing material things around at another moment? The obvious answer is that this is impossible, since only matter can interact with matter.

And that obvious answer is wrong, completely backward from the truth known to spiritual science. Mind is able to affect matter because the essence of both is *non-material*—Spirit. By the same token, God effortlessly guides and communicates with everything in existence because that "everything" is made of Himself—Spirit. When you decide to raise your hand, and instantly do so, this is an infinitesimal reflection of the divine power which created this universe. How difficult is it to tell yourself to do something, and then do it? One aspect of you—the mind—is telling another aspect of you—a hand—to move. There is nothing complex or mysterious about this. Again, it is a weak echo of God's omnipotence and omnipresence. Within unity, there are no barriers to action; no time and no space; no resistance of any sort. Ultimate reality is not divided against itself.

> *When we say that God is not in the heavens, we do not mean that He is not in the heavens. What we mean is that the heavens do not encompass Him, but he encompasses them. He has a connection to the heavens, but it is ineffable and inscrutable, just as He has established an ineffable and inscrutable connection with you. Everything is in the hand of His Power. Everything is the locus of His Self-manifestation and under His control.*
>
> —Rumi[14]

Spiritual geography

Now we turn to another question: if Spirit is the thread that connects God-as-God and God-in-Creation—heaven and earth—what else is strung along this cosmic necklace? People often say, "When I die, I hope to go to heaven." Does this mean that there only are two floors in the structure of creation, the physical universe and heaven? What about hell? If this exists, is it "up" or "down" from earth? Or do these terms have any meaning in regards to spiritual geography?

Previously it was noted that mysticism does not put much emphasis on a travel guide approach to spiritual science. That is, rather than reading and hearing about higher domains of consciousness, it is much better to actually go to these regions. When you are freezing in the middle of a snowy winter, which of these options would you prefer? One, go to the library and check out a book about tropical islands. Read it while a blizzard whistles around you. Two, consult a travel agent and buy an airplane ticket to Hawaii. Sit on the beach and bask in the warm sun. Well, let's see. Speaking for myself, I'd choose the second option. Who wouldn't?

Perfect mystics are the travel agents of mysticism, and much more besides. They provide us with a brief description of the heavens above—an easily understood "brochure," so to speak—sufficient to provide added impetus for us to begin our spiritual journey. Certainly the misery we encounter in this physical universe should be reason enough to want to return to the bliss of our original home, just as freezing weather leads us to think about a vacation in more temperate climes. But if a person in the northern latitudes did not know that the tropics existed, he would be inclined to stay at home and put a few more logs on the fire. Similarly, hearing about the levels of creation helps us to realize that the path back to God and Ultimate Reality *can* be traversed. It is not a fantasy. That royal road of Spirit is more real than all we see around us.

Descriptions of higher levels of consciousness also serve this purpose: we learn what can, and what cannot, be taken on our journey back to God. Our travel agent, the perfect mystic, informs us that the conveyance of Spirit has an exceedingly strict baggage limit. No body, no mind, no ego can go with us all the way to our final destination. "Hey, wait a minute!," is the alarmed reaction of many when they hear this news. "What's the point of making this journey if no part of me will be there at the end of the trip?" "Calm down," soothes our mystic guide. "You will not need these coverings over your soul where you are going. Further, they weigh you down to such an extent that you cannot board the vehicle of Spirit with them." This makes sense. People do not take fur coats and heavy boots to Hawaii. And only a few pieces of luggage are accepted at the check-in of a passenger plane flight. Your grand piano must stay at home.

In addition, having an overview of the stages through which one must pass before reaching God provides reassurance to the first-time traveler. Leaving New York, how do I know that the plane I'm on *really* is going to Hawaii? Maybe it has been hijacked, or the pilot is lost, and I will end up somewhere I do not want to be. If I am aware that the airplane will pass over the Great Plains, Rocky Mountains, and the Pacific Ocean before reaching Honolulu, then seeing these milestones from my window seat allows me to relax and enjoy the ride. Similarly, perfect mystics describe just enough about the sights and sounds of each higher region of existence for the spiritual scientist to confirm that he indeed is on the path that leads to God.

A seeker of Ultimate Reality does not want to stop part way. Mysticism teaches that each spiritual domain reached in the course of contemplative mediation appears to be the final truth. Lekh Raj Puri writes, "In this scheme of creation, each stage controls and gives energy to all the planes below it. Thus each stage seems to be the last and final. . . . "[15] This is why it is absolutely essential for students of mysticism to learn this science from a perfect mystic who knows each stage of the inner journey. Otherwise, it is all too easy to be deluded into believing that one is having a vision of God, whereas actually that experience is taking place on a plane of consciousness barely above this crude physical universe.

God's creation is much vaster and more majestic than we can imagine. In the course of recorded history, relatively few people have traversed the Royal Road of God-realization from start to finish while in the human body. These are the perfect mystics, the saints and prophets. Others start that journey while in the physical form, and are able to continue it after death. These are the less-than-perfect mystics, who by far constitute the majority of spiritual teachers. So the *source* of information about higher regions of creation is all-important. An imperfect spiritual scientist is able to provide a less-than-complete description of what stages of consciousness lie between the everyday waking state and God. This is not because of any willful intent to deceive or mislead, but is simply a result of ignorance about what lies beyond the stage he has reached.

Five levels of God's creation

The following pages contain a broad overview of the levels of creation. It must be recognized that not only is it impossible for saints to describe the true nature of these non-symbolic domains of existence in words, but it is difficult to convey even a sense of what distinguishes the different levels. Thus in various contexts mysticism teaches that there are three, four, five, eight, or eighteen stages of consciousness—and these do not exhaust the ways of counting the uncountable. Such seeming inconsistencies arise because undivided wholeness can be cut up conceptually in many ways. Reality itself, of course, is not affected by all of this mental dissecting.

To someone standing outside on the sidewalk, a building appears to consist of five stories. But from the inside, landings are observed in between the floors. Are these to be counted as separate levels—giving ten stories—or as parts of a single floor? Or should we speak of only the first floor and what lies above ground level? Five, ten, or two: no matter how many divisions are ascribed to that building, it remains what it is, a single structure. So for the sake of continuity I will remain with the five-level model of creation cited in previous chapters.[16]

Fifth region

Beyond all is God-as-God. Beyond anything created there is no time or space as we know them. There are no divisions at all in the formlessness of unity. In the words of the perfect mystic Seth Shiv Dayal Singh:

> How far can I say? No one was there; the creation of the four realms had not taken place. What there was do I tell thee now: Wondrous Wonder 'twas all in Himself; Wonder, Wonder, Wonder. *Wonder then took on a form.* In Himself He ever remains; that state doth He ever retain. His Being doth no one know; He himself doth tell us of Himself. . . . He was Himself, and no other was there. Then arose his Will and came the manifestation of [the highest region].[17]

Level five is that top of creation, the domain of Universal Spirit. Perfect mystics call it the unfathomable, nameless, and invisible region, because this essence of God cannot be understood, spoken, or seen by any faculty other than direct perception by the soul. One's mind and body are left far behind before the gates of this region are passed. Julian Johnson writes, "It is inhabited by countless multitudes of pure souls, who know no stain of imperfection, no sorrow and no death . . . one single soul living there radiates a light equal to that of sixteen times the total light capacity of our sun."[18]

This often is called the home of the Supreme Being. Though omnipresent in His creation, in this highest domain of consciousness He is revealed most completely. The radiance of God is such that "one hair" of His is said to excel the brightness of trillions of physical suns and moons. Jagat Singh says that "millions of suns pale into insignificance before the effulgence of even one atom of this region."[19] Since this final truth is the conscious ground of all being—physical, mental, and spiritual—actually it is not a stage *within* creation, but the Ultimate Reality of every aspect of existence. Hence Seth Shiv Dayal Singh says, "This is the beginning and the end of everything and circumscribes all. The love and energy of this region vibrate at every place, in [each] part of the whole."[20]

Fourth region

Level four was the first stop of Spirit as it formed additional domains of creation. This also is a realm of pure spirituality, untainted by mind or matter, but the lower portion of this region is the source of those cruder constituents of existence: the creative energy of mind, and the inert consciousness of matter. Hence in its descent the unified force of Spirit divides here into two sub-forces. By the same token, spiritual science teaches that when the soul is able to rise to this domain, it understands the true meaning of the mystic saying, "I am That." For consciousness now leaves behind the covering of mind, which up to this point has accompanied the soul on its upward journey. Lekh Raj Puri writes, "Here the soul perceives that it is in essence the same as the Absolute Lord, and then it proceeds to merge in and become one with Him."[21]

It is impossible to visualize the immensity of this region. Insofar as our usual conception of space can be applied to levels of higher consciousness, perfect mystics say that this plane is seventy times as large as the whole of what lies below it: levels one, two, and three of creation. The mind boggles at such a statement, because it is most difficult to conceive of the vastness of even our physical universe—which constitutes a tiny portion of the "three worlds" that *together* would fit into one-seventieth of the fourth level of existence. Fortunately for seekers of Ultimate Reality, spiritual faculties more subtle than the mind are able to perceive directly the majestic geography of creation.

> *Speech is an astrolabe in its reckoning.*
> *How much does it really know of the sky and the sun?*
> *Or of that Sky which holds this heaven as a speck;*
> *and that Sun which shows this sun to be a grain of sand?*
> —Rumi[22]

Third region

Level three is known in spiritual science as the "causal" region, because it is the effective cause of everything that lies below. Thus in a sense Universal Mind—the dominant force in this domain of existence—is the power that guides our physical universe. But from a higher perspective mind, whether universal or personal, merely is a link between ethereal Spirit and crude matter. Just as our mind requires a brain to mediate the neuro-chemical linkages between thoughts and physical sensations/actions, so does the soul need a mental body to interact with the mental-material worlds. In the words of Julian Johnson, "Owing to the extreme fineness of spirit, it cannot contact the coarser worlds without an intermediate instrument. Hence it is obliged to clothe itself in some sort of medium of contact."[23]

In other words, here the soul puts on a covering of mind so it can function in a denser domain of existence. From our perspective, mind is ethereal. And it is, compared to physical matter. But mysticism regards mind as a subtle form of matter which manifests relatively little energy and no innate consciousness. Spirit is the power behind the mind, just as mind is the power behind matter. To exercise its delegated control over the lower regions of creation, Universal Mind manifests various subsidiary forces: these include three attributes, and five subtle forms, of materiality. As John Davidson puts it, these are "the pre-cursors or blueprint of the dense material out of which our physical universe and body are constructed . . . the blueprint of the blueprint, so to speak, of our gross physical universe."[24] Hence, says Davidson:

> Mind is the power which is primarily responsible for the apparent or illusory division of the One Word [Spirit] into the myriad forms and rhythms we experience in the physical creation. It is the creator of the physical, astral and causal domains, and our individual human mind is only one small part of all that goes to make up this greater Mind. . . . The Mind is both the architect and the administrator of these regions, the highest level within these realms of the Mind being termed the Universal Mind.[25]

Thus the third region often is considered—explicitly or implicitly—to be the final truth, since creation of the lower domains starts from here. However, even though Universal Mind is the creator of our physical universe, this power only is a mid-level manager in the grand organization of existence. Also, that executive carries out his assigned duties through karmic laws of cause and effect, not unconditional love—which is the "law" of pure Spirit in higher spheres of consciousness. Still, Julian Johnson says that "Of course, when compared with man, he [Universal Mind] is very exalted, full of light, goodness, wisdom and power. It is only when compared with the Positive Power that his lesser light becomes manifest."[26] So perfect mystics stress that devotion to mind—whether universal or personal—cannot lead to Ultimate Reality.

Seth Shiv Dayal Singh says that God has created myriads of aggregates of three worlds (material-astral-causal), and that each aggregate is overseen by a universal mind.[27] Thus not only is mind a middling official subordinate to the edicts of Spirit, this minion is merely one of many "branch managers" of lower realms of creation. Sawan Singh writes that we know that "this earth, with the moon and the planets, is revolving around the sun. The sun, like other solar systems, is revolving around another luminous entity, far brighter than itself. Similarly, this three-tiered universe, with its Lord [Universal Mind] is revolving around that True Being [of the fifth region] and that Immaculate One in its turn is going round its source, the Eternal Immaculate One beyond time and timelessness, form and formlessness."[28]

Second region

Level *two* commonly is called the astral region, because here the soul puts on another covering—the astral body, which is said to sparkle with millions of little particles resembling star dust.[29] This domain lies just above the physical universe. "Above," of course, is in reference to the vibrational energy of consciousness, and not to anything material. It does no good to pray with the eyes uplifted, for heaven is not in that direction. The anonymous medieval mystic who wrote *The Cloud of Unknowing* humorously observed that when some disciples hear "that men should lift up their hearts unto God, at once they are star-gazing as if they wanted to get past the moon, and listening to hear an angel sing out of heaven. In their mental fantasies they penetrate to the planets, and make a hole in the firmament, and look through!"[30] Spiritual science teaches that, in truth, the sky which must be pierced to enter non-physical planes of existence is within, not without.

This "sky" is visible to everyone. Close your eyes and notice what is within yourself. Do not imagine or think about anything, nor try to use your physical eyes to do the seeing. That darkness—which most people observe—is what separates us from the astral region. John Davidson says that this inner vacuum is linked into our human consciousness, and it "prevents most of us from leaving the physical body during our lifetime and traveling in the higher regions. It is the inner prison wall. . . . "[31] By means of contemplative meditation, or at the time of death, the mind and soul pass through this inner sky and enter a realm which often mistakenly is considered to be heaven.

As beautiful as it may be in comparison to earth, this domain is far removed from the regions of pure Spirit. It is better thought of as a gateway than a destination. In the words of Julian Johnson:

> This region constitutes the negative part of all the superphysical zones. That is, it lies most distant from the positive pole of creation. . . . Lying nearest to the physical universe, it forms the port of entry for all the higher regions. . . . The great majority of human souls at the time of death pass to some sub-plane of this region. . . . This section of creation is not immortal or imperishable. Neither are its inhabitants.[32]

Recall that both spiritual and material science consider final truth to be unchanging. Ultimate reality is permanent. Even if an entity, or law, lasts for hundreds of billions of years—or trillions upon trillions of years—it is not true if one day it ceases to exist. Only the highest spiritual region, level five, is eternal. Since the lower domains of creation are the projection of God's Spirit, they are subject to being reabsorbed, or dissolved, when He chooses. In the parlance of mysticism, a *simple dissolution* reaches up to the third region, and a *grand dissolution* extends up to the fourth region. Johnson notes that "of course, both of these dissolutions include the entire physical universe, every sun, moon, and planet in it."[33] Using the terms of material science, this dissolution is the Big Crunch which correctly is hypothesized to be the eventual endpoint of the Big Bang.

However, perfect mystics assure us that our part of creation will continue to exist for a long time. And so long as it does, it will be sustained by the all-pervading conscious energy of Spirit that passes through the "transforming stations" of the various planes. Starting from the fifth level and moving downwards, mysticism teaches that each region is marked by diminished consciousness and energy. In reality all still is one with God, but the illusion of separateness increases as the light of Spirit is dimmed by coverings of mind and matter. Julian Johnson says that for God to project Himself as this physical universe ". . . His vibrations are lowered to the material plane in order that he may manifest upon that plane."[34]

> *We are all darkness and God is light;*
> *this house receives its brightness from the Sun.*
> *The light here is mixed with shadow—*
> *if you want light, come out of the house onto the roof.*
> —Rumi[35]

This subtle form of physical energy—or, if you like, crude form of spiritual energy—is described as "ether" by perfect mystics. Substituting this word for a non-English term used by Seth Shiv Dayal Singh, he tells us that "The entire creation below [the astral plane] derives its life and vitality from the manifesting power of the ether of this region; that is, this ether vitalizes all the creation below it."[36] Thus while material science recognizes only the horizontal types of energy that are evident on this material plane of existence, spiritual science knows that all forms of physical and mental energy are part of a vertical energy spectrum that extends far beyond the reaches of this universe. In John Davidson's words:

From our human point of view, looking horizontally or outwardly at the physical world from within a physical body, we can provide descriptions at the physical level. From a universal point of view, looking vertically or inwards, we find that the physical universe is a reflection downwards of energy vibrations from more subtle worlds, which are, in turn, reflections of more inner or subtle worlds or energy fields. Creation, or existence, then, is a stepping down or a progression outwards from the Source [of Spirit].[37]

First region—the physical universe

Level one is material universe—the bottom floor of creation, the final step taken by Spirit in its downward course. Here that all-pervading conscious energy takes on its crudest manifestation, physical matter. Because mind and matter are vibrating at such a low rate, perfect mystics say that this world appears "frozen" in comparison to higher domains of consciousness. While Mother Earth may seem to possess a certain grace and beauty, this is a weak reflection of the glory in those realms of existence which enjoy a closer connection to God.

Spiritual science warns: do not become attached to the outward forms of nature. No matter how beautiful and enticing these may appear, every material entity is cold and dark in comparison to the warmth and brightness of God's essence. Take care that you do not cling for life to the ice of this physical world, for the temperate breeze of Spirit can cause materiality to vanish in an instant. For a universe, this is dissolution. For a person, this is death. What then will be your support?

> *There is no doubt that this world is midwinter. Why are inanimate objects called "solid"? Because they are all "frozen." These rocks, mountains, and other coverings that garb this world are all "frozen." If the world is not midwinter, why is it frozen? The concept of the world is simple and cannot be seen, but through effect one can know that there are such things as wind and cold. . . . When that "divine" breeze comes, the mountains of this world will begin to melt and turn to water. Just as the heat of midsummer causes all frozen things to melt. . . .*
>
> —Rumi[38]

The simple truth of Spirit

In the opening verses of the Gospel according to St. John, we find: "In the beginning was the Word, and the Word was with God, and the Word was God. The same was in the beginning with God. All things were made by him; and without him was not any thing made that was made." This Word is Spirit, God-in-Action—not a written or spoken word. It is the creator and sustainer of this universe and every other realm of existence. It is the only Power. All other forces—spiritual, mental, or physical—are emanations of Spirit, stepped-down in energy and consciousness commensurate with the tasks assigned to them by God.

The mind wonders: *Could truth be so simple?* Certainly. Charan Singh says: "Teachings for realizing the Lord are so simple, so very simple that we do not understand them. It is the habit of the human mind to create problems and then try to solve them, and take pleasure in solving them. We cannot accept simple thinking in a simple way. Truths are so simple that we cannot accept them."[39]

The appearance of complexity is inevitable when the mind comes into play, for the microcosm—*us*—is a reflection of the macrocosm—*cosmos*. Just as the unitary power of Spirit divides when it forms the region of Universal Mind, so does our soul consciousness become fractured through sensory perceptions and mental cognition. Thinking keeps us ignorant of what lies beyond thought, and each sight or sound of materiality makes us blind and deaf to spirituality. In the course of discussing parts and wholes it was noted that at every moment of our life we have a choice: Face toward Ultimate Reality, or away from it? Move closer to, or further from, final truth?

> *The more awake one is to the material world,*
> *the more one is asleep to spirit.*
> —Rumi[40]

There is no room for equivocating here, no middle ground on which the seeker of absolute truth can stand. The path to God is as sharp as a razor's edge. You are either on it, or off it. The essence of spirituality is contained in that very term: *Spirit*uality. Spirit leads to God and absolute knowledge. Everything else leads to mind and matter. Logic argues that if the all-pervading conscious energy of Spirit branched out as it descended from the unified trunk of the highest region, through the limbs of the various domains of existence, to end in the twigs and leaves of this physical universe—then it should be possible to retrace the course of creation by moving backward from the leaves, to the twigs, to the branches, to the trunk.

This is the hope of those who believe that the door to spirituality is opened by passing through the antechamber of materiality. Matter leads to mind, and mind to God. This sounds reasonable. But perfect mystics say that it is false, for logic pertains to parts, not the whole. The reality of Spirit may be likened to the sap, the lifeblood of a tree, that runs through every part of it. It is not found by remaining on the outside of the myriad branches of existence, no matter how far one travels along the twigs and leaves of mind and matter. The root of existence never can be reached that way. Become the sap, become Spirit, and the mysteries of the entire tree of creation are known.

Kabir, a perfect mystic, posed these questions:

Is the creator of the world greater, or the One who created him?
Is knowledge greater, O Lord, or the source of all knowledge?
Is this mind greater, or the place where mind finds lasting rest?
Is the lord of the three worlds greater, or the one who knows him?[41]

Returning to the Source

Mysticism teaches that the heart of the matter is this—in truth, there is only God and Spirit, which is God-in-Action. Nothing else exists, nothing. Charan Singh writes, "Except for the Lord no other power has created anything in this entire creation. Everything is His own projection. . . . God is that Power which is permeating in every particle of His creation. . . ."[42] However, with every additional level of creation the power of Spirit is increasingly hidden. First, behind a veil of mind. Then, behind a denser covering of subtle matter. And finally, behind a thick wall of crude physical matter.

Enough is enough. Anyone who longs for union with the reality of God must begin uncreating what stands between them and Him. Creativity has become an unnatural obsession with us. We strive to be creative in arts and crafts, music and literature, personal and business affairs. Those who actively create things through thinking and physical activity are admired as productive members of society. Those who sit quietly in contemplative meditation and seek to know the source of mind and matter are considered to be out of touch with the reality of this world. It indeed is strange that actions which further divide the unbroken wholeness of existence—by adding more parts to the myriad entities which already are present in the universe—are so revered.

Mystics view things differently. Lao Tzu says, "In the pursuit of learning, every day something is acquired. In the pursuit of Tao [Spirit], every day something is dropped. . . . Returning to the source is stillness, which is the way of nature."[43] Rumi tells us, "Most of the people of paradise are simpletons. . . . The true possessors of intellects have sent their intellects to that side; the fool has remained on this side, where the Beloved cannot be found."[44] We need not be afraid of becoming less by leaving behind our mind and body. Rather, we will become much more—Spirit, the unseen essence of everything in existence.

Echoes in the new physics.

Physics and mysticism both are devoted to the pursuit of unity. However, the nature of the oneness being sought distinguishes these sciences. As we have seen, spiritual science seeks to unify mind and matter in the all-pervading conscious energy of Spirit. This occurs at the level of the fourth region of creation, where those two sub-forces are transformed—or left behind—and the soul enters realms of pure spirituality. Similarly, the goal of the new physics is to arrive at a Theory of Everything which explains the disparate phenomena of our universe in terms of a single underlying law, or force.

Waves are not the ocean

But this "everything" actually is only the crude energy and matter which make up the bulk of this lowest region of existence. Even mind is not part of physicists' unified theories, with soul and Spirit far beyond their present understanding. While spiritual and material science share a common goal, knowledge of Ultimate Reality, only mysticism is able to reach that end. Quantum theory and the rest of the new physics is concerned with only the densest manifestation of Spirit—physical matter and energy. These entities are unconscious consciousness, lacking a soul. Hence until material scientists learn to study *conscious consciousness* through the research tool of contemplative meditation, their attempts to unify the forces of nature will be woefully incomplete. Physicist Arthur Eddington cogently described the problem:

> Interpreting the term material (or more strictly, physical) in the broadest sense as that with which we can become acquainted through sensory experiences of the external world, we recognise now that it corresponds to the waves, not to the water of the ocean of reality. My answer does not deny the existence of the physical world, any more than the answer that the ocean is made of water denies the existence of ocean waves; only we do not get down to the intrinsic nature of things that way. Like the symbolic world of physics, a wave is a conception which is hollow enough to hold almost anything; we can have waves of water, of air, of aether, and (in quantum theory) waves of probability. So after physics has shown us the waves, we have still to determine the content of the waves by some other avenue of knowledge.[45]

Shadows of reality

Spiritual science is that avenue. It teaches that the content of every wave in existence is, in essence, Spirit. This fact is not evident because we live in what Rumi termed a "frozen" world. The physical universe possesses much less energy and consciousness than higher domains of creation. Here Spirit is a dim reflection of God's light, a weak echo of His voice. But that divine unity expresses itself in each and every part of existence. Oneness always is oneness, sometimes more hidden, sometimes more revealed.

As was briefly noted in the previous section, material science finds this fact reflected in fractals—the visible traces of complex deterministic systems.

John Briggs writes, "Fractals are images of the way things fold and unfold, feeding back into each other and themselves. The study of fractals has confirmed many of the chaologist's insights into chaos, and has uncovered some unexpected secrets of nature's dynamical movements as well . . . [such as] *self-similarity*. This means that as viewers peer deeper into the fractal image, they notice that the shapes seen at one scale are similar to the shapes seen in the detail at another scale."[46]

Looking out my window, I see an oak tree in our backyard. It is a single trunk for about one-third of its height, then divides into four large limbs. Each of the limbs has numerous branches, and each branch splits into many twigs. This tree is a fractal, because every part of it resembles—to a remarkable degree—the whole. If I look at only one of the four main limbs, that limb is an almost perfect down-sized replica of the entire tree. Likewise, all of the branches look like tiny oak trees. Even a single leaf maintains the pattern of the whole, as it too has a main trunk and branching veins. We would, of course, not be able to take shelter from a rainstorm under a leaf. And Ultimate Reality cannot be known through any of the parts of creation, even though they reflect the nature of higher truths.

The unified theories of the new physics are attempts to understand one twig of one limb of a branch from the main trunk of existence. As such, they are useful to a seeker of final knowledge only as a metaphor, so to speak, of the genuine unification of mind and matter which occurs through contemplative meditation. These theories bring Spirit down-to-earth, providing our minds with a faint echo of what otherwise could not be heard at all. Recall this quotation from Rumi: "His rules are manifest in all creation, because all things are the shadow of God, and the shadow is like the person. . . . [yet] Not everything that is in the person shows in the shadow, only certain things. So not all the attributes of God show in this shadow. . . ."

Unification of the material forces of nature is a shadow of the soul's union with God. But what exactly is meant by those "forces"? We are accustomed to looking at the world as being composed of matter and energy, two separate entities. For example, the sun is made of matter, and gives off energy. There certainly appears to be a difference between the elements in the sun's core—matter—and the electromagnetic radiation—energy—released when nuclear reactions in the core convert hydrogen into helium. Quantum theory, however, erased this apparent distinction between matter and energy, particles and forces. Heinz Pagels says:

> Previously, physicists imagined the world was divided into matter and energy. The matter resided in particles and the energy in fields that interacted with the particles, causing them to move. Now [with quantum theory] a unified view was established. The dualisms of energy and matter,

particle and field, were dissolved, and everything could be seen to be interacting quantum fields. There isn't anything to material reality except the transformation and organization of field quanta—that is all there is.[47]

This assertion that field quanta is "all there is" to materiality may strike you as unsatisfying, especially since most physicists believe there is no non-material reality. What does it mean to reduce existence to the interactions of quantum fields? The new physics has no definitive answer, in part because it does not know what these fields are made of. If they are neither energy, nor matter—but the essence underlying both—then what *is* this foundation of material reality? Physics can tell us what fields *do*, but not what they *are*.

Physicist Arthur Eddington wrote these words over fifty years ago, and they still ring true: "If today you ask a physicist what he has finally made out the aether or the electron to be, the answer will not be a description in terms of billiard balls or fly-wheels or anything concrete; he will point instead to a number of symbols and a set of mathematical equations which they satisfy. What do the symbols stand for? The mysterious reply is given that physics is indifferent to that; it has no means of probing beneath the symbolism."[48] This is because the new physics has reached the boundary between the symbolic material world, and non-symbolic higher domains of existence. Bound by limited reason and intellect, the human mind only can perceive *shadows* of reality in this transition zone.

Quantum fields—shadows of Spirit

Physicist Paul Davies notes that because the uncertainty principle makes it impossible to simultaneously attribute a definite position as well as a definite motion to an atom, "Atoms and subatomic particles inhabit a shadowy world of half-existence. Then there are still more abstract entities such as fields. The gravitational field of a body certainly exists, but you cannot kick it, let alone see or smell it. Quantum fields are still more nebulous, consisting of quivering patterns of invisible energy."[49]

Quivering patterns of invisible energy, this is what physics tells us material reality is made of. Why not recognize this entity for what it is? A low-energy manifestation of Spirit. However, there is nothing wrong with continuing to call it a "quantum field." The name literally is immaterial. What is important is to realize that something subtle and ethereal, yet powerful and objectively real, creates and sustains physical existence. Fritjof Capra says, "The quantum field is seen as the fundamental physical entity: a continuous medium which is present everywhere in space. Particles are merely local condensations of the field; concentrations of energy which come and go, thereby losing their individual character and dissolving into the underlying field."[50]

Forces, then, are inseparable from particles. In the new physics, forces are considered to be interactions between particles, mediated through fields. Do not be concerned if this fails to be crystal clear to you. For our purposes it is not necessary to understand the details of how physics considers the four forces of nature—electromagnetism, the weak and strong nuclear forces, and gravity—arise through the exchange of particles. John Barrow sums up the situation when he writes, ". . . we see that the forces of Nature are deeply entwined with the elementary particles of Nature. They cannot be considered independently."[51]

Unifying the four forces of nature

So when physicists theorize about unifying forces, particles come along for the ride, so to speak. This permits the new physics to view a Theory of Everything—which generally is expressed as a unification of the four forces of nature—as being capable of explaining *all* of material existence: energy and matter, fields and particles. Progress toward this end began in the nineteenth century, when Michael Faraday demonstrated that electricity and magnetism were not two separate forces, but one—electromagnetism. He found that a moving or changing electric field generates a magnetic field, and a moving or changing magnetic field generates an electric field (this is how dynamos produce electricity). So electricity and magnetism came to be recognized as different forms of a single electromagnetic field.[52]

Unifying additional forces of nature turned out to be much more difficult. It was not until 1967 that Steven Weinberg and Abdus Salam, working independently, were able to explain how the electromagnetic and weak nuclear forces could be united. Steven Hawking explains that the Weinberg-Salam theory exhibits a property known as spontaneous symmetry breaking:

> This means that what appear to be a number of completely different particles at low energies are in fact found to be all the same type of particle, only in different states. At high energies all these particles behave similarly. The effect is rather like the behavior of a roulette ball on a roulette wheel. At high energies (when the ball is spun quickly) the ball behaves in essentially only one way—it rolls round and round. But as the wheel slows, the energy of the ball decreases, and eventually the ball drops into one of the thirty-seven slots in the wheel. In other words, at low energies there are thirty-seven different states in which the ball can exist. If, for some reason, we could only observe the ball at low energies, we would then think that there were thirty-seven different types of ball![53]

Anthony Zee likens this symmetry-breaking to identical triplets being separated at birth and forgetting their kinship. The two particles which carry the weak force are massive in comparison to the massless photon—which mediates the electromagnetic force. However, at much higher energies than are evident in the everyday world those "weak" particles also become effectively massless. Thus, says Zee (speaking as one of those particles), "It is just that, at low energies, we are dragged down by our masses, so you think we are weak."[54] The energy at which the electromagnetic and weak forces merge into a unified *electroweak* force is about ninety proton masses[55] (in particle physics, energy and mass are equivalent, so one can be expressed in terms of the other). A proton mass is the energy needed to create one proton, which is the same as one GeV, or one giga-electron volt.

Four become three

Since the electroweak force manifests at ninety proton masses, or ninety GeV, and existing particle accelerators can reach energies of about one hundred GeV, it has been possible to confirm the theory developed by Weinberg-Salam. So the new physics now recognizes three elemental forces of nature: electroweak, strong nuclear, and gravity. Is it possible for our understanding to go further, and reduce this number to two? Most physicists believe that this can be done, but it entails taking a *big step* up to the next rung of the energy ladder of unification. Paul Davies says that the onset of Grand Unified Theories in particle physics—where the strong force is united with the electroweak force—is believed to occur at about 10^{14} GeV.[56]

Imagine that one GeV is equivalent to one second of time. That is, every second one GeV is added to a hypothetical pool of energy, like drops steadily dripping out of a faucet. It takes just ninety seconds, or one-and-a-half minutes, to produce enough energy to reach the ninety GeV electroweak level. But to accumulate 10^{14} GeV would take over thirty-one million years at the rate of one GeV per second. This indicates how far removed are Grand Unified Theories from the low energies available to mankind. Even the most advanced particle accelerators being planned would reach energies of only a few thousand GeV, which in terms of the previous metaphor is less than a hour along that thirty-one million year journey to the Grand Unified Theory energy level. Hawking wryly notes that "a machine that was powerful enough to accelerate particles to the grand unification energy would have to be as big as the Solar System—and would be unlikely to be funded in the present economic climate. Thus it is impossible to test grand unified theories directly in the laboratory."[57]

Three are theorized to be two

This does not stop physicists from developing theories about grand unification, and seeking indirect evidence for their hypotheses, but no candidate for a Grand Unified Theory is close to being proven. In fact, John Boslough says that "the one prediction of the various grand unified theories that did seem to be testable—that is, that every proton in the universe eventually would decay—has led to a number of ongoing experiments that have utterly failed to turn up a single shard of evidence that the models were correct."[58] So the new physics' countdown toward a confirmable Theory of Everything, which describes the *one* elemental force of nature, moved from "four" to "three" in the 1960s and has been on hold at that point since.

As was noted previously, particle physics has found it necessary to substitute abstract mathematics for hard experimental evidence. Some popular books about the new physics describe Grand Unified Theories in rich detail, yet fail to mention that no solid support exists for these hypotheses. David Lindley observes that "Grand unification, regarded as a general sentiment rather than a specific physical theory, is taken to have solved the problem of where matter in the universe comes from. . . . Grand unification, which is in physical terms an entirely speculative and wholly unverified theory, is nevertheless taken by cosmologists as a done deal, something they can teach to their graduate students."[59]

Once again, we find that out of necessity the new physics has become less committed to experimental verification than mysticism. Lacking any means of directly contacting the high energy at which grand unification is believed to occur, material scientists are forced to concoct symbolic representations of that reality. Spiritual science possesses the research tool of contemplative meditation—which permits the seeker of Ultimate Reality to unify his own consciousness with the all-pervading conscious energy of Spirit. That divine energy is contacted at a level far above even the GUT energy. This is reflected in the final theoretical rung of physics' unification ladder, the so-called Theory of Everything which unites gravity with the electroweak and strong forces.

Two become one—the theorized superforce

The leading hypothesis for a Theory of Everything will be discussed under the final principle of spiritual science. Any candidate for such a theory must explain the inconceivably high energy possessed by our universe a tiny fraction of a second after the Big Bang. More accurately, the universe did not "possess" this immense energy, it *was* that energy. For with a Theory of Everything the countdown moves from the three forces of electroweak/strong/gravity, past the two forces of electroweak-strong/gravity (grand unification), to the one superforce which created the physical universe.

What is the energy of this superforce? Paul Davies writes that it is the same as the Planck energy (mentioned in a previous section): 10^{19} proton masses, or GeV. This, he says, represents "the ultimate scale of energy at which all of physics comes together in spectacular simplicity."[60] Recall that if one GeV per second "drips" into a pool, the grand unification level is reached in a bit over thirty-one million years. By contrast, the superforce level is reached only after three trillion years, which is over two hundred times longer than the fifteen billion years our universe has existed. This Planck energy is a hundred trillion times higher than the energy that would have been available at the Superconducting Supercollider[61]—which the Congress of the United States stopped funding in 1993.

According to Davies, it would take a particle accelerator as long as our Milky Way galaxy—one hundred-thousand light years—to attain the Planck energy of the primal superforce. If this could be done, he says, it would "enable us to manipulate the greatest power in the universe, for the superforce is ultimately responsible for generating all forces and all physical structures. It is the fountain-head of all existence. With the superforce unleashed, we could change the structure of space and time, tie our own knots in nothingness, and build matter to order. . . . Truly we should be lords of the universe."[62]

One only can be thankful that material science is *unable* to replicate that superforce. (I am content with letting God determine the structure of space and time. Simply living in—and trying to understand—the existence He has created is challenging enough.) Mysticism teaches that it is no accident that the conscious energy which formed this universe is so far outside of our ability to control it. The purpose of creation is not that we become separate little gods, but rather that we should unite our soul consciousness with the One Supreme God. True, the superforce of Spirit is all-powerful. But by seeking to lay hold of that force and use it for self-centered ends, we miss the central teaching of spiritual science: Ultimate reality is an unbroken wholeness—to realize this final truth we must become it, not stand outside of it as a distinct possessor and manipulator of that power.

A central problem facing both material and spiritual scientists is this: the energy of the superforce, or Spirit, is available but inaccessible. Steven Weinberg writes, "It is not that the energy itself is unavailable—the Planck energy is roughly the same as the chemical energy in a full automobile gasoline tank. The difficult problem is to concentrate all that energy on a single proton or electron."[63] Thus, *focus* is the key to unlocking the mysteries of the universe. This is the heart of contemplative meditation, which serves as the "consciousness accelerator" for mystics seeking to contact the superforce of Spirit. Our attention *is* Spirit, just as electromagnetism and the other three forces of nature *are* the superforce of the Big Bang.

Frozen energy

However, in both the inner (mental) and outer (physical) spheres of existence, that all-pervading conscious energy has become frozen. Ice appears completely different from liquid water, yet has the same atomic structure. Their illusory dissimilarity arises because atoms of ice vibrate at a lower, or cooler, energy level than do atoms of liquid water. Physicist James Trefil says:

> In exactly the same way, theorists tell us that the fundamental forces look different to us because we live in a universe that has been expanding and cooling for 15 billion years. If we go back far enough in time, the argument goes, we could come to a temperature where the differences would disappear. Just as the ice melts so that it's basic similarity to water becomes manifest, so do the forces "melt" when the temperature is high enough. This melting process is what we have called unification.[64]

In much the same way as physicists seek to understand how the low-energy forces of nature are created and subsumed by the superforce, so do mystics strive to unite their consciousness with the all-pervading Sun of Spirit. After ego, mind and body "melt," only the essence of Ultimate Reality remains.

> *The world is snow and ice, and Thou art the burning summer—*
> *no trace of it remains, oh King, when Thy traces appear. . . .*
> *Like the flame of a candle next to the sun—it is not,*
> *but it is when you consider: The candle's essence exists,*
> *for if you place some cotton upon the flame, it will be consumed.*
> *But the flame does not exist: It gives you no light—the sun has annihilated it.*
> *   —Rumi*[65]

Such is love, losing one's self in the being of another. Love is the keynote of mysticism, yet is not at all "mystical." In the same fashion, unification is the goal of the new physics, yet is not at all "physical." Both spiritual and material science seek an objective reality which is beyond time and space as we know it. This reality—call it Spirit or superforce—is the creator and sustainer of material existence.

Paul Davies writes: ". . . these investigations [for a unified theory] point toward a compelling idea, that all nature is ultimately controlled by the activities of a single *superforce*. The superforce would have the power to bring the universe into being and to furnish it with light, energy, matter, and structure. But the superforce would amount to more than just a creative agency. It would represent an amalgamation of matter, spacetime, and force into an integrated and harmonious framework that bestows upon the universe a hitherto unsuspected unity."[66] Remember, this is a physicist talking—but if you substitute "Spirit" for "superforce" in that quotation, the words are indistinguishable from those of a mystic.

Sometimes this objection to mysticism is raised: "If the all-pervading conscious energy of Spirit is so powerful, why is it not observed?" Well, ask a physicist the same question: "If the superforce is so powerful, why is *it* not observed?" The answer likely will be: "Because we live in a low-energy world, where that unimaginably potent force cannot be manifested. Give me a particle accelerator as large as a galaxy, and I will show the material reality of the superforce to you." Similarly, a perfect mystic would respond: "Shift your attention away from mind and matter, for Spirit is hidden in these entities. Focus the now-diffuse energy of your consciousness into a single point. Then you will be able to contact the non–material reality of Spirit."

Gravity, the superforce's first-born child

Even though both the superforce and Spirit are veiled in the everyday world, gravity is the force of nature which has the closest kinship to the superforce. It is believed to have been the first force to separate from the original unitary energy of the cosmos. Hazen and Trefil say that 10^{-43} second after the Big Bang "all four forces were unified and things were as beautiful and simple and elegant as they could be."[67] By 10^{-33} second, gravity had made its appearance in the newborn universe, while the strong nuclear force still was united with the electroweak force. So of the four forces of nature presently evident in the universe, gravity is the firstborn child of the superforce. Understanding a few facts about gravity thus can help us comprehend the nature of that superforce, and by implication, Spirit—which is the essence of both.

Gravity apparently is by far the weakest physical force. The entire mass of the earth exerts a gravitational force upon our body, yet it is quite easy for us to pick up our feet from the ground and walk. A tiny magnet uses electromagnetic force to overcome gravity and lift iron shavings from a tabletop. Anthony Zee writes that "gravity is by an absurd margin the most feeble force in Nature. . . . It is only because macroscopic objects contain such an enormous number of particles that gravity makes its presence known."[68] Every particle in the earth, he says, is pulled toward every particle in your body—and indeed, every particle of anything is pulled toward every particle of everything else in the universe.

Gravity always is attractive. It has no opposite, no countervailing force as does electromagnetism with its positive and negative charges. As was noted previously, in a poetic sense gravity is a manifestation of love. It ceaselessly seeks to bring entities together, not to force them apart. This is what makes gravity as powerful as it is. Though the gravitational force between any two atoms is vanishingly small, add up all of the atoms that comprise Mother Earth and she is able to hold all but the most powerful rockets within her grasp. Further, at high energies gravity is the most potent force in existence. In a black hole, for example, *nothing* escapes from its hold. Mass and energy are equivalent in physics, so either a large mass or high energy produces strong gravity.

In the vastly energetic instants immediately after the Big Bang, gravity was king. Zee writes that there was an epoch in the history of the universe "when gravity was the most powerful force in the universe. . . . Gravity is a mysterious interloper from an energy scale far beyond our experience."[69] Why is it so weak today? Because if it were not, materiality could not exist. If gravity now had anywhere near the strength it did at the birth of the universe, nothing would be as we know it. Paul Davies notes that if the gravitational force was altered by as little as one part in 10^{40} stars like the sun could not exist.[70] According to Anthony Zee, ". . . a force that acts on every particle had better be extremely weak, given the enormous number of particles abroad in the universe. Gravity could not be much stronger without the universe's looking completely different."[71]

Similarly, the all-pervading conscious energy of Spirit must be stepped-down, reduced in energy and intensity, so that mind and matter may continue to exist as separate entities in this "frozen" domain of God's creation. When God wills this divine power to act, universes burst in and out of existence like foam tossed by a breaking wave.

> *You think the shadow is the substance:*
> *so to you the substance has become a cheap toy.*
> *Wait until the day when that substance*
> *freely unfolds its wings.*
> *Then you will see the mountains become soft as wool,*
> *and this earth of heat and ice become as nothing;*
> *you will see neither the sky nor the stars,*
> *nor any existence but God—the One, the Living, the Loving.*
> *—*Rumi[72]

Coverings of the superforce

The true nature of gravity and the superforce, like that of Spirit, is hidden behind an increasingly dense series of veils. Mysticism teaches that the coverings of Spirit are mind and matter, which accrue as the soul descends from higher spiritual realms of existence. Physics teaches that the coverings of the superforce are the four material forces of nature, which manifested one by one as the universe cooled after the Big Bang. Hazen and Trefil call those manifestations "freezings," for with the appearance of each force "the fabric of the universe changed in a fundamental way, much as water changes when it freezes into ice."[73]

So by combining the teachings of spiritual and material science, we can understand why the all-pervading conscious energy of Spirit is difficult to perceive on this plane of reality. Starting with the formless and nameless state of God-as-God, mysticism recognizes four primary freezings which reduced the vibratory energy of Spirit: creation of the fourth, third, second, and first regions of existence. Then physics tells us that following the creation of this physical universe—the first region of mysticism—at least four secondary freezings took place: manifestation of the forces of gravity, strong nuclear, weak nuclear, and electromagnetism (the exact number of freezings is open to question; Hazen and Trefil count six, three of matter and three of forces).

How then could the essence of Ultimate Reality *not* be hidden behind all this frozen mind and matter? Freezings have piled upon freezings as acts of creation cascaded from God's will, "Be!"

> *God most High created the world by a word. His command, when He desires a thing, is to say to it, 'Be", and it is.*
>
> —Rumi[74]

Mercifully, that Voice of "Be" is still with us. Hidden, yes, but very much part of the fabric of material and non-material spheres of existence. More accurately, it is not part of the fabric, but is the very warp and woof of creation. Spirit is both the weaver and the weaved, the clothier and the cloth. God's whisper never has been separate from creation's thunder. One *is* the other. The preceding discussion of physics' unified theories helps us understand how such is possible. At the instant of the Big Bang there was one fundamental force. Now there are four, and a host of particles related to those forces. Four have emerged from one, but how could those forces ever be different from the original unified superforce? Where could any "differentness" come from when physics teaches that everything in this universe can be traced to a single source?

And that source, say both physicists and mystics, continues to sustain existence. Like an infinitely vast ocean of conscious energy, Spirit—or call it superforce if you like—flows through all of creation, sometimes appearing as a powerful torrent, sometimes as a stagnant pool. But always it is the same water of life, always. There is nothing else in existence. Where is this sea of energy? All around us. Similarly, where are radio and television signals? All around us. We do not realize this until we turn on a device which is precisely tuned to receive and decode the information contained in those electromagnetic waves. However, even without a radio or television, a signal strength meter makes it possible to be aware that those waves exist—but such a relatively crude device cannot show you the picture and sound being carried by the waves, as can a television set.

Mysteries of the quantum vacuum

In somewhat the same manner, physics knows that the *vacuum* contains virtually infinite energy, even though the exact nature of that all-pervading force is unknown to material scientists. We are accustomed to thinking of a vacuum as being empty, like a jar with all of the air expelled from it. However, the new physics views it as anything but that. Danah Zohar says that "the quantum vacuum is very inappropriately named because it is not empty. Rather, it is the basic, fundamental, and underlying reality of which everything in this universe—including ourselves—is an expression. . . . By analogy, if we lived in a world of sound, the vacuum could be conceived as a drum skin and the sounds it makes as vibrations of that skin. The vacuum is the *substrate* of all that is."[75]

Similarly, mathematician Roger Penrose asks how one can quantify the amount of matter in existence—which naturally involves coming to grips with what "matter" is and where it can be located. He concludes that "we seem to be driven to deduce that if this mass-energy is to be located at all, it must be in this *flat empty space*—a region completely free of matter or fields of any kind. In these curious circumstances, our 'quantity of matter' is now either *there*, in the emptiest of empty regions, or it is nowhere at all!"[76]

Recall that under the second principle of spiritual science it was noted that many physicists believe that a vacuum fluctuation was the event which produced the Big Bang and our physical universe. So whatever the vacuum is, material science tells us that it contains enough power to be the creator and sustainer of everything in existence. From this perspective, physical matter and energy are inconsequential waves upon the all-pervading sea of the vacuum. In the words of physicists David Bohm and F. David Peat:

> Current quantum field theory implies that what appears to be empty space contains an immense "zero point energy," [i.e., it remains even at temperatures of absolute zero, when thermal energy oscillations are absent] coming from all the quantum fields that are contained in this space. Matter is then a relatively small wave or disturbance on top of this "ocean" of energy. Using reasonable assumptions, the energy of one cubic centimeter of space is far greater than would be available from the nuclear disintegration of all the matter in the known universe! Matter is therefore a "small ripple" on this ocean of energy. But since we, too, are constituted of this matter, we no more see the "ocean" than probably does a fish swimming in the ocean see the water.[77]

John Davidson likens the situation to a tug-of-war.[78] When two evenly matched teams pull on each end of a strong rope with all of their might, the rope does not move. But though no motion is evident, it is full of potential energy (if the rope snaps, that energy probably will make members of the two teams fall on their backsides). Similarly, we do not observe the vacuum energy because it is "locked in" to the structure of physical reality. Only minute surface quiverings of that immense force—known to physics as short-lived "virtual" particles—are apparent. Physicist Larry Abbott observes: "By convention, energies are often measured in relation to the vacuum. When it is defined in this way, the vacuum automatically has zero energy in relation to itself."[79]

But as noted by Bohm and Peat, when attempts are made to calculate the actual energy of the vacuum (or *cosmological constant*, a term used in relativity theory), that energy ends up being essentially infinite. Perhaps the greatest unsolved mystery in physics and cosmology is why this stupendous power does not produce the physical effects—such as curved space—that are predicted by current theories. In the words of Anthony Zee, "The naturally expected value of the cosmological constant [or vacuum energy] comes out to be about 10^{123} times the maximum value that the astronomers tell us the cosmological constant possibly can have. This discrepancy of a factor of 10^{123} between observation and theoretical expectation has been called the biggest error in the history of physics."[80]

Infinite energy pervading the universe

So physics knows that the superforce is still with us in the unimaginably vast zero-point energy of the vacuum, but cannot measure or experience it directly. The ancient conception of an all-pervasive yet invisible "ether" has been resurrected by modern physics—even though most physicists would reject that term (if not the concept). Hans Christian Von Baeyer writes:

> A vacuum that is utterly dark is nonetheless suffused with an electro-magnetic field that fluctuates in gentle random waves of all wavelengths, each wavelength with its own zero point energy. What's more, since there is an infinity of these vacuum fluctuations, the sum total of all zero point energies, even in a compact volume of, say, a cubic inch, must be infinite. The impossibly dense ether has been replaced by an infinite energy density pervading the entire universe.[81]

Infinite energy pervading the entire universe—this is how the vacuum is viewed by the new physics. Once again findings of material science have brought us back to the beginning and end of all existence: Spirit, the all-pervading conscious energy which—being God's essence—manifests His attributes of omnipresence, omniscience, and omnipotence. The laws of quantum physics imply that the "nothingness" of the physical vacuum contains much more energy than all of the matter in the universe. In fact, that nothing can be viewed as the source of everything in this domain of creation—the forces of nature, atomic particles, the structure of space itself.

The workshop and treasure of God is in nonexistence: You are deceived by existence, so how should you know about nonexistence.
                                                        —Rumi[82]

In like fashion, mysticism teaches that the seeming emptiness of the inner sky seen within ourselves when we close our eyes actually is the gateway to absolute truth, knowledge, and bliss. We fail to recognize that all of the thoughts and images of materiality which ceaselessly fly about in the mind distract us from realizing the essence of pure consciousness: Spirit. That nothing experienced in the depths of contemplative meditation actually is the ground of everything in existence.

As we shall find in the next principle of spiritual science, the all-pervading conscious energy of Spirit is contacted directly by spiritual scientists. What price would a theoretical physicist pay in order to hear and see the superforce which created this physical universe? In his wildest dreams, can he imagine being able to unite his very consciousness with that power, thus knowing it completely? This would seem to be utter fantasy, but for serious students of mysticism it becomes a concrete reality through the research tool of contemplative meditation.

Principle 7.
Spirit appears as audible spiritual vibration.

*Bring the sky beneath your feet
and listen to celestial music everywhere.*[1]
—Rumi

*We have heard these melodies in Paradise;
Though earth and water have cast their veil upon us,
We retain faint remembrances of those heavenly songs.*[2]
—Rumi

Rumi advises us to "bring the sky beneath your feet." The lower reaches of that inner sky appear as the darkness initially perceived in meditation. By bringing our attention beyond that point, we are able to reach a higher level of consciousness where Spirit—the essence of Ultimate Reality—can be heard and seen by subtle faculties of the soul. "Impossible!," says the skeptical material scientist. "This is all wishful thinking, self-fulfilling imagination. Mystics want to believe in God so strongly that their minds conjure up hallucinations which they take to be spiritual realities." Not true. Spiritual science prevents such subjectivity by eliminating all personal thoughts and emotions through the rigorous experimental methods of contemplative meditation. This is no different from the research approach of material science.

When astronomers want to capture faint light from distant stars, they place powerful telescopes on top of high mountains—or in space—to avoid the contamination of atmospheric effects. The scientists in charge of the microwave antenna which first detected the weak radiation remaining from the Big Bang had to assure that pigeon droppings or other extraneous influences were not creating the signal they received. Experiments designed to detect a theorized—but not yet observed—decay of protons are being conducted with tanks of ultra pure water buried deep underground to avoid interference from cosmic rays (this has been termed "Zen physics": sitting and waiting for something that may never happen.)[3]

Contemplative meditation—the ultimate scientific experiment

My point is this: material scientists ably use their minds to develop research designs which enable them to focus precisely on the particular physical phenomenon being studied. Sophisticated technological and mathematical tools do a remarkably good job of sorting out pertinent and extraneous data, valid and invalid observations. Small irregularities in the cosmic microwave radiation—which heretofore had appeared completely smooth—have been discovered by scientists who painstakingly separated those true background fluctuations from the many false microwave sources coming from our own galaxy. And even though purposeful frauds and careless errors are found in material science—as in spiritual science—these are exceptions, not the rule.

So if a physicist can use his mind to find ways of eliminating influences which may confound perceptions of material reality, why cannot a mystic transform his consciousness into a tool which clearly perceives non-material reality? If material science is able to control for subjectivism and invalid observations, what prevents spiritual science from doing the same? Nothing. All that is required is the will to follow the "experimental design" developed by perfect mystics, as this leads to direct knowledge of Ultimate Reality. Follow that design—the core of which is the previously discussed techniques of contemplative meditation—and one discovers the objective truth of spiritual realms of existence. Stray from it, and one runs a large risk of confusing what is personal with what is universal.

I doubt that these arguments have convinced those readers who are confirmed materialists. All right, let me join your camp for a moment. To make an argument, I will assume that Ultimate Reality is *material* in nature, as most physicists believe. As we have seen, this strongly implies that the mind also has a material origin, since a monistic world view is favored by almost all scientists. Then why could not that material mind be made into an instrument for contacting the essence of material reality? Since both are the same substance, only in different forms, this would seem to be feasible. Like attracts like.

> *Each kind attracts its own kind—*
> *no cow ever went before a fierce lion. . . .*
> *Parts travel to their wholes. . . .*
> *The mother seeks her child,*
> *principles seek out their derivatives. . . .*
> *Every kind bursts its chain to go to its own kind.*
> *Whose kind am I, caught here in the snare?*
> —Rumi[4]

Even if one believes in materialism, contemplative meditation seems equally well-suited for conducting experiments into the nature of Ultimate Reality. For if the mind—or whatever you believe it is that does your thinking and perceiving—indeed is composed of a material substance, then focusing that entity's attention upon a single point should produce some interesting results. Who can say that it is impossible to concentrate the energy of consciousness—be it of material or spiritual origin—onto the equivalent of a proton? It is a reasonable hypothesis that this focused energy would bring one closer to a unified domain of reality, that "infinite energy pervading the entire universe" recognized by physics.

After all, physicists theorize that a suitable particle accelerator could transform low-energy protons into the virtually infinite power of the superforce. Indeed, Steven Weinberg has been quoted as saying that the energy of the superforce is equal to that of a full tank of gasoline—*if* the energy contained in that tank could be concentrated upon a single proton. Presently this seems to be technically impossible, but even if it could be accomplished by a technological breakthrough, another problem faces material scientists: having reached the unified energy of the superforce, *they* remain separate from it. What sort of unification is this?

The big hole in a Theory of Everything

Let us dream a bit. Imagine the excitement as the world's leading physicists cluster in the control center of the newly constructed Absolutely Super Supercollider. Two beams of protons are accelerated to speeds, and energies, which far surpass any previously attained. Suddenly the beams are brought together in a head-on collision, and for a tiny fraction of a second the primal power of the superforce which produced our universe is re-created on earth. This is an undeniable fact. The Absolutely Super Supercollider's counters, dials, computer monitors, and mathematical models all confirm that such has occurred. A loud cheer erupts from the assembled scientists. Champagne bottles pop open. Visions of Nobel prizes dance through heads. But when the excitement dies down. . . .

Nothing really has changed. The superforce was there for a moment, and then it was gone. It has been *reflected* in computer printouts and meter readings, yet those dry-as-dust measurements are not the power of the universe itself. Further, the consciousness of those who designed and carried out this grand experiment has not been altered. These material scientists are no closer to a truly unified Theory of Everything. When "everything" leaves out the human mind which is attempting to grasp the nature of existence, a gaping hole remains in one's understanding of Ultimate Reality.

Further, imagine the crowded press conference where humanity is told about this historic attainment of the superforce energy—an event whose significance makes the first manned moon landing pale in comparison. Questions like these likely would be asked: "Since physics has replicated the force which created our universe, have you found God?" "Is the superforce a conscious entity?" "Why did creation occur?" Could any member of the Absolutely Super Supercollider team answer such questions? No. In effect they would have to reply, "Well, we saw the needle on the collider's Energyometer go up into the 'SUPERFORCE!' range. Whether that energy is conscious or not is a metaphysical question beyond the domain of physics. And of course we have no way of knowing whether creation was a random, or purposeful, event."

This scenario is completely hypothetical. But it helps us understand just how limited a Theory of Everything would be even if it could be confirmed experimentally by material science. Such a theory could not answer the *big* questions: Does God exist? What is the meaning of creation? We have seen that physics and mysticism already agree that an all-pervading energy is the foundation of material existence. In my opinion, the single substantial point of disagreement between these sciences is whether that power is conscious, or unconscious. Whether this essence of Ultimate Reality is material or non-material actually is of lesser importance. For even though physics pays lip service to being a "material" science, we have seen that extremely little materiality can be found at the heart of quantum physics and Big Bang cosmology. There the all-pervading vacuum force— devoid of energy or particles as we know them—holds center stage.

To know Ultimate Reality, become it

Spiritual science offers a simple means of approaching this problem: to know whether or not the all-pervading energy of the cosmos is conscious, *ask it.* That is, it is impossible to resolve conclusively this question of consciousness by relying on external observations of the material universe. Even when those observations strongly imply that a conscious being was responsible for the remarkable degree of regularity and order in existence, counter-arguments always can be made. That seeming order could be a random occurrence, or an illusion produced by the human mind.

Mysticism teaches that the only way to know every secret of the superforce—or Spirit—is to become it through rigorous spiritual experimentation. If this power is superconscious, then anyone who is able to unite himself with it will become superconscious. And if this all-pervading power is just material energy, or incapable of being contacted by the human mind, then something else—or nothing—will happen in meditation. Spiritual scientists, however, have confirmed many times that our personal consciousness is able to merge with God's universal consciousness. In this way, Ultimate Reality is heard and seen directly by the ears and eyes of the soul.

Charan Singh says, "The higher and higher we go [in meditation], the more space we are covering to look down upon. Rather, we are expanding ourselves; from a part, we are becoming the whole. When we merge into the Lord, we become the Lord. . . . To know Him is to merge back into Him. That is the only way to know Him, and not intellectually."[5] This knowing, teaches mysticism, is by direct perception of the soul. And the dominant faculties of the soul are the same as the primary sense organs of the physical body: vision and hearing.

Seeing and hearing Spirit

Physicist Nick Herbert notes that "there is a close parallel between the senses of vision and hearing because both involve sensing the frequencies of certain vibrations."[6] He says that physical vision is "a subjective appreciation of electromagnetic vibrations possessing wavelengths between 400 and 700 nanometers [billionths of a meter], otherwise known as 'light' . . . the ear is sensitive to sound frequencies between 20 cycles and 20,000 cycles per second."[7] The human body is incapable of sensing vibrations outside of these ranges.

Spirit is a non-material vibration of God-in-Action, and so cannot be seen or heard by the physical eyes or ears—no matter how sensitive those organs might be. Indeed, perceptions of material phenomena pull our consciousness outward and downward, away from the point where Spirit can be contacted. We are not aware of this soul-power because our attention is diffused, rather than concentrated. Just as the energy within a tank of gasoline would be equivalent to the superforce if it could be focused upon a single proton, so is it possible for our consciousness to become one with Spirit if it could be withdrawn to a single point. As Jesus said, "if therefore thine eye be single, thy whole body shall be full of light."[8] And sound.

Sawan Singh writes that "The Word [of Spirit] is ringing in every atom. We do not hear it because we are not in touch with it within ourselves."[9] What prevents us from contacting this immanent power? Lack of concentration. Perfect knowledge, bliss, and love is within us, not without. Yet virtually all of our attention is scattered outside, in material sensations, thoughts, images, imagination, and emotions. Inner realms remain unknown, *terra incognita*. Even if we manage to close our eyes and forget the outside world for a moment, awareness of our physical body remains. This too keeps us bound to limited reality.

In the words of Charan Singh, "Spirit is even now in our body. The soul is only a ray of that Spirit and the soul is spread in the whole body . . . we have to withdraw that consciousness back to the eye center to be attracted to that Spirit, then it will pull the soul upwards. Spirit is everywhere, but you have to withdraw your consciousness to that stage where the Spirit can pull the soul like a magnet."[10] In this realm of material existence soul and Spirit appear as two separate entities, even though the essence of each is the same. Charan Singh says, "And what is the difference between soul and Spirit? It is the level of consciousness."[11] A physicist would give much the same answer if asked about the difference between electromagnetism and gravity: "it is the level of energy." At high energy levels, separate forces and particles merge into one entity. This applies to both material and spiritual domains of existence.

The soul's homeward journey

Contemplative meditation elevates the soul to a plane of consciousness where it unites with the all-pervading conscious energy of Spirit. This bears some resemblance to the launch of a space shuttle. Consider *soul* to be the command vehicle which is to be lifted into space. *Mind* is the powerful rocket on top of which that vehicle sits. Our *body* is the launching pad and gantries that support the various components of the space shuttle. The mission, the over-arching goal, is to place the command vehicle—soul—in a high "orbit."

The launching pad of our body plays an important role in preparing for this mission. Our physical senses, after all, are the means by which we learn about meditation and the other research methods of spiritual science. But just as the gantries which supply the space shuttle drop away in the final seconds before lift-off, so must we become detached from materiality—including our body—before mystical transport into higher domains of consciousness occurs. The power for this transport initially comes from mind, which serves as the engine for overcoming the pull of the physical senses and thoughts about this world.

As has been noted previously, in contemplative meditation the spiritual scientist repeats words associated with non-material planes of existence. This gradually pulls the mind away from this lowest domain of creation, just as the engines of the shuttle's booster rockets cause it to rise above the launch pad—barely moving at first, then more and more rapidly until it disappears into the clouds. The command vehicle of the soul is controlling the engine of the mind, but cannot rise up without its power. Those two entities, soul and mind, are linked at this point: the conscious astronauts cannot complete their mission without the aid of unconscious booster rockets.

However, after reaching a certain height those rockets drop away, and the command vehicle travels on under its own power. Similarly, we have seen that the soul leaves the mind at the top of the third region of creation. Even earlier, at a point in the second region, Spirit becomes the motive force for mystical transport. This force is audible as sound, and visible as light. It is the divine dynamo which energizes every part of creation. Charan Singh says, ". . . this Sound not only leads us but actually takes us back to the Father. First we follow it; then as we make internal progress, we merge into it and ride, or ascend, to our home by means of the Sound, the Word. It is constantly pulling us inside like a magnet and attracting us homeward."[12]

Physicists use electromagnetic energy to power particle accelerators because this is the only force available to them in this low energy world of ours. The ultimate purpose of these sophisticated accelerators is to understand the nature of the unified superforce. It is obvious that if the superforce could be tapped directly, sufficient energy would exist to test a Theory of Everything. But then there would be no need for such accelerators, because the questions being posed of them would have been answered. No one digs a well next to a huge lake of pure water.

In much the same way, mystics make use of the mind's energy to reach the level of consciousness where the "pulling power" of Spirit can be utilized for further, and much more rapid, progress. If it were possible to contact God-in-Action through our everyday consciousness, this would not be necessary. But just as the space shuttle is not able to go into orbit without the aid of booster rockets, our soul is enfeebled by its material surroundings and needs mind to propel it into spiritual regions. However, the conscious force of Spirit always is controlling this process of mystical transport, just as rockets do not fire on their own: their unconscious power is guided by intelligent personnel in mission control.

> *The poor body will not move until the spirit moves:*
> *Until the horse goes forward, the saddlebag stands still.*
> —Rumi[13]

Enjoying the sound of Ultimate Reality

So anyone who wishes to make the journey to final truth must ride upon the wave of Spirit, which appears as audible vibration—the sound of Ultimate Reality. This sound is heard not by the physical ears, but by a faculty of the soul. Sawan Singh says that it "is heard with the ears of the soul. . . . The Sound is in reality God-in-Action. He projects Himself into everything and revels in this play. . . . It is the unstruck music that resounds within. What we hear within is its reverberation, by gaining which the mind becomes still."[14] Through concentration, the energy of our consciousness is raised to a level where it can be attracted by the power of Spirit.

The soul then takes great pleasure in hearing what has been called Divine Music, or the "music of the spheres." One interpretation of this latter phrase—associated with Pythagoras—is that it reflects a pre-scientific belief that the planets, stars and other heavenly bodies emit an inaudible tone as they move through an invisible ether. However, a deeper understanding views that music as the all-pervading vibratory energy of Spirit. Being the essence of both external materiality (including the stars and planets) *and* our internal reality, it can be perceived inwardly. Julian Johnson writes that "in truth it is God himself that vibrates all through infinite space. . . . When he speaks, everything in existence vibrates, and that is the Sound . . . it is the only way in which the universal Spirit can manifest itself to human consciousness."[15]

The audible vibrations of Spirit have been described by spiritual scientists from many different religions, countries, and times. How could it be otherwise? The essence of Ultimate Reality will be perceived by anyone who knows how to contact it. Mystics do, of course, differ in how they describe the music of Spirit, for this is conditioned by their culture and circumstances. A fourteenth century Christian mystic, Richard Rolle, described his experience of the Holy Spirit in this fashion:

> This peace experienced by the spirit is very sweet. A divine and dulcet melody comes down to fill it with joy. The mind is ravished with this sublime and effortless music and it sings the joy of everlasting love . . . [I felt] an infusion and apprehension of heavenly spiritual sound which belonged to the song of eternal praise and to the sweetness of a melody inaccessible to normal hearing. These sounds cannot be known or heard by anybody but the one who receives them and he has to keep himself pure and separate from the world. . . . Nobody who is absorbed in worldly matters knows anything about it. . . .[16]

And here are the words of a twentieth-century Sufi mystic—Inayat Khan—who lived almost six hundred years and half a world away from the English mystic Rolle. Their essential message, however, is the same.

> Abstract sound is called *saut-e sarmad* by the Sufis; all space is filled with it. . . . The knower of the mystery of sound knows the mystery of the whole universe . . . the sound of the abstract is always going on within, around, and about man. As a rule, one does not hear it because one's consciousness is entirely centered in material existence. . . . Those who are able to hear the *saut-e sarmad* and meditate on it are relieved from all worries, anxieties, sorrows, fears and diseases; and the soul is freed from captivity in the senses and in the physical body. The soul of the listener becomes the all-pervading consciousness. . . .[17]

The Chinese Taoists taught that Tao, or Spirit, could be perceived as sound. Livia Kohn says that "in the cosmology of Taoist mystical philosophy, one may imagine the Tao as a tone of a certain wavelength that pervades and encompasses all there is. Or as the Taoists themselves have it, a certain quality of *qi* [cosmic energy] that underlies and furnishes all existence."[18] Plotinus, a mystic who came from Egypt and taught philosophy in Rome during the third century, wrote that "energy runs through the Universe and there is no extremity at which it dwindles out."[19] Peter Gorman notes that "Plotinus often speaks of the cosmos as a harmony, but the real abode of the music of the gods is the intelligible world beyond the three-dimensional cosmos. In describing the mystical journey to that world Plotinus bids the initiate wait until he hears musical sounds proceeding from the intelligible:"

> If, for instance, someone were waiting to hear a desired sound, he would withdraw from other sounds and rouse his ear for the time when that paragon among auditory sensations should approach; so too on earth he should forgo listening to perceptible sounds, unless it is strictly necessary, and preserve the psychic faculty of apprehension pure and prepared to hear tones from on high.
>
> —Plotinus[20]

Many more examples could be given of how conscious communion with the Word of God, or Spirit, is the common denominator of every deep religion and mystical discipline. The experiment of contemplative meditation has been replicated many times, over many centuries, in many cultures, and the results reported by serious investigators of Ultimate Reality always are the same. The all-pervading conscious energy of Spirit, Tao, Saut-e Sarmad, Holy Ghost—the name is unimportant—is perceived as audible spiritual vibration.

Spiritual light and sound are one

And yes, as divine light. God's light is not separate from His sound. Sawan Singh says that "the Word gives out both, light and sound. At this end, in the physical plane, the light and sound are lost in gross matter. On the finer planes—astral, causal, and spiritual, sound is audible and light is visible. At the upper end the Sound is the finest music, unheard by human ears—and the Light is of millions of suns and moons in one ray."[21] Even though the power of Spirit combines both light and sound, often perfect mystics emphasize the audible manifestation of God-in-Action. The Book of Genesis tells us: "And God said, Let there be light: and there was light."[22] This implies that the saying, the voice of God, preceded His light. Sound also tends to be the first attribute of Spirit perceived by the beginning student of contemplative meditation.

Still, both sound and light accompany the spiritual scientist on his journey through higher domains of consciousness to the reality of God. The sound, according to perfect mystics, comes from the light, and the light from the sound. Electromagnetism behaves similarly: as Hazen and Trefil put it, "Electricity and magnetism are two inseparable aspects of one phenomenon: you cannot have one without the other."[23] In the same fashion, Spirit appears in two guises to guide the soul Homeward. Charan

Singh writes, "The Word combines both light and sound. The sound is meant to determine the direction from which it comes, and the light to enable us to travel toward it."[24]

Spirit sometimes is compared to a watercourse which makes various sounds as it descends from high in the mountains to the ocean. Near its source in pure snow fields, the water cascades as a swiftly running creek, bouncing over boulders and falls. Midway in its journey, it changes to a smoothly flowing river. And as the water approaches the sea, it spreads out into a stagnant and almost motionless estuary. No alteration ever occurs in its molecular composition: it remains H_2O in every place, just as Spirit is forever One. But depending upon the environment through which the watercourse passes, the sound it makes differs.

One Truth, One Way

In much the same fashion, Seth Shiv Dayal Singh explains that there are five spiritual sounds corresponding to the five previously described regions of creation: "Each region has its own distinctive Sound and its own characteristic secret. . . . It is via the Sound of each region that the soul can, by degrees, ascend from one region to another, up to the highest stage. The ascent of the spirit is absolutely impossible in any other way. . . . "[25] That final statement may cause a visceral reaction in some readers, for many people dislike the idea that there is only one path to God. Why cannot each individual choose the religion, faith, or belief system which makes the most sense to *them*? Well, they can. In the bazaar of subjective thinking—which includes theology and philosophy—there is no way of conclusively determining which product is genuine, and which counterfeit.

However, in the marketplace of objective science—which includes mysticism—there are means of sorting out truth from falsehood. Several chapters have been devoted to this topic, and I do not wish to cover that ground again. In brief, truth is lasting, and final truth is eternal; limited reality is composed of many parts, and Ultimate Reality is an unbroken wholeness. Therefore, from the standpoint of absolute knowledge whatever leads to that unchanging Unity of God is true, and what does not, is false. When perfect mystics tell us that no way other than the path of Spirit leads to God, they are speaking from the perspective of objective science—not subjective philosophy.

This is no different from a physicist who teaches that in the course of attaining the energy of the superforce, the four forces of nature become increasingly unified—from four, they become three; from three, two; from two, one. Is it narrow-minded to say that there is only one way by which those disparate forces merge into the superforce? No, it is a fact. In order to go from many to one—whether that unity be the limited reality of the superforce which created this physical universe, or the Ultimate Reality which formed all of creation—at some point all paths must converge upon that summit of final truth.

Charan Singh writes, "God is one, and the way to gain Him is also one. No second way exists. The way to gain Him consists in withdrawing our consciousness from the body and joining it with Spirit."[26] This occurs, however, by degrees. Since there are five levels of creation, the soul must progress through five manifestations of Spirit before complete union with God is attained. At each stage, as was pointed out in the quotation above, the sound of God-in-Action takes on a different tone and His light a unique cast. These attributes of Spirit cannot be described accurately in terms of physical sounds and lights, but imperfect approximations are possible.

The sounds of Spirit

In terms of the levels of creation described previously, the characteristic sound of the fifth, or highest, region is said to be like that of the *vina* (a stringed instrument); of the fourth, like the *flute* or *harp;* of the third, like *thunder* or the beating of *drums*; of the second, like a big resounding *bell.*[27] What about the first, or lowest, region of creation—this physical universe? Unfortunately, when our consciousness is confined to this realm of existence Spirit is barely audible to the ear of the soul. The reason, as was noted previously, is that the crudity of mind and matter obscure the pristine vibratory resonance of God-in-Action. At this level we are able to discern only the vaguest echoes of His voice. Lekh Raj Puri says:

> When we, after initiation by a Perfect Master, begin practising the Word we listen to internal sounds. First we hear confused sounds like those of a flowing river, a running train or of showers of rainfall. Gradually the sounds change into those of the insect cicada, and [that] of small tinkling bells (metallic ring) comes in. Finally when our concentration has reached a high pitch, we hear the clear sound of a big bell, resounding and reverberating. That is the first real sound of the Word. It has a great power of attraction. . . . From plane to plane, the sound changes, till the soul reaches the Fifth Stage which is our true home. . . . [28]

Could the voice of God be ringing in your ears?

It is important to realize that not only do students of contemplative meditation hear the above-mentioned "confused" sounds of Spirit, but others as well—many of whom are completely unaware of what is being experienced. Charan Singh notes that "some even run to the doctors, thinking that there is something wrong with their ears."[29] Indeed, so-called ringing in the ears—termed *tinnitus* by medical science—is one of the most common health concerns. I am not implying that everyone who hears an unexplained sound in their head is listening to the voice of God. If you just set off a loud firecracker, or attended a raucous rock concert, one of these experiences is almost certainly the likely cause of the ringing in your ears. Certain diseases and medications also may produce tinnitus symptoms. But a great many cases of tinnitus, most in fact, baffle physicians.

In a general household survey of twenty-three thousand individuals in Great Britain, "15 per cent of the sample reported hearing noises in the head or ears of a kind not due to known external stimuli. . . . Two people in every 100 interviewed reported that they heard noises in their head or ears *continuously*. Roughly two-thirds of those with tinnitus said that there was nothing specific that brought it on—its appearance was quite unpredictable."[30] What is heard by people with tinnitus? Interestingly, sounds quite similar to those described by spiritual science as being the initial manifestations of Spirit. Leslie Sheppard and Audrey Hawkridge provide a listing of some of the sounds reported by those with tinnitus.[31] These include: running water; swarm of bees; buzzing; low throbbing hum; fizzing; constant whistling. And, as reported by the American Tinnitus Association, the following were the most common descriptions from a survey of thirteen thousand people with tinnitus: ringing, hissing, crickets, the sound of a high tension wire.

If you are bothered by an unexplained sound in your ears or head, do not hesitate to seek medical advice. According to the American Tinnitus Association survey, only thirty-one percent of people with this condition have tried any form of treatment for it. And of those who *have* tried some treatment, masking—which disguises but does not eliminate the sound— was the most effective measure. But only forty percent of those treated with a masking device found any benefit from it. So by all accounts tinnitus is a prevalent, but little understood, condition.

As a student of spiritual science, I find it interesting—and even a bit amusing—that so many people are trying to get rid of sounds in their head which bear such a close resemblance to sounds *welcomed* by students of contemplative meditation. Mystics say that early in the course of the soul's journey back to God, Spirit appears as the sound of a flowing river, running train, rainfall, the insect cicada, or a metallic ring. Those with tinnitus say that they hear such sounds as running water, low throbbing hum, crickets, or ringing. There is little doubt that some tinnitus "sufferers" are experiencing a spiritual phenomenon well known to mystics. Spiritual scientists realize that this inner sound is a manifestation of Spirit, while others do not know what to make of it. One person enjoys this echo of God-in-Action, and another fears that he is afflicted by an unknown disease.

Ultimate objective truth is **within us, not without**

Mysticism teaches that Spirit actually is the cure for all our ills. What could make us happier and healthier than listening to the voice of God, the sound of Ultimate Reality? Conversely, why does hearing and seeing absolute truth in the depths of our consciousness seem like a fanciful—even bizarre—notion to some people? How is this different from an astronomer looking through the eyepiece of a telescope at some limited truth about material reality? Consciousness, or the mind, is doing the observing in both cases. It comes down to this: we have much more faith in the existence of an objective *external* reality, than in the existence of an objective *internal* reality.

As was noted in the chapter concerning the four states of being, this attitude reflects an unfounded assumption: that the only substantive reality is in the public world of physical objects and shared symbols—words, numbers, and the like. So when a perfect mystic says that he has heard and seen Ultimate Reality in contemplative meditation, the reaction of many is to say: "That is impossible. Inside our heads there is only a subjective reality. We must look outside of ourselves to find objective reality."

I hope that by now the reader realizes the glaring flaws in this argument. First, we have seen that both physics and mysticism agree that whatever Ultimate Reality is, it is an unbroken wholeness. Second, spiritual science teaches—and material science is coming to accept—that consciousness is the fundamental ground of existence. Third, given the unity which underlies the seeming separateness of this material universe, no part ever can fully understand the whole. These premises lead to a simple conclusion: if absolute truth is a whole, and the human mind—or consciousness—is a part trying to understand that oneness, then somehow one must merge into the other. Since a whole obviously already is itself, and every part of itself, this means that it is up to the part to do the merging. And this is what occurs in contemplative meditation.

Mysticism teaches that the process of unifying the soul with Spirit proceeds in stages, with each stage being marked by certain sounds and sights. What is so strange about this? It would be much stranger if this low-energy physical universe was the only part of creation with any attributes, the rest of existence being silent, dark, and featureless. Thankfully, such is not the case. Mystics tell us that there is much to see and hear in higher domains of creation. However, these phenomena are but way-stations which the spiritual scientist passes on his way to the final goal: union with God. Lekh Raj Puri provides a hint of the wonders which lie within:

> In one pulse of universal consciousness the mystic sees thousands of worlds being created and dissolved, and thousands of worlds sweeping through the infinity of space and rolling in the eternity of Time; and then he looks

beyond all Time and Space, beyond Heaven and Hell, beyond cosmos and chaos, beyond past, present, and future; and beholdeth the Divine Light of Absolute Naked Truth, and entereth the supreme Essence of the ultimate transcendent Being, and becometh one with Him. . . . Thus [mystic] transport is not nothingness, nor a delusion or creation of our own imagination. It is a transcendent experience of objective Reality.[32]

And such is attained by uniting one's consciousness with Spirit, the all-pervading conscious energy which comprises everything in existence. How simple. To know all, become all. Sawan Singh summarizes the ageless mystical wisdom concerning Spirit:

It is a current of consciousness or divine Sound. When there is motion in the consciousness, the Sound appears. We call this the Name or Word of God. This Name or Sound is the sustainer of the entire universe. Wherever there is creation, there is divine Sound or Spirit. Creation cannot be carried out without the divine Sound. This Sound is all-pervading. No place is without It. It is, however, manifest at some places, and unmanifest at others. It is resounding from head to foot. It is the essence and real substance of the universe.[33]

Echoes in the new physics.

Throughout these pages we have seen that spiritual and material science agree about many aspects of the nature of reality, and differ on other issues—some central, some peripheral. For example, mysticism holds that an all-pervading *conscious* energy is the foundation of existence, while the mainstream of physics concurs that this substrate is "all-pervading" and "energy"—but is reluctant to admit that it is conscious. As has been noted previously, this view implies that somehow a part of the physical universe—the human brain—has been able to manifest a quality which is not present within the whole.

Mysticism teaches that the essence of Ultimate Reality, Spirit, indeed is conscious, and appears as audible spiritual vibration. Since most physicists deny that consciousness—in the sense of an entity separate from the mind and brain—even exists, the "audible spiritual" portion of this principle of spiritual science would find few adherents among the ranks of material scientists. However, the "vibration" aspect is a core tenet of the new physics. So the crux of the matter is this: mysticism says that it is possible for a person's consciousness to become attuned to the frequency of the all-pervading conscious energy which resonates throughout the cosmos. Then the sound of Spirit, God-in-Action, can be heard by a faculty of the soul. In essence, a drop of consciousness comes to vibrate in concert with the ocean of consciousness.

Waves of vibrating energy

As shall be shown in this section, material science also believes that vibrating energy is the root of existence. Whether it might be possible to hear, or otherwise perceive, this energy directly is not a subject that physicists seem to spend much time considering; consequently this issue rarely is raised in other books about the new physics. However, material scientists make many allusions to sound and music when describing the subatomic realm. This is because physical sounds are produced by waves, and waves of various kinds are the fundamental constituent of quantum reality.

Physicist Fritjof Capra writes that "sound is a wave with a certain frequency which changes when the sound does . . . particles—the modern equivalent of the old concept of atoms—are also waves with frequencies proportional to their energies. According to [quantum] field theory, each particle does indeed 'perpetually sing its song' producing rhythmic patterns of energy (the virtual particles) in 'dense and subtle forms.'"[34] In this quotation Capra is comparing findings of the new physics to the philosophy expressed by a Tibetan mystic, who said that "All beings, all things, even those things that appear to be inanimate, emit sounds . . . these beings and things are aggregates of atoms that dance and by their movements produce sounds."[35]

As Capra notes, this indeed is in tune with modern physics, which "pictures matter not at all as passive and inert, but as being in a continuous dancing and vibrating motion whose rhythmic patterns are determined by the molecular, atomic and nuclear structures."[36] Similarly, P.C.W. Davies and J.R. Brown observe that electrons circle the atomic nucleus in stationary or standing wave patterns: "Much as a cavity can be made to resonate at different discrete musical notes, so the electron waves vibrate with certain well-defined energy patterns . . . quantum theory predicts the existence of discrete non-radiating energy levels in which the wave associated with the electron just 'fits' around the nucleus, forming standing wave patterns reminiscent of the notes on a musical instrument."[37]

Recall that the quantum wave function determines the probability that a subatomic particle is located in a certain place. The Schrödinger equation describes how those wave functions change over time. Steven Weinberg says that "the Schrödinger equation is mathematically the same sort of equation (known as a partial differential equation) that had been used since the nineteenth century to study waves of sound or light. . . . Some of the solutions of the Schrödinger equation for electrons in atoms simply oscillate at a single pure frequency, like the sound wave produced by a perfect tuning fork."[38]

So at the heart of quantum reality the new physics finds vibrating waves which behave much like sound waves. But knowing how they act is not the same as knowing what they are. As was previously discussed, the mainstream perspective is that these are some sort of mathematical entity, as opposed to something "real." Weinberg says that "electron waves are not waves *of* anything; their significance is simply that the value of the wave function at any point tells us the probability that the electron is at or near that point."[39] However, Gary Zukav raises an obvious question: ". . . how can wave functions predict *anything* when they are defined as completely unrelated to physical reality? This is a scientific version of the philosophical question, 'How can mind influence matter?' "[40]

Is a quantum wave conscious?

In other words, physics has found that something which can be precisely described by mathematical equations directs all of the matter and energy in the universe, for physical existence is built up out of the myriad subatomic particles that are under the sway of quantum wave functions. But there is no physical sign of that something. It leaves no tracks, apart from its quantum effects. Most physicists conclude, therefore, that those waves are *nothing*. "If we cannot sense them," goes the argument, "then even though they act like actual waves of sound or light, they must be part of mathematical reality—not real reality." A minority of physicists find this reasoning unsatisfying.

After all, how could a non-material entity be so effective in guiding material particles? This is like the aforementioned ghost who can glide through solid walls, but also knock over lamps. If something is not physical, how does it interact with physical objects? Physics seems to have embraced a dualism which is at odds with its overarching monistic philosophy of materialism. Monism argues, "If the wave function exists, it must be composed of the same substance as the rest of physical reality. Physics says this is matter, and mysticism says that it is Spirit. Regardless, those waves must be *something*, not nothing at all."

Physicist David Bohm developed a version of quantum theory which resolved this conceptual difficulty. A central feature of Bohm's interpretation of quantum physics is the *pilot wave*. Mathematician John Casti describes how Bohm views this wave:

> In the pilot wave picture, every quantum object is a real particle possessing definite attributes at all times. Associated with each such object is a pilot wave that is also real but undetectable other than through its effects on the particle. This wave is called the quantum potential, and serves the function of "reading" the environment and reporting its findings back to the particle. Let me emphasize here that this is a real wave and should not be

confused with the quantum wave function, a purely mathematical gadget for making predictions. The particle then acts in accordance with the information provided by its associated pilot waves.[41]

Clearly Bohm has a radically different view of the quantum realm. What is all this talk about waves and particles "reading the environment," "reporting findings," and "acting in accordance with information"? Does Bohm actually believe that subatomic particles, and their pilot waves, are *conscious*? Yes, it appears that he does. In the course of an interview with Renée Weber, Bohm said that "the wave function is a kind of mental side to the electron, the information content determining its nature . . . the electron is doing much the same as what we are doing when we react to a situation. . . . It is gathering information about us, about the whole universe. . . . Therefore it is observing, if you take that in its literal sense."[42]

If one accepts Bohm's interpretation of quantum physics, many perplexing questions are resolved. Why does a photon of light behave like a particle in some experimental situations, and like a wave in others? Because the pilot wave "tells" the photon how to adjust to different configurations of a researcher's apparatus: "One slit is open; you must go through it and act like a particle. Now two slits are open; go through both of them like a wave."

John Casti writes, ". . . the quantum potential senses the presence of a measuring apparatus of a certain type and immediately notifies the particle, which then adjusts its behavior to accommodate to the kind of attribute the device is designed to measure."[43] As I noted previously, mysticism would see this as the workings of the universal law of karma. If a researcher desires to observe a photon as a wave, nature obliges. Bohm's pilot wave thus can be viewed as a manifestation of Spirit—God's Word—which acts in accord with laws of cause and effect in this domain of creation.

> *God speaks words of power to souls—*
> *To things of naught, without eyes or ears,*
> *And at these words they all spring into motion;*
> *At His words of power these nothings arise quickly,*
> *And strong impulse urges them into existence.*
> *Again, He speaks other spells to these creatures,*
> *And swiftly drives them back again into Not-being.*
> —Rumi[44]

So in David Bohm's interpretation of quantum physics, consciousness is very much part of the vibrating energy evident in the subatomic realm. And that realm turns out to be completely deterministic. The probabilities found in the traditional Copenhagen interpretation are considered to be, in physicist David Albert's words, " . . . a matter of ignorance and not a matter of the operations of any irreducible element of chance in the fundamental laws of the world. Nevertheless, this theory [of Bohm's] entails that some such ignorance exists for us, as a matter of principle . . . the act of measurement unavoidably gets in the way of what is being measured."[45] We have, then, a vision of physical reality which is in accord with principles of spiritual science: causes and effects proceed deterministically under the guidance of conscious laws, but the inner workings of those laws ordinarily cannot be known completely.

The strange world of superstrings

Having learned that a well-respected—if not universally accepted—theory of the new physics holds that consciousness permeates the subatomic world, let us return to the matter of vibrating energy. Paul Davies and John Gribbin summarize modern quantum field theory:

> . . . it paints a picture in which solid matter dissolves away, to be replaced by weird excitations and vibrations of invisible field energy. In this theory, little distinction remains between material substance and apparently empty space, which itself seethes with ephemeral quantum activity. The culmination of these ideas is the so-called superstring theory, which seeks to unite space, time and matter, and to build all of them from the vibrations of submicroscopic loops of invisible string inhabiting a ten-dimensional imaginary universe.[46]

This superstring theory currently is the leading candidate for the new physics' Theory of Everything. Richard Morris says, "Superstring theory, after all, is what we sometimes like to call 'the only game in town.' Though scientists have explored every possible idea, no one has found any other plausible way of unifying the four forces [of nature] within one theory, and as we have seen, the unification of the forces is the theoretical physicist's Holy Grail."[47] In the course of discussing the previous principle of spiritual science, *Spirit creates and sustains every level of creation*, we saw that the new physics is confident that those four forces were one—the superforce—at the instant of the Big Bang. It also appears clear that energy levels are the key to unifying those now-disparate forces: at higher energies forces merge; at lower energies, they separate.

But knowing these broad outlines of a Theory of Everything is a far cry from explaining exactly *how* unification takes place. This is why superstrings have captured the fancy of so many theoretical physicists. Though the mathematics of superstring theory is extremely difficult, almost intractable, the fundamental concept underlying these equations is simple: in the words of Anthony Zee, "A fundamental particle is represented as a bit of vibrating string. If the bit of string is much shorter than the resolution of our detection instruments, it will look like a point particle. The remarkable feature of string theory is that by vibrating in different ways, the string can appear to us as different particles. . . . Thus, string theory holds out the promise of a truly grand unification, in which gravity is tied intrinsically to the grand unified interaction."[48]

Well and good. But what *are* these pieces of "string?" First, they are not matter, energy, time, or space as we know them—because superstring theory holds that each of these aspects of physical existence is produced through the vibrations of those strings. Further, superstrings are believed to possess a length, but no other dimension, and are incredibly small. F. David Peat says that "if we could imagine shrinking down from our own size to that of a single elementary particle, then we would then have to perform an equally powerful act of the imagination to shrink down to the size of a string."[49] Thus they exist at the utmost edge of physical existence, seemingly as close as a quasi-material entity could be to whatever lies beyond.

A string is massless, and its ends vibrate at the speed of light. "Yet," notes Peat, "despite this absence of mass, it is still possible for this string to represent massive particles. The reason is that because the string vibrates and rotates, it will have a series of energy levels, just as a violin string has a series of notes. Since Einstein has shown that energy and matter are related by the equation $E=mc^2$, these energy levels also have associated masses."[50] This budding theory of superstrings is quite elegant. Infinitesimal building blocks of creation vibrate at varying energy levels, thereby producing out of a single substance the myriad variety of matter and forces in our universe. David Freedman says that "at high energies they are their own bizarre entities. At lower energies their vibrations give them the characteristics of gravity or a quark or any of the other forces or fundamental particles."[51]

The mathematics of superstring theory implies a huge gap between those higher and lower vibrational energies. That is, the "octaves" of superstrings take a tremendous jump from that which produces all of the phenomena of materiality, and their natural energy—where they exist as themselves. Dennis Overbye sums up this musical aspect of superstring theory:

> You could think of the elementary particles as notes played on God's ten-dimensional guitar. All our ears could hear in this cold lonely universe in the era of galaxies and protons was the lowest bottom octave. All the particles known to physics . . . belonged to this lowest octave, in which their masses were negligible (to a string's way of thinking). The next octave and every octave after that would raise the mass-energy of these notes by nineteen orders of magnitude. . . . So the full power of

superstrings was reserved for a crack of a moment at the beginning of time, when God's accelerator was fully charged, a golden second when the symmetry of nature must have been more dazzling than a diamond with a thousand faces. Physics itself was somehow a relic of that ancient era.[52]

Hidden higher dimensions of reality

We have come, then, to the most fundamental—though still speculative—theory of the new physics: the core reality of our universe is vibrating bits of what Steven Weinberg terms "tiny one-dimensional rips in the smooth fabric of space."[53] As noted in the quotation above, superstrings are considered to exist in a ten-dimensional space, only four of which are evident in our everyday world (time, plus height, width, and breadth). So it does make sense to view them as "rips" in the fabric of physical time and space, for superstrings are seen as links between this realm of existence and some unknown dimensions. In superstring theory those other dimensions, though unseen, are considered to be part and parcel of the reality in which we live.

F. David Peat writes, ". . . although the six compactified [or hidden] dimensions of superstring space are invisible to us, their effects pervade our world at every level. They are responsible for the appearance and the number of generations of the elementary particles and for the existence of the forces of nature."[54] We have reached a point where it is exceedingly difficult to distinguish mysticism and physics. Theories of material science now are telling us that six higher dimensions, which vibrate at levels almost infinitely more energetic than the physical universe, produce all phenomena. This is in accord with teachings of spiritual science, for we have seen that Spirit creates and sustains four higher domains of creation before reaching our level of existence.

Further, superstrings are considered to possess both an infinite capacity to create, and are eternal. How much nearer to the all-pervading conscious energy of Spirit can the new physics come? Steven Weinberg says,

> Each string can be found in any one of an infinite number of possible states (or *modes*) of vibration, much like the various overtones produced by a vibrating tuning fork or violin string. The vibrations of ordinary violin strings die down with time because the energy of vibration of a violin string tends to be converted into a random motion of the atoms of which the violin string is composed, a motion we observe as heat. In contrast, the strings which concern us here are truly fundamental and keep vibrating forever; they are not composed of atoms or anything else, and there is no place for their energy of vibration to go.[55]

Similarly, mysticism makes a distinction between "struck" and "unstruck" sounds. All outer sounds are of the first variety, since they are produced by some cause. Eventually struck sounds end, though they may last billions of years. On the other hand, Sawan Singh says there "is a continuous Sound which is not perishable. . . . The Lord is without form. He manifests as Unending Sound. . . . This Unstruck Music is resounding in the sky of every mind. . . . The souls of God-men leave their bodies and become absorbed in the unending Music."[56]

This may seem remarkable. Even unbelievable. But it is simply the truth of Ultimate Reality as known to spiritual science, which is completely compatible with the findings of material science. Where physics leaves off, mysticism continues. We have found that findings of the new physics have reached the utmost limit of physical existence, penetrating right to the edge of materiality. What is found there? Vibrating energy—an energy which some physicists believe is conscious. The significant difference between spiritual and material science concerns this question: *Can that vibrating energy be perceived directly in one's consciousness?*

Mysticism unequivocally says, "Yes." The all-pervading conscious energy of Spirit makes up everything in existence. This includes us: our physical body, our mind, our soul. We are conscious because a drop of the ocean of God, the soul, is our true innermost self. Veiled by coverings of mind and matter, the spiritual faculties of the soul—hearing and seeing—are dormant. When we close our eyes and look within the inner sky of our consciousness, we generally perceive only silence and darkness. But through contemplative meditation it is possible to raise one's attention into non-material regions of reality.

The superstring theory of physics postulates six (or more) dimensions beyond the time and space of our universe. Mysticism teaches that there are four (or more, depending upon how one views them) higher domains of consciousness lying above the everyday waking state. Many physicists call those superstring dimensions "imaginary" mathematical entities, even though they are theorized to underlie all physical phenomena. Every perfect mystic says that the higher regions are objectively real, more real than anything on earth, and can be entered by those who devote themselves assiduously to the practice of contemplative meditation.

Jump into the ocean of Spirit

Again: human beings are formed from the same substance which comprises the material universe, and everything else in existence: Spirit. So when we reach the deepest essence of ourselves, we also are in touch with the foundation of Ultimate Reality—the final Theory of Everything. This Truth is an all-pervading conscious energy which is audible as spiritual sound, and visible as spiritual light.

> *How could the fish not jump*
> *Immediately from dry land into water*
> *When the sound of water from the ocean*
> *Of fresh waves springs to his ear?*
>   —Rumi[57]

Andrew Harvey interprets the deep meaning of these poetic words uttered by the great spiritual scientist, Jalaluddin Rumi:

That sound is a real mystical sound heard in meditation by yogis, Sufis, and lovers of God. That sound is the sound of the "water" that is drenching the universe at every moment. That water is the Divine Light, at every moment drenching, soaking, and saturating all things. Nothing else is going on but this rain of blissful Light that is soaking creation at all moments. When you begin to awaken, you hear that rain of Light as water. The sound of water—because God is kind and gives us sounds that we can be familiar with—alerts us to the secret endless ocean of Light that is the universe, "the ocean of fresh waves". . . . When you begin to hear that sound, how could you not want to "leap from the dry land" of the ego, desire, and ignorance, the dry land of this terrestrial, banal, separative reality, into the sea of gnosis and rapture, "the ocean of fresh waves"?[58]

Such is the truth which lies at the heart of both the new physics and ageless mysticism—the ocean of God's Spirit is the fountainhead of creation. Whether concealed as the dry land of materiality, or revealed as the fresh waves of spirituality, God-in-Action always is unbroken wholeness. Ocean waves ceaselessly break upon the shoreline, while the depths of the sea are calm. Similarly, Spirit's conscious energy vibrates continually in every domain of existence, created and sustained by the eternal, boundless, unchanging presence of God. If the force of Spirit was absent from our universe for an instant, nothing would remain. In the same fashion, when the divine energy of soul is withdrawn from your body, you will die.

It behooves us to become acquainted with this power behind all other powers while we are alive. For Spirit will be your only true Friend when you leave everything familiar behind at the time of death. Material scientists seek to understand this "superforce" through mathematical equations and conceptual models, even though whatever created everything in our universe obviously preceded numbers and words as we know them—and is the reality that remains when ideas about reality fade away. Physics has been able to gain some insights into the subtle nature of existence, but these merely are reflections of reflections of final non-symbolic truth. Mysticism urges: "A genuine scientist is not content with such weak echoes and blurry images. Your consciousness *is* the clear Voice and bright Light of God. Learn how to focus that attention through the discipline of contemplative meditation, and perceive Him directly."

Epilogue

Writing these final pages, I feel like a lecturer who has been speaking for hours in a darkened auditorium. Finally he comes to the conclusion of his talk. Peering into the shadows beyond the stage spotlights, he is unable to discern how many people remain in their seats. "Perhaps by now I am talking largely to myself," he thinks. Indeed, some members of my original audience undoubtedly took a break from reading these pages and failed to return from that intermission. Others, like you, have been persistent enough to reach this epilogue. But having been exposed to the message of *God's Whisper, Creation's Thunder* does not mean that you are convinced by it. Many will regard this book as mildly interesting and informative, nothing more.

If this describes you—someone who is curious about mysticism, yet skeptical about the legitimacy of spiritual science—I only can hope that you will choose to read other books about pure mysticism and not close the door on this approach to knowing the truth about Ultimate Reality and your innermost self. There are many ways of describing the indescribable teachings of perfect mystics. I have emphasized the scientific tenor of mysticism, and this may not appeal to those who approach spirituality with a different bent. Perhaps reading this book has strengthened your commitment to whatever philosophy or metaphysics you *do* believe in. Tread with confidence that chosen path to truth. But remember the way of spiritual science if you ever encounter an immovable roadblock or impassable dead end on that path.

Turn faith into reality

Those who do not agree with the specific principles of mysticism discussed previously still may find that applying the general scientific method to spirituality will bring new vitality to their chosen faith. By this I mean: whatever you believe, make it more than a belief. Turn your faith into a vibrant reality. If you are reasonably sure that your religion, philosophy, or metaphysics is true, act in accord with that reason so that wispy abstract concepts become solid experiential facts. And do this *now*, not after death. What if your faith is sincere but misguided? Once you die, it will be too late to live your life over. It is much better to confirm here and now those truths which are being counted on to support you for eternity.

Ask yourself, "How do I know that what I believe is true?" If this question cannot be answered, then you have become content with faith, not facts; with reason, not reality; with supposition, not science; with philosophy, not proof. It never fails to amaze me that among the myriad diverse creeds, religions, and other forms of metaphysical belief, so many claim to possess the final truth about existence. This is impossible if Ultimate Reality is something other than a subjective mirage with no substance apart from appearances. The material sciences—such as physics—have proven that objective laws of nature exist independent of personal perceptions or beliefs. I have argued that the same is true of spiritual science. Thus if there are a thousand different ways of making perfect sense of the cosmos, only *one* of them can be correct. Are you absolutely sure that what you believe is that single penultimate truth? If not, keep searching for the evidence that will convince you.

Let me address myself now to those readers who sense an unmistakable ring of truth in the tenets of spiritual science described in this book. This sensation, let me assure you, has nothing to do with the author's means of expression, and everything to do with his subject matter: truth. Truth cannot be hidden from those who seek it sincerely and tenaciously. Being the all-pervading conscious essence of God, Spirit's truth finds those whom truth wants to be found by. The title of a book on Zen expresses this idea nicely: *That Which You are Seeking is Causing You to Seek.*[1]

In a movie about the life of C.S. Lewis, *Shadowlands*, a student tells Lewis, "We read to know that we are not alone." Reader, I hope that this has been the case for you. None of us is alone. We need to be reminded continually of this fundamental fact of existence. At this very moment everyone is enfolded in the loving arms of Spirit. If this Unity which underlies every aspect of creation only could be realized, we never would feel anxious, worried, isolated, confused, or fearful. But it is not enough to think about, or believe in, the presence of that oneness. We must cease existing as illusory separate parts, and become the reality of the whole.

This is the goal of contemplative meditation, which serves as the experimental laboratory of spiritual science. Within that laboratory of our consciousness, all the truths of existence can—and will—be revealed to those who long for that knowledge with every fiber of their being. This is a promise which unfailingly is kept by perfect mystics, the professors of mysticism. If you truly want to learn what they know, find such a teacher. Let go of everything else in this world if necessary, but catch hold of a perfect mystic. The beauty is that as you look for Him, He is finding you. Sincere seekers of Ultimate Reality are not searching on their own. Though the journey toward final truth often seems lonely and long, every step one takes in that direction is being guided. Believe this with all your heart.

Spirituality and the scientific method

Some readers already may be pursuing a discipline of contemplative meditation aimed at merging one's consciousness with Spirit. I hope that this book has strengthened your appreciation of the scientific nature of the practice in which you are engaged. Knowledge of God is attained by those who meditate with the rigor and exactitude demanded of scientific research, and the passion and love required of spiritual devotion. When a student of mysticism is able to manifest both of these qualities, barriers to truth crumble. One or the other is not sufficient. This is why mysticism is termed *spiritual science*. To achieve success in contemplative meditation, one should conduct his experiments in the laboratory of consciousness with the precision of a physicist and the selflessness of a saint.

Let me speak now to those readers who still are wedded to the promises of material science: I understand and respect your commitment to the scientific method. You are absolutely correct in demanding demonstrable proof of theorized laws of existence, whether those laws be of a physical or spiritual nature. However, I hope that you will remain open to a broader definition of "proof." Continue to consider the possibility that there may be objective, yet non-symbolic, realms of existence. What kind of proof could be provided of such domains? I have argued that evidence for a higher state of consciousness cannot be found in a lower state, just as someone dreaming is unable to demonstrate the reality of wakefulness.

If this is true, then entering a higher domain of consciousness is the only way to prove that such a state exists. Through contemplative meditation the everyday mental activities of observing, thinking, and feeling are unified into a fourth state of being—*contemplating*, which leads one to regions of consciousness "above" the material universe. While some find this fantastic and unbelievable, spiritual science describes the precise methods by which the experiment of God-realization is conducted. If you are skeptical about whether the principles of mysticism described in this book are valid, then by all means seek to refute them. But to do so, you must replicate the research approach taught by perfect mystics.

Several years ago there was a flurry of excitement concerning the possibility of generating virtually limitless amounts of energy through cold fusion. But no one was able to replicate the initial experiment which supposedly resulted in this type of fusion. Hence, that energy source remains theoretically possible, but unproven. Material scientists could repeat the methods which purportedly resulted in cold fusion, but not the claimed results. The same scientific method—attempted replication of experimental results—must be used by those who seek to disprove the assertions of mysticism described in these pages.

Around the globe, hundreds of thousands of students of spiritual science meditate for several hours a day, every day. In addition, they follow certain codes of behavior—including abstinence from alcohol and mind-altering drugs—designed to assure that the laboratory of their consciousness is fit for this research. I could describe other actions that are necessary, or valuable, adjuncts in the study of mysticism, but my purpose is simply to provide some idea of what is required to properly practice contemplative meditation. The results of this practice are evident only within the student's consciousness. As has been discussed throughout this book, the objective non-symbolic reality of higher domains of consciousness cannot be reflected in symbolic words, numbers, or images. By the same token, the *non-existence* of such realms also cannot be proven by such means.

Given this fact, I will be a devil's advocate and suggest how someone who doubts the validity of spiritual science could substantiate his position: engage wholeheartedly in the disciplines associated with contemplative meditation for, say, ten years. Follow every precept enjoined by a perfect mystic; rise early each day and meditate as instructed for two or three hours; empty your consciousness of all images, thoughts, and emotions during that time. Do this for a decade, then write an exposé about your experiences. But be warned: if you do everything I have suggested, the results of your research will be different from what you expect. Still, give it a try. Attempt to debunk mysticism by becoming a mystic.

Real science is direct vision, not mere talk

My ploy is obvious, of course. For I know that anyone who drinks even one drop of the ocean of Spirit will be changed forever. All who dive into the sea of that divine consciousness are drenched with God's presence. It makes no difference whether they believe in Him or not. Experience transcends belief. I am not asking any reader to have blind faith in the reality of Spirit—only to open the eye of your consciousness and look within. Consider to be true what is directly seen and heard there. Do not trust dry ideas and barren concepts, whether they be your own or anyone else's. Such lifeless abstractions block our awareness of the living current of Spirit which ceaselessly flows within us.

In 1994 an exciting event occurred in the world of physics: the first experimental evidence for the "top" quark. Recall that quarks, along with electrons, are believed to be the fundamental building blocks of matter. Theories predict that several varieties of quarks—including the top—existed only in the high energy of the universe's earliest moments, and are absent from ordinary matter. However, a particle accelerator was able to reproduce that energy (which still was far less than that of the Big Bang itself) for a fraction of a second, and convert it into the previously unobserved top quark.

This was front-page news. But seekers of Ultimate Reality should consider some sobering facts about the top quark's discovery, for these illustrate how far material science is able to journey toward final truth. First, physicists must put forth great effort to lay bare the secrets of existence—which also is the case in mysticism. It took 440 physicists from 34 countries 17 years to reveal the top quark.[2] This is a tremendous expenditure of time, money, and mental energy. I admire the determination shown by these scientists in their hunt for knowledge about the universe, especially given the nature of their quarry—and what they were able to capture.

This massive effort, after all, was directed at finding evidence for a constituent of matter which does not even exist in the present-day universe. Of course, it could be argued that this also is the goal of mysticism: to contact the hidden essence of existence. However, there are large differences between physicists' search for the top quark, and mystics' quest for God's Spirit. For what did the material scientists actually find? This variety of quark was not sensed by any instruments, nor were its tracks detected. After lasting for a hundred billionth of a trillionth of a second, the top quark decays into lighter particles. These remnants are what physicists detect of that quark, or as science writer Sharon Begley puts it, "sometimes the decay products of its decay products."[3]

Further, even this knowledge is based on probabilities, and hence uncertain. Out of sixteen million collisions between protons and antiprotons, physicists detected *twelve* which resulted in decay products characteristic of the top quark. Something else could have produced these effects, but statistical calculations indicated such to be unlikely. So this is how the top quark was "discovered." Not seen, not heard, not sensed in any way, nor proven to exist beyond a shadow of a doubt, but implied. Strongly implied, to be sure, but implied all the same. Indications, reflections, probabilities: these are what the new physics is able to provide seekers of final truth. Is statistical evidence that evanescent decay products probably came from an imperceptible top quark sufficient for you, or do you yearn for more direct and substantial knowledge about the essence of reality?

If so, consider becoming a student of spiritual science. Continue to learn about material science also, if you like, for these disciplines are complementary. However, the beauty is that while a physicist needs to know facts and formulas, methodologies and mathematics—the mystic forgets all that he has learned about the material world, at least during the time he is engaged in the experiment of contemplative meditation. This allows spiritual science to be practiced by anybody, anywhere. Still, mysticism is no easier to master than physics; indeed, it is much more difficult, going as it does against the grain of social convention. "Acquire more information," we are told from an early age by our teachers and parents, "and you will be a wiser person." Spiritual science says: "Your ideas are what veil you from the essence of reality, which is pure consciousness. Stop *thinking* about what is true, and *become* Truth."

Rumi, as always, says it best.

> *In truth everything and everyone is a shadow of the Beloved,*
> *and our seeking is His seeking and our words are His words. . . .*
> *We search for Him here and there, while looking right at Him.*
> *Sitting by his side, we ask: "Oh Beloved, where is the Beloved?"*
> *Enough with such questions.*
> *Let silence take you to the core of life.*
> *All your talk is worthless when compared*
> *with one whisper of the Beloved.*
>
> —Rumi[4]

Listen for that whisper. It is the only voice truly worth attending to. Every mystery of the universe is contained within the Word of God which ceaselessly resounds in each particle of creation. More than that: the Creator *is* His creation. The mind trembles at the wonder and glory of this. We are not separate from the fabric of existence. The reality of God is closer to us than we can imagine. That very imagination veils our awareness of Spirit. Rumi says, "Know real science is seeing the fire directly, not mere talk, inferring the fire from the smoke."

Equations, words, thoughts, concepts, beliefs, theories, intuitions, expectations, dreams. All of these are smoke, too wispy to support you. This book also is smoke, but it points in the direction of Spirit's fire. Burn up your fears and uncertainties in that divine flame. Drown your ignorance and misconceptions in that holy water. This is why you were born—to die and be born again, this time as the eternal essence of Ultimate Reality.

> *Leave aside your body and become spirit! Dance to that world!*
> *Flee not, even if for now dying is tumult and commotion. . . .*
> *Why should we flee the spirit?*
> *We find true spirit when we surrender the spirit.*
> *Why should we flee the mine? Dying is a mine of gold!*
> *Once you have been delivered from this cage,*
> *your home will be the rosegarden.*
> *Once you have broken the shell,*
> *dying will be like the pearl.*
> *When God calls you and pulls you to Himself,*
> *going is like paradise, dying like the pool of heaven.*
>
> —Rumi[5]

Notes

Prologue
1. John Barrow, *Theories of Everything*, p.1.
2. E. H. Whinfield, *Teachings of Rumi: The Masnavi*, p.122.

Is Ultimate Reality Real, and Who Can Tell?
1. William Chittick, *The Sufi Path of Love*, p.204.
2. William Chittick, *The Sufi Path of Love*, p.201.
3. Martin Gardner, *The Whys of a Philosophical Scrivener*, p.11.
4. Ibid, p.18.
5. Bryan Appleyard, *Understanding the Present*, p.10.
6. Daniel Dennett, *Consciousness Explained*, pp. 431, 433, 435.
7. Ibid, p.455.
8. W. M. Thackston, Jr., *Signs of the Unseen*, p.194.
9. Joe Rosen, *The Capricious Cosmos*, p.4.
10. Ibid, p.6.
11. Ken Wilber, *Quantum Questions*, p.153.
12. Paul Davies, *The Mind of God*, p.103.
13. John Boslough, *Masters of Time*, p.224.
14. Ibid.
15. Douglas Hofstadter, *Gödel, Escher, Bach*, p.75.
16. Ibid, p.77.
17. Bryan Appleyard, *Understanding the Present*, p.233.
18. E. H. Whinfield, *Teachings of Rumi: The Masnavi*, p.306.

Divisions of Consciousness
1. Ken Wilber, *Quantum Questions*, p.18.
2. Bryan Appleyard, *Understanding the Present*, p.140.
3. Paul Davies, *The Mind of God*, p.84
4. Paul Davies, *Superforce*, p.51.

5. David Lindley, *The End of Physics*, p.4.
6. W. M. Thackston, Jr., *Signs of the Unseen*, p.203.

7. Paul Davies, *God and the New Physics*, p.151.
8. Anthony Zee, *Fearful Symmetry*, p.132.
9. Nevit Orguz Ergin, *Magnificent One*, p.28.
10. Nick Herbert, *Elemental Mind*, p.74.
11. Joseph Traub and Henryk Wozniakowski, "Breaking Intractibility," *Scientific American*, p.107.
12. Ibid, pp 102, 107.
13. H.L. Mencken, *A New Dictionary of Quotations*, p.1192.
14. Robert Hutchins, editor, *Great Books of the Western World, vol. 8*, p.295.
15. Robert March, *Physics for Poets*, p.9.
16. Ibid, p.10.
17. Robert Hazen and James Trefil, *Science Matters*, p.xvii.
18. A.J. Arberry, *Mystical Poems of Rumi*, p.62.
19. M.O'C. Walshe, *Meister Eckhart*, pp. 66, 118, 126.
20. William Chittick, *The Sufi Path of Love*, p.178.

The Spiritual Science of Mysticism
1. Anthony Zee, *Fearful Symmetry*, p.280.
2. Deepak Chopra, *Quantum Healing*, pp. 45, 67.
3. Camille and Kabir Helminski, *Rumi—Daylight*, p.151.
4. King James Version, St. John 14:2.
5. Ken Wilber, *Eye to Eye*, p.34.
6. Sawan Singh, *Philosophy of the Masters, series iv*, p.lxii.
7. Robert Hutchins, editor, *Great Books of the Western World, vol. 19*, p.35.
8. Gopal Singh, *Sri Guru Granth Sahib, vol 1*, p.165.
9. Kabir Helminski, *Love is a Stranger*, p.6.
10. William Chittick, *The Sufi Path of Love*, p.2.
11. Camille and Kabir Helminski, *Rumi—Daylight*, p.10.
12. Jonathan Star and Shahram Shiva, *A Garden Beyond Paradise*, p.xii.

13. Thomas Matus quoted in Fritjof Capra and David Steindl-Rast, *Belonging to the Universe*, p.49
14. William Chittick, *The Sufi Path of Love*, p.291.
15. Ibid, p.127.
16. Ibid, p.128.
17. Ibid, p.222.

The Material Science of the New Physics

1. John Barrow, *Theories of Everything*, p.11.
2. Fritjof Capra, *The Tao of Physics*, p.6.
3. Ibid, p.23.
4. Raymond Moody, *Reflections on Life After Life*, p.86.
5. Joe Rosen, *The Capricious Cosmos*, p.19.
6. Ibid, p.24.
7. Ian Stewart, *Does God Play Dice?*, p.293.
8. Joe Rosen, *The Capricious Cosmos*, p.28.
9. Ibid, p.21.
10. Ibid, p.23.
11. Neil McAleer, *The Cosmic Mind-Boggling Book*, p.xii.
12. Brian Walker, *Hua Hu Ching*, p.56.
13. Kabir Helminski, *Love is a Stranger*, p.72.
14. William Chittick, *The Sufi Path of Love*, p.112.
15. Joe Rosen, *The Capricious Cosmos*, p.25.
16. James Trefil, *Reading the Mind of God*, p.198.
17. Paul Davies, *God and the New Physics*, p.159.
18. Gopal Singh, *Sri Guru Granth Sahib, vol 1*, p.5.
19. King James Version, St. John 1:1
20. Ahmed Ali, *Al-Qur'an*, 2:117.
21. King James Version, Genesis 1:3
22. Gia-Fu Feng and Jane English, *Tao Te Ching*, ch. 1.
23. Raimuindo Panikkar, *The Vedic Experience*, p.111.
24. Joe Rosen, *The Capricious Cosmos*, p.26.
25. Anthony Zee, *Fearful Symmetry*, p.3.
26. Ibid, p.5.
27. Arthur Arberry, *Discourses of Rumi*, p.163.

Theory of Everything, or God?

1. John Barrow, *Theories of Everything*, p.10.

2. Lekh Raj Puri, *Mysticism—The Spiritual Path, vol. II*, p.4.
3. Coleman Barks, *We Are Three*, p.11.
4. Paul Davies, *The Mind of God*, p.82.
5. Ibid.
6. B. Alan Wallace, *Choosing Reality*, p.68.
7. W. M. Thackston, Jr., *Signs of the Unseen*, p.147.
8. Sawan Singh, *Spiritual Gems*, p.29.
9. Stephen Hawking, *A Brief History of Time*, p.12.
10. William Chittick, *The Sufi Path of Love*, p.224.
11. Anthony Zee, *Fearful Symmetry*, p.9.
12. Robert Bly, *The Kabir Book*, p.17.
13. Nick Herbert, *Elemental Mind*, p.26.
14. Ibid, p.27.

Matter's Magic Act—Now You See It, Now You Don't

1. Robert Hazen and James Trefil, *Science Matters*, p.165.
2. Heinz Pagels, *The Cosmic Code*, p.21.
3. Ibid, p.181.
4. Fritjof Capra, *The Tao of Physics*, p.62.
5. Nick Herbert, *Faster Than Light*, p.108.
6. Fritjof Capra, *The Tao of Physics*, p.58.
7. Edward Harrison, *Masks of the Universe*, p.125.
8. F. David Peat, *Einstein's Moon*, p.28.
9. Nick Herbert, *Quantum Reality*, p.159.
10. Ibid, p.55.
11. Paul Davies, *Superforce*, p.133.
12. Fritjof Capra, *The Tao of Physics*, p.62.
13. Gary Zukav, *The Dancing Wu Li Masters*, p.32.
14. Heinz Pagels, *The Cosmic Code*, p.203.
15. Paul Davies and John Gribbin, *The Matter Myth*, p.18.
16. Ibid.
17. Ken Wilber, *Quantum Questions*, p.132.
18. Ibid, p.129.
19. William Chittick, *The Sufi Path of Love*, p.133.
20. David Lindley, *The End of Physics*, p.19.
21. Stephen Hawking, *A Brief History of Time*, p.134.
22. Ibid, p.139.
23. David Lindley, *The End of Physics*, p.20.
24. Willem Drees, *Beyond the Big Bang*, p.72.

25. Stephen Hawking, *A Brief History of Time*, p.174.
26. Joe Rosen, *The Capricious Cosmos*, pp. xii, 126, 128.
27. Paul Davies, *The Mind of God*, pp. 226, 231.
28. John D. Barrow, *Theories of Everything*, p.209.
29. Bryan Appleyard, *Understanding the Present*, p.216.
30. David Lindley, *The End of Physics*, p.255.
31. Camille and Kabir Helminski, *Rumi—Daylight*, p.204.

Universities of Ultimate Reality

1. Steven Weinberg, *Dreams of a Final Theory*, p.254.
2. Ibid, p.257.
3. William Chittick, *The Sufi Path of Love*, p.149.
4. Brian Walker, *Hua Hu Ching*, p.24.
5. Aldous Huxley, *The Perennial Philosophy*, p.vii.
6. Ibid, p.viii.
7. Ibid, pp. ix, xi.
8. Sawan Singh, *Spiritual Gems*, p.322.
9. Lekh Raj Puri, *Mysticism—The Spiritual Path, vol. I*, p.134.
10. King James Version, St. John: 10:33, 34, 38.
11. William Chittick, *The Sufi Path of Love*, p.125.
12. King James Version, St. John: 10:16.
13. William Chittick, *The Sufi Path of Love*, p.246.
14. Ibid, p.122.
15. Ibid, p.123.
16. V.K. Sethi, *Kabir, the Weaver of God's Name*, pp. 210-211.
17. Ken Wilber, *Eye to Eye*, p.278.
18. Charan Singh, *The Master Answers*, p.204.
19. William Chittick, *The Sufi Path of Love*, p.344.
20. Ibid, p.275.
21. Charan Singh, *Spiritual Discourses*, p.98.
22. William Chittick, *The Sufi Path of Love*, p.276.
23. Ibid, p.277.
24. King James Version, Jeremiah: 5:21.

25. Lekh Raj Puri, *Mysticism—The Spiritual Path, vol. I*, pp. 89, 92-93.
26. David Bruning, "Fixing Hubble," *Astronomy*, p.37.
27. William Chittick, *The Sufi Path of Love*, p.177.
28. William Chittick, *The Sufi Path of Love*, p.136.
29. Lekh Raj Puri, *Mysticism—The Spiritual Path, vol. I*, p.167.
30. Stephen Hawking, *A Brief History of Time*, p.9.
31. Ibid, p.10.
32. William Chittick, *The Sufi Path of Love*, p.125.
33. Stephen Hawking, *A Brief History of Time*, p.11.
34. E.H. Whinfield, *Teachings of Rumi: The Masnavi*, p.52

Meaning and Form

1. Steven Weinberg, *The First Three Minutes*, pp. 154-55.
2. Gerald Schroeder, *Genesis and the Big Bang*, pp. 172-73.
3. William Chittick, *The Sufi Path of Love*, p.20.
4. Ibid, p.21.
5. Ken Wilber, *Quantum Questions*, p.171.
6. Ibid, p.174.
7. Charan Singh, *The Master Answers*, p.322.
8. Charan Singh, *Spiritual Discourses*, pp. 33, 41, 114, 167.
9. William Chittick, *The Sufi Path of Love*, p.203.
10. Camille and Kabir Helminski, *Rumi—Daylight*, p.109.
11. David L. Chandler, "Here's another for the book of records," *Oregonian*.
12. Dennis Overbye, *Lonely Hearts of the Cosmos*, p.411.
13. Neil McAleer, *The Cosmic Mind-Boggling Book*, p.155.
14. John Updike, *Roger's Version*, pp. 10, 20.
15. Ibid, p.21.
16. William Chittick, *The Sufi Path of Love*, p.59.
17. Arthur J. Arberry, *Discourses of Rumi*, p.235.
18. Seth Shiv Dayal Singh, *Sar Bachan*, p.27.

19. Camille and Kabir Helminski, *Rumi—Daylight*, p.66.
20. Tony Rothman, "This is the Way the World Ends," *Discover*, p.81.
21. Camille and Kabir Helminski, *Rumi—Daylight*, p.55.
22. Andrew Harvey, *The Way of Passion*, p.101.
23. Julian Johnson, *The Path of the Masters*, pp. 436, 438.
24. Camille and Kabir Helminski, *Rumi—Daylight*, p.90.
25. Arthur Arberry, *Discourses of Rumi*, p.238.
26. John Polkinghorne, *Reason and Reality*, p.39.
27. William Chittick, *The Sufi Path of Love*, p.215.

Principle 1.

1. James Cowan, *Where Two Oceans Meet*, p.79.
2. William Chittick, *The Sufi Path of Love*, p.70.
3. Sawan Singh, *Philosophy of the Masters, series iv*, p.39.
4. William Chittick, *The Sufi Path of Love*, p.183.
5. Kabir Helminski, *Love is a Stranger*, p.50.
6. Charan Singh, *Quest for Light*, p.185.
7 F. David Peat, *Einstein's Moon*, p.113.
8. Ibid, p.76.
9. Ibid, p.69.
10. Heinz Pagels, *The Cosmic Code*, p.149.
11. Nick Herbert, *Faster Than Light*, p.159.
12. Paul Davies and John Gribbin, *The Matter Myth*, p.224.
13. Nick Herbert, *Quantum Reality*, p.241.
14. Nick Herbert, *Faster Than Light*, p.179.
15. Nick Herbert, *Quantum Reality*, p.229.
16. Ibid, p.162.
17. P.C.W. Davies and J.R. Brown, *The Ghost in the Atom*, p.50.
18. Nick Herbert, *Quantum Reality*, p.230.

Principle 2.

1. Coleman Barks, *Feeling the Shoulder of the Lion*, p.39.
2. Charan Singh, *St. John, the Great Mystic*, p.5.

3. W.M. Thackston, Jr., *Signs of the Unseen*, p.85.
4. Sawan Singh, *Philosophy of the Masters, series iv*, p.119.
5. William Chittick, *The Sufi Path of Love*, p.43.
6. Ibid, p.48.
7. Charan Singh, *Divine Light*, p.266.
8. E.H. Whinfield, *Teachings of Rumi: The Masnavi*, p.263.
9. John Boslough, *Masters of Time*, p.80.
10. Anthony Zee, *An Old Man's Toy*, p.64.
11. F. David Peat, *Superstrings and the Search for The Theory of Everything*, p.6.
12. Paul Davies, *The Mind of God*, p.63.
13. Paul Davies, *Superforce*, p.203.
14. Joe Rosen, *The Capricious Cosmos*, pp. 139-140.
15. Stephen Hawking, *A Brief History of Time*, p.1.
16. Heinz Pagels, *The Cosmic Code*, pp. 243, 244, 247.
17. William Chittick, *The Sufi Path of Love*, pp. 176,177.
18. Willem Drees, *Beyond the Big Bang*, p.192.
19. Ivar Ekeland, *The Broken Dice and Other Mathematical Tales of Chance*, p.46.
20. Heinz Pagels, *The Cosmic Code*, p.284.
21. Paul Davies, *Superforce*, p.185.
22. Edward Harrison, *Masks of the Universe*, p.176.
23. Neil McAleer, *The Cosmic Mind-Boggling Book*, p.173.
24. John Boslough, *Masters of Time*, p.80.
25. Joe Rosen, *The Capricious Cosmos*, p.50.
26. David Lindley, *The End of Physics*, p.182.
27. Henrietta McCall, *Mesopotamian Myths*, p.52.
28. Joseph Campbell, *The Masks of God: Oriental Mythology*, p.130.
29. Jagat Singh, *Science of the Soul*, p.22.

Principle 3.

1. William Chittick, *The Sufi Path of Love*, p.66.
2. Camille and Kabir Helminski, *Rumi—Daylight*, p.186.
3. Francis Crick, *The Astonishing Hypothesis*, p.21.

4. J.R. Puri, *Guru Nanak—His Mystic Teachings*, p.59.
5. Sawan Singh, *Philosophy of the Masters, series iv*, p.10.
6. William Chittick, *The Sufi Path of Love*, p.48.
7. Charan Singh, *The Master Answers*, p.359.
8. Lekh Raj Puri, *Mysticism—The Spiritual Path, vol. II*, p.202.
9. Arthur Arberry, *Tales from the Masnavi*, p.277 (all "chickpea" quotations are from this source).
10. Sawan Singh, *Philosophy of the Masters, series iv*, pp. 119-120, 158-160.
11. Sawan Singh, *My Submission*, p.27.
12. Julian Johnson, *The Path of the Masters*, pp. 324-325.
13. Jonathan Star and Shahram Shiva, *A Garden Beyond Paradise*, p.149.
14. Neil McAleer, *The Cosmic Mind-Boggling Book*, p.9.
15. Dennis Overbye, *Lonely Hearts of the Cosmos*, pp. 42-43.
16. Gerald Schroeder, *Genesis and the Big Bang*, p.57.
17. Steven Weinberg, *The First Three Minutes*, p.146.
18. Anthony Zee, *An Old Man's Toy*, p.xxii.
19. Roger Penrose, *The Emperor's New Mind*, p.338.
20. Paul Davies, *Superforce*, p.173.
21. Roger Penrose, *The Emperor's New Mind*, p.318.
22. King James Version, St. Matthew 4:4.
23. Roger Penrose, *The Emperor's New Mind*, p.340.
24. Paul Davies, *God and the New Physics*, p.168.
25. Camille and Kabir Helminski, *Rumi—Daylight*, p.113.
26. Paul Davies, *The Mind of God*, p.135.
27. Ibid, p.220.
28. Sawan Singh, *Philosophy of the Masters, series iv*, p.10.
29. Ken Wilber, *Quantum Questions*, pp. 132, 141, 143.
30. Steven Weinberg, *Dreams of a Final Theory*, p.220.
31. Willem Drees, *Beyond the Big Bang*, p.83.
32. Ibid, p.82.

33. David Lindley, *The End of Physics*, p.247.
34. Bryan Appleyard, *Understanding the Present*, p.175.
35. David Lindley, *The End of Physics*, p.247.
36. Menas Kafatos and Robert Nadeau, *The Conscious Universe*, p.170.

Principle 4.
1. Arthur Arberry, *Discourses of Rumi*, p.62.
2. Ibid, p.92.
3. King James Version, Exodus 21:24.
4. Charan Singh, *The Master Answers*, pp. 323-24.
5. Ibid, p.355.
6. Arthur Arberry, *Discourses of Rumi*, p.47.
7. William Chittick, *The Sufi Path of Love*, p.226.
8. Ibid, pp. 154, 174.
9. Arthur Arberry, *Discourses of Rumi*, p.111.
10. Charan Singh, *Spiritual Discourses*, p.161.
11 Camille and Kabir Helminski, *Rumi—Daylight*, p.54.
12. Julian Johnson, *The Path of the Masters*, p.357.
13. Charan Singh, *Quest for Light*, p.200.
14. Charan Singh, *Die to Live*, p.176.
15. W.M. Thackston, Jr., *Signs of the Unseen*, p.26.
16. Robert Hazen and James Trefil, *Science Matters*, p.24.
17. Julian Johnson, *The Path of the Masters*, p.446.
18. William Chittick, *The Sufi Path of Love*, p.178.
19. Menas Kafatos and Robert Nadeau, *The Conscious Universe*, p.13.
20. Ibid.
21. John Horgan, "Particle Metaphysics," *Scientific American*, p.106.
22. Ibid, p.98.
23. Menas Kafatos and Robert Nadeau, *The Conscious Universe*, p.171.
24. Ibid, p.14.
25. Ibid, p.121.
26. Nick Herbert, *Quantum Reality*, p.114.
27. Fritjof Capra, *The Tao of Physics*, p.138.
28. Ibid.
29. Menas Kafatos and Robert Nadeau, *The Conscious Universe*, p.43.
30. Ibid, p.180.

31. Heinz Pagels, *The Cosmic Code*, p.11.
32. Ibid, p.12.
33. Stephen Hawking, *A Brief History of Time*, p.54.
34. Ibid, p.55.
35. Menas Kafatos and Robert Nadeau, *The Conscious Universe*, p.48.
36. Nick Herbert, *Quantum Reality*, p.68.
37. Heinz Pagels, *The Cosmic Code*, p.72.
38. Menas Kafatos and Robert Nadeau, *The Conscious Universe*, p.55.
39. Nick Herbert, *Quantum Reality*, p.124.
40. Ibid, p.194.
41. W.Y. Evans-Wentz, *Tibetan Yoga and Secret Doctrines*, p.162.
42. Clifton Wolters, *The Cloud of Unknowing and Other Works*, pp. 67-68.
43. Menas Kafatos and Robert Nadeau, *The Conscious Universe*, p.118.
44. Ibid, p.50.
45. Ibid, p.176.
46. Ibid, pp. 164-165.
47. Ibid, pp. 179-180.
48. Camille and Kabir Helminski, *Rumi— Daylight*, p.165.
49. Menas Kafatos and Robert Nadeau, *The Conscious Universe*, p.113.
50. Clifton Wolters, *The Cloud of Unknowing and Other Works*, pp. 144-145.
51. Ibid, p.111.
52. Ibid, pp. 142-143.
53. Menas Kafatos and Robert Nadeau, *The Conscious Universe*, pp. 121-122.

Principle 5.

1. William Chittick, *The Sufi Path of Love*, p.184.
2. Camille and Kabir Helminski, *Rumi— Daylight*, p.124.
3. King James Version: Galatians 6:7.
4. King James Version: St. Matthew 7:20.
5. Sawan Singh, *Philosophy of the Masters, series iii*, p.66.
6. Sawan Singh, *Spiritual Gems*, p.108.
7. John Davidson, *The Secret of the Creative Vacuum*, p.45.
8. Sawan Singh, *Spiritual Gems*, p.274.
9. John Davidson, *Subtle Energy*, p.152.
10. Julian Johnson, *The Path of the Masters*, pp. 321, 339.

11. Albert Einstein quoted in Larry Dossey, *Recovering the Soul*, p.147.
12. W.M. Thackston, Jr., *Signs of the Unseen*, p.208.
13. William Chittick, *The Sufi Path of Love*, p.254.
14. Ibid.
15. Sawan Singh, *Spiritual Gems*, p.28.
16. E.H. Whinfield, *Teachings of Rumi: The Masnavi*, p.219.
17. Lekh Raj Puri, *Mysticism—The Spiritual Path, vol. I*, p.147.
18. Charan Singh, *The Master Answers*, pp. 382-384.
19. Charan Singh, *Light on St. Matthew*, p.135.
20. Sawan Singh, *Spiritual Gems*, p.275.
21. Julian Johnson, *The Path of the Masters*, p.363.
22. Brian Weiss, *Through Time Into Healing*, p.55.
23. Sawan Singh, *Spiritual Gems*, p.28.
24. W.M. Thackston, Jr., *Signs of the Unseen*, p.70.
25. Sawan Singh, *Tales of the Mystic East*, p.182.
26. William Chittick, *The Sufi Path of Love*, p.217.
27. W.M. Thackston, Jr., *Signs of the Unseen*, p.159.
28. Arthur Arberry, *Discourses of Rumi*, p.111.
29. Shanti Sethi, *Message Divine*, p.75.
30. Sawan Singh, *Spiritual Gems*, p.31.
31. Charan Singh, *Spiritual Heritage*, p.174.
32. William Chittick, *The Sufi Path of Love*, p.243.
33. Robert March, *Physics for Poets*, p.76.
34. Heinz Pagels, *The Cosmic Code*, p.68.
35. Ibid.
36. B. Alan Wallace, *Choosing Reality*, p.63.
37. Heinz Pagels, *The Cosmic Code*, 87.
38. Ibid, p, 94.
39. Ivar Ekeland, *The Broken Dice and Other Mathematical Tales of Chance*, pp. 49-50.
40. Ian Stewart, *Does God Play Dice?*, p.2.
41. Heinz Pagels, *The Cosmic Code*, p.85.
42. Ibid, p.86.
43. B. Alan Wallace, *Choosing Reality*, p.64.
44. David Bohm and F. David Peat, *Science, Order, and Creativity*, p.131.

31. Heinz Pagels, *The Cosmic Code*, p.11.
32. Ibid, p.12.
33. Stephen Hawking, *A Brief History of Time*, p.54.
34. Ibid, p.55.
35. Menas Kafatos and Robert Nadeau, *The Conscious Universe*, p.48.
36. Nick Herbert, *Quantum Reality*, p.68.
37. Heinz Pagels, *The Cosmic Code*, p.72.
38. Menas Kafatos and Robert Nadeau, *The Conscious Universe*, p.55.
39. Nick Herbert, *Quantum Reality*, p.124.
40. Ibid, p.194.
41. W.Y. Evans-Wentz, *Tibetan Yoga and Secret Doctrines*, p.162.
42. Clifton Wolters, *The Cloud of Unknowing and Other Works*, pp. 67-68.
43. Menas Kafatos and Robert Nadeau, *The Conscious Universe*, p.118.
44. Ibid, p.50.
45. Ibid, p.176.
46. Ibid, pp. 164-165.
47. Ibid, pp. 179-180.
48. Camille and Kabir Helminski, *Rumi—Daylight*, p.165.
49. Menas Kafatos and Robert Nadeau, *The Conscious Universe*, p.113.
50. Clifton Wolters, *The Cloud of Unknowing and Other Works*, pp. 144-145.
51. Ibid, p.111.
52. Ibid, pp. 142-143.
53. Menas Kafatos and Robert Nadeau, *The Conscious Universe*, pp. 121-122.

Principle 5.
1. William Chittick, *The Sufi Path of Love*, p.184.
2. Camille and Kabir Helminski, *Rumi—Daylight*, p.124.
3. King James Version: Galatians 6:7.
4. King James Version: St. Matthew 7:20.
5. Sawan Singh, *Philosophy of the Masters, series iii*, p.66.
6. Sawan Singh, *Spiritual Gems*, p.108.
7. John Davidson, *The Secret of the Creative Vacuum*, p.45.
8. Sawan Singh, *Spiritual Gems*, p.274.
9. John Davidson, *Subtle Energy*, p.152.
10. Julian Johnson, *The Path of the Masters*, pp. 321, 339.

11. Albert Einstein quoted in Larry Dossey, *Recovering the Soul*, p.147.
12. W.M. Thackston, Jr., *Signs of the Unseen*, p.208.
13. William Chittick, *The Sufi Path of Love*, p.254.
14. Ibid.
15. Sawan Singh, *Spiritual Gems*, p.28.
16. E.H. Whinfield, *Teachings of Rumi: The Masnavi*, p.219.
17. Lekh Raj Puri, *Mysticism—The Spiritual Path, vol. I*, p.147.
18. Charan Singh, *The Master Answers*, pp. 382-384.
19. Charan Singh, *Light on St. Matthew*, p.135.
20. Sawan Singh, *Spiritual Gems*, p.275.
21. Julian Johnson, *The Path of the Masters*, p.363.
22. Charan Singh, *Divine Light*, p. 321.
23. Sawan Singh, *Spiritual Gems*, p.28.
24. W.M. Thackston, Jr., *Signs of the Unseen*, p.70.
25. Sawan Singh, *Tales of the Mystic East*, p.182.
26. William Chittick, *The Sufi Path of Love*, p.217.
27. W.M. Thackston, Jr., *Signs of the Unseen*, p.159.
28. Arthur Arberry, *Discourses of Rumi*, p.111.
29. Shanti Sethi, *Message Divine*, p.75.
30. Sawan Singh, *Spiritual Gems*, p.31.
31. Charan Singh, *Spiritual Heritage*, p.174.
32. William Chittick, *The Sufi Path of Love*, p.243.
33. Robert March, *Physics for Poets*, p.76.
34. Heinz Pagels, *The Cosmic Code*, p.68.
35. Ibid.
36. B. Alan Wallace, *Choosing Reality*, p.63.
37. Heinz Pagels, *The Cosmic Code*, 87.
38. Ibid, p, 94.
39. Ivar Ekeland, *The Broken Dice and Other Mathematical Tales of Chance*, pp. 49-50.
40. Ian Stewart, *Does God Play Dice?*, p.2.
41. Heinz Pagels, *The Cosmic Code*, p.85.
42. Ibid, p.86.
43. B. Alan Wallace, *Choosing Reality*, p.64.
44. David Bohm and F. David Peat, *Science, Order, and Creativity*, p.131.
45. Ibid, p.149.

28. Sawan Singh, *Philosophy of the Masters,* series iv, p.29.

29. Julian Johnson, *The Path of the Masters,* p.317.

30. Clifton Wolters, *The Cloud of Unknowing and Other Works,* p.129.

31. John Davidson, *The Secret of the Creative Vacuum,* p.71.

32. Julian Johnson, *The Path of the Masters,* p.267.

33. Ibid, p.268.

34. Ibid, p.96.

35. William Chittick, *The Sufi Path of Love,* p.22.

36. Seth Shiv Dayal Singh, *Sar Bachan,* p.34.

37. John Davidson, *Subtle Energy,* p.27.

38. W.M. Thackston, Jr., *Signs of the Unseen,* p.59.

39. Charan Singh, *The Master Answers,* p.260.

40. Camille and Kabir Helminski, *Rumi— Daylight,* p.25.

41. V.K. Sethi, *Kabir, the Weaver of God's Name,* p.513.

42. Charan Singh, *Quest for Light,* p.211.

43. Gia-Fu Feng and Jane English, trans., *Tao Te Ching,* ch. 16, 48.

44. William Chittick, *The Sufi Path of Love,* p.224.

45. Ken Wilber, *Quantum Questions,* p.180.

46. John Briggs, *Fractals: the Patterns of Chaos,* p.23.

47. Heinz Pagels, *The Cosmic Code,* p.239.

48. Ken Wilber, *Quantum Questions,* p.181.

49. Paul Davies, *The Mind of God,* p.85.

50. Fritjof Capra, *The Tao of Physics,* p.196.

51. John Barrow, *Theories of Everything,* p.74.

52. Robert March, *Physics for Poets,* p.73.

53. Steven Hawking, *A Brief History of Time,* p.71.

54. Anthony Zee, *Fearful Symmetry,* p.227.

55. Paul Davies, *Superforce,* p.168.

56. Ibid.

57. Steven Hawking, *A Brief History of Time,* p.74.

58. John Boslough, *Masters of Time,* p.6.

59. David Lindley, *The End of Physics,* p.167.

60. Paul Davies, *Superforce,* p.168.

61. Steven Weinberg, *Dreams of a Final Theory,* p.234.

62. Paul Davies, *Superforce,* p.168.

63. Steven Weinberg, *Dreams of a Final Theory,* p.234.

64. James Trefil, *Reading the Mind of God,* p.200.

65. William Chittick, *The Sufi Path of Love,* p.180.

66. Paul Davies, *Superforce,* p.5.

67. Robert Hazen and James Trefil, *Science Matters,* p.151.

68. Anthony Zee, *An Old Man's Toy,* pp. 34-35.

69. Ibid, pp. 199, 205.

70. Paul Davies, *God and the New Physics,* p.188.

71. Anthony Zee, *An Old Man's Toy,* p.34.

72. Camille and Kabir Helminski, *Rumi— Daylight,* p.127.

73. Robert Hazen and James Trefil, *Science Matters,* p.150.

74. Arthur Arberry, *Discourses of Rumi,* p.87.

75. Danah Zohar, *The Quantum Self,* p.225.

76. Roger Penrose, *The Emperor's New Mind,* p.221.

77. David Bohm and F. David Peat, *Science, Order, and Creativity,* p.199.

78. John Davidson, *The Secret of the Creative Vacuum,* p.31.

79. Larry Abbott, "The Mystery of the Cosmological Constant," *Scientific American,* p.106.

80. Anthony Zee, *An Old Man's Toy,* p.238.

81. Hans Christian Von Baeyer, "Vacuum Matters," *Discover,* p.111.

82. William Chittick, *The Sufi Path of Love,* p.178.

Principle 7.

1. Camille and Kabir Helminski, *Rumi— Daylight,* p.150.

2. E.H. Whinfield, *Teachings of Rumi: The Masnavi,* p.182.

3. Dennis Overbye, *Lonely Hearts of the Cosmos,* p.234.

4. William Chittick, *The Sufi Path of Love,* pp. 97, 98.

5. Charan Singh, *The Master Answers,* pp. 86, 201.

6. Nick Herbert, *Elemental Mind,* p.65.

7. Ibid, pp. 62, 64.

8. King James Version, St. Matthew 6:22.

9. Sawan Singh, *The Dawn of Light,* p.191.

10. Charan Singh, *Thus Saith the Master*, p.385.
11. Charan Singh, *Spiritual Heritage*, p.164.
12. Charan Singh, *Light on Saint John*, p.141.
13. William Chittick, *The Sufi Path of Love*, p.28.
14. Sawan Singh, *Philosophy of the Masters, series i*, p.93.
15. Julian Johnson, *The Path of the Masters*, pp. 475, 477.
16. Karen Armstrong, *The English Mystics of the Fourteenth Century*, pp. 33, 39, 58.
17. Hazrat Inayat Khan, *The Music of Life*, pp. 25-26.
18. Livia Kohn, *Taoist Mystical Philosophy*, p.104.
19. Robert Hutchins, editor, *Great Books of the Western World, vol. 17*, p.131.
20. Peter Gorman, *Pythagoras: A Life*, p.201.
21. Sawan Singh, *Philosophy of the Masters, series i*, p.157.
22. King James Version, Genesis 1:3.
23. Robert Hazen and James Trefil, *Science Matters*, p.39.
24. Charan Singh, *Spiritual Discourses*, p.29.
25. Seth Shiv Dayal Singh, *Sar Bachan*, p.34.
26. Charan Singh, *Spiritual Discourses*, p.85.
27. Lekh Raj Puri, *Radha Swami Teachings*, pp. 177-192; Sawan Singh, *Philosophy of the Masters, series i*, p.94.
28. Lekh Raj Puri, *Teachings of the Gurus*, p.159.
29. Charan Singh, *Die to Live*, p.231.
30. Robert Slater and Mark Terry, *Tinnitus: A Guide for Sufferers and Professionals*, p.92.
31. Leslie Sheppard and Audrey Hawkridge, *Tinnitus: Learning to Live With It*, p.57.
32. Lekh Raj Puri, *Mysticism—The Spiritual Path, vol. I*, p.174.
33. Sawan Singh, *Philosophy of the Masters, series iv*, p.211.
34. Fritjof Capra, *The Tao of Physics*, p.229.
35. Alexandra David-Neel, *Tibetan Journey*, p.186.
36. Fritjof Capra, *The Tao of Physics*, p.180.
37. P.C.W. Davies and J.R. Brown, *The Ghost in the Atom*, p.3.
38. Steven Weinberg, *Dreams of a Final Theory*, p.70.
39. Ibid, p.72.
40. Gary Zukav, *The Dancing Wu Li Masters*, p.259.
41. John Casti, *Paradigms Lost*, p.462.
42. Renée Weber, *Dialogues with Scientists and Sages*, pp. 113, 120.
43. John Casti, *Paradigms Lost*, p.463.
44. E.H. Whinfield, *Teachings of Rumi: The Masnavi*, p.25.
45. David Albert, "Bohm's Alternative to Quantum Mechanics," *Scientific American*, p.63.
46. Paul Davies and John Gribbin, *The Matter Myth*, p.14.
47. Richard Morris, *The Edges of Science*, p.166.
48. Anthony Zee, *Fearful Symmetry*, p.272.
49. F. David Peat, *Superstrings and the Search for The Theory of Everything*, p.3.
50. Ibid, p.108.
51. David Freedman, "The New Theory of Everything," *Discover*, p.59.
52. Dennis Overbye, *Lonely Hearts of the Cosmos*, p.372.
53. Steven Weinberg, *Dreams of a Final Theory*, p.213.
54. F. David Peat, *Superstrings and the Search for The Theory of Everything*, p.161.
55. Steven Weinberg, *Dreams of a Final Theory*, p.214.
56. Sawan Singh, *Philosophy of the Masters, series iv*, p.171-175.
57. Andrew Harvey, *The Way of Passion*, p.221.
58. Ibid.

Epilogue
1. Published by: A Center for the Practice of Zen Buddhist Meditation, 1990.
2. Sharon Begley, "How Many Scientists Does It Take to Screw in a Quark," *Newsweek*, p.54.
3. Ibid.
4. Andrew Harvey, *The Way of Passion*, pp. 109-110.
5. William Chittick, *The Sufi Path of Love*, p.186.

Bibliography

Abbott, Larry. "The Mystery of the Cosmological Constant." *Scientific American,* May 1988.

Albert, David. "Bohm's Alternative to Quantum Mechanics." *Scientific American,* May 1994.

Ali, Ahmed, trans. *Al-Qur'an.* Princeton: Princeton University Press, 1990.

Appleyard, Bryan. *Understanding the Present.* New York: Doubleday, 1992.

Arberry, A. J., trans. *Mystical Poems of Rumi.* Chicago: University of Chicago Press, 1968.

Arberry, Arthur J., trans. *Discourses of Rumi.* Richmond: Curzon Press, 1993.

Arberry, Arthur J., trans. *Tales from the Masnavi.* Richmond: Curzon Press, 1993.

Armstrong, Karen, trans. *The English Mystics of the Fourteenth Century.* London: Kyle Cathie, 1991.

Barks, Coleman, trans. *Feeling the Shoulder of the Lion.* Putney: Threshold, 1991.

Barks, Coleman, trans. *We Are Three.* Athens, Georgia: Maypop, 1987.

Barrow, John D. *Theories of Everything.* Oxford: Clarendon Press, 1991.

Begley, Sharon. "How Many Scientists Does It Take to Screw in a Quark." *Newsweek,* May 9, 1994.

Bly, Robert. *The Kabir Book.* Boston: Beacon Press, 1977.

Bohm, David and F. David Peat. *Science, Order, and Creativity.* New York: Bantam Books, 1987.

Boslough, John. *Masters of Time.* Addison Wesley Publishing Co., 1992.

Briggs, John. *Fractals: the Patterns of Chaos.* New York: Simon & Schuster, 1992.

Bruning, David. "Fixing Hubble." *Astronomy,* January 1994.

Campbell, Joseph. *The Masks of God: Oriental Mythology.* New York: Viking Press, 1962.

Capra, Fritjof. *The Tao of Physics.* Bantam Books, 1977.

Capra, Fritjof and David Steindl-Rast, *Belonging to the Universe.* Harper San Francisco, 1991.

Casti, John L. *Paradigms Lost.* New York: Avon Books, 1990.

Chittick, William. *The Sufi Path of Love.* Albany: State University of New York Press, 1983.

Chopra, Deepak. *Quantum Healing.* New York: Bantam Books, 1989.

Cowan, James G. *Where Two Oceans Meet.* Rockport: Element, 1992.

Crick, Francis. *The Astonishing Hypothesis.* New York: Charles Scribner's Sons, 1994.

David-Neel, Alexandra. *Tibetan Journey.* London: John Lane The Bodley Head, 1936.

Davidson, John. *Subtle Energy.* Essex: The C.W. Daniel Company Limited, 1988.

Davidson, John. *The Secret of the Creative Vacuum.* Essex: The C.W. Daniel Company Limited, 1989.

Davidson, John. *The Robe of Glory.* Shaftesbury: Element Books, 1992.

Davies, P.C.W. and J.R. Brown, editors. *The Ghost in the Atom.* Cambridge: Cambridge University Press, 1991.

Davies, Paul. *God and the New Physics.* New York: Simon & Schuster, 1983.

Davies, Paul. *Superforce.* New York: Simon & Schuster, 1984.

Davies, Paul. *The Mind of God.* New York: Simon & Schuster, 1992.

Davies, Paul and John Gribbin. *The Matter Myth.* New York: Simon & Schuster, 1992.

Dennett, Daniel C. *Consciousness Explained.* Boston: Little, Brown and Company, 1991.

Dossey, Larry. *Recovering the Soul.* New York: Bantam Books, 1989.

Drees, Willem B. *Beyond the Big Bang.* La Salle: Open Court Publishing Co., 1990.

Ekeland, Ivar. *The Broken Dice and Other Mathematical Tales of Chance.* Chicago: The University of Chicago Press, 1993.

Ergin, Nevit Orguz., trans. *Magnificent One.* Larson Publications: 1993.

Evans-Wentz, W.Y. *Tibetan Yoga and Secret Doctrines.* New York: Oxford University Press, 1968.

Feng, Gia-Fu and Jane English, trans. *Tao Te Ching.* New York: Vintage Books, 1972.

Freedman, David. "The New Theory of Everything." *Discover,* August 1991.

Gardner, Martin. *The Whys of a Philosophical Scrivener.* New York: William Morrow and Company, Inc., 1983.

Gleick, James. *Chaos*. New York: Viking, 1987.

Gorman, Peter. *Pythagoras: A Life*. London: Routledge & Kegan Paul, 1979.

Goswami, Amit. *The Self-Aware Universe*. New York: G.P. Putnam's Sons, 1993.

Harrison, Edward. *Masks of the Universe*. New York: Macmillan Publishing Co., 1986.

Harvey, Andrew. *The Way of Passion*. Berkeley: Frog, Ltd., 1994.

Hawking, Stephen. *A Brief History of Time*. New York: Bantam Books, 1988.

Hazen, Robert M. and James Trefil. *Science Matters*. New York: Doubleday, 1991.

Herbert, Nick. *Quantum Reality*. New York: Doubleday, 1987.

Herbert, Nick. *Faster Than Light*. New York: A Plume Book, 1989.

Herbert, Nick. *Elemental Mind*. New York: Dutton, 1993.

Helminski, Camille and Kabir, trans. *Rumi—Daylight*. Putney: Threshold, 1990.

Helminski, Kabir, trans. *Love is a Stranger*. Putney: Threshold, 1993.

Hofstadter, Douglas R. *Gödel, Escher, Bach*. New York: Vintage Books, 1980.

Horgan, John. "Particle Metaphysics." *Scientific American*, February 1994.

Hutchins, Robert, editor. *Great Books of the Western World, vol. 8*. Chicago: Encyclopedia Britannica, Inc., 1952.

Hutchins, Robert, editor. *Great Books of the Western World, vol. 17*. Chicago: Encyclopedia Britannica, Inc., 1952.

Hutchins, Robert, editor. *Great Books of the Western World, vol. 19*. Chicago: Encyclopedia Britannica, Inc., 1952.

Huxley, Aldous. *The Perennial Philosophy*. New York: Harper & Row, 1970.

Johnson, Julian p. *The Path of the Masters*. Punjab: Radha Soami Satsang Beas, 1980.

Kafatos, Menas and Robert Nadeau. *The Conscious Universe*. New York: Springer- Verlag, 1990.

Khan, Inayat Hazrat. *The Music of Life*. New Lebanon: Omega Press, 1988.

Kohn, Livia. *Taoist Mystical Philosophy*. Albany: State University of New York Press, 1991.

Lindley, David. *The End of Physics*. New York: BasicBooks, 1993.

McAleer, Neil. *The Cosmic Mind-Boggling Book*. New York: Warner Books, 1982.

McCall, Henrietta. *Mesopotamian Myths*. Austin: University of Texas Press, 1990.

March, Robert H. *Physics for Poets*. McGraw Hill, Inc., 1992.

Mencken, H.L., editor. *A New Dictionary of Quotations*. New York: Alfred A. Knopf, 1960.

Moody, Raymond A. *Reflections on Life After Life*. New York: Bantam Books, 1989.

Morris, Richard. *The Edges of Science*. New York: Prentice Hall Press, 1990.

Overbye, Dennis. *Lonely Hearts of the Cosmos*. New York: HarperCollins, 1991.

Pagels, Heinz. *The Cosmic Code*. New York: Bantam Books, 1990.

Panikkar, Raimundo. *The Vedic Experience*. Berkeley: University of California Press, 1977.

Peat, F. David. *Superstrings and the Search for The Theory of Everything*. Chicago: Contemporary Books Inc., 1988.

Peat, F. David. *Einstein's Moon*. Chicago: Contemporary Books Inc., 1990.

Penrose, Roger. *The Emperor's New Mind*. New York: Oxford University Press, 1989.

Peterson, Ivars and Carol Ezzell. "Crazy Rhythms." *Science News*, September 5, 1992.

Polkinghorne, John. *Reason and Reality*. Philadelphia: Trinity Press International, 1991.

Puri, J.R. *Guru Nanak—His Mystic Teachings*, Punjab: Radha Soami Satsang Beas, 1982.

Puri, Lekh Raj. *Radha Swami Teachings*. Punjab: Radha Soami Satsang Beas, 1972.

Puri, Lekh Raj. *Mysticism—The Spiritual Path, vol. I*. Punjab: Radha Soami Satsang Beas, 1986.

Puri, Lekh Raj. *Teachings of the Gurus*. Punjab: Radha Soami Satsang Beas, 1987.

Puri, Lekh Raj. *Mysticism—The Spiritual Path, vol. II*. Punjab: Radha Soami Satsang Beas, 1988.

Rosen, Joe. *The Capricious Cosmos*. New York: Macmillan Publishing Co., 1991.

Rothman, Tony. "This is the Way the World Ends." *Discover*, July 1987.

Schroeder, Gerald L. *Genesis and the Big Bang*. New York: Bantam Books, 1990.

Sethi, Shanti. *Message Divine*. Punjab: Radha Soami Satsang Beas, 1987.

Sethi, V.K. *Kabir, the Weaver of God's Name*. Punjab: Radha Soami Satsang Beas, 1984.

Shah, Idries. *The Way of the Sufi*. New York: E.P. Dutton, 1970.

Sheppard, Leslie & Audrey Hawkridge. *Tinnitus: Learning to Live With It*. Bath: Ashgrove Press, 1989.

Singh, Charan. *The Master Answers*. Punjab: Radha Soami Satsang Beas, 1969.

Singh, Charan. *Quest for Light*. Punjab: Radha Soami Satsang Beas, 1972.

Singh, Charan. *Spiritual Discourses*. Punjab: Radha Soami Satsang Beas, 1974.

Singh, Charan. *Thus Saith the Master*. Punjab: Radha Soami Satsang Beas, 1974.

Singh, Charan. *Divine Light*. Punjab: Radha Soami Satsang Beas, 1976.

Singh, Charan. *St. John, the Great Mystic*. Punjab: Radha Soami Satsang Beas, 1978.

Singh, Charan. *Die to Live*. Punjab: Radha Soami Satsang Beas, 1979.

Singh, Charan. *Light on St. Matthew*. Punjab: Radha Soami Satsang Beas, 1979.

Singh, Charan. *Spiritual Heritage*. Punjab: Radha Soami Satsang Beas, 1983.

Singh, Charan. *Light on Saint John*. Punjab: Radha Soami Satsang Beas, 1985.

Singh, Gopal. *Sri Guru Granth Sahib, vol 1*. New Delhi: World Book Centre, 1989.

Singh, Jagat. *Science of the Soul*. Punjab: Radha Soami Satsang Beas, 1982.

Singh, Sawan. *Philosophy of the Masters, series i*. Punjab: Radha Soami Satsang Beas, 1963.

Singh, Sawan. *Spiritual Gems*. Punjab: Radha Soami Satsang Beas, 1965.

Singh, Sawan. *Philosophy of the Masters, series iii*. Punjab: Radha Soami Satsang Beas, 1965.

Singh, Sawan. *My Submission*. Punjab: Radha Soami Satsang Beas, 1980.

Singh, Sawan. *The Dawn of Light*. Punjab: Radha Soami Satsang Beas, 1985.

Singh, Sawan. *Tales of the Mystic East*. Punjab: Radha Soami Satsang Beas, 1983.

Singh, Sawan. *Philosophy of the Masters, series iv*. Punjab: Radha Soami Satsang Beas, 1989.

Singh, Seth Shiv Dayal. *Sar Bachan*. Punjab: Radha Soami Satsang Beas, 1971.

Slater, Robert and Mark Terry. *Tinnitus: A Guide for Sufferers and Professionals*. London: Croom Helm, 1988.

Star, Jonathan and Shahram Shiva, *A Garden Beyond Paradise*. New York: Bantam Books, 1992.

Stewart, Ian. *Does God Play Dice?*. Cambridge: Blackwell Publishers, 1992.

Talbot, Michael. *The Holographic Universe*. New York: HarperCollins, 1991.

Thackston, Jr., W.M. *Signs of the Unseen*. Putney: Threshold Books, 1994.

Traub, Joseph F. and Henryk Wozniakowski. "Breaking Intractibility." *Scientific American*, January 1994.

Trefil, James. *Reading the Mind of God*. New York: Anchor Books, 1989.

Updike, John. *Roger's Version*. New York: Alfred A. Knopf, 1986.

Von Baeyer, Hans Christian. "Vacuum Matters." *Discover*, March 1992.

Walker, Brian. *Hua Hu Ching*. Livingston: Clark City Press, 1992.

Wallace, B. Alan. *Choosing Reality*. Boston: New Science Library, 1989.

Walshe, M.O'C. *Meister Eckhart*. Longmead: Element Books, 1987.

Weber, Renée. *Dialogues with Scientists and Sages*. London: Arkana, 1990.

Weinberg, Steven. *The First Three Minutes*. Basic Books: 1988.

Weinberg, Steven. *Dreams of a Final Theory*. New York: Pantheon Books, 1992.

Weiss, Brian. *Through Time Into Healing*. New York: Simon & Schuster, 1992.

Whinfield, E.H., trans. *Teachings of Rumi: The Masnavi*. London: The Octagon Press, 1989.

Wilber, Ken, editor. *Quantum Questions*. Boston: New Science Library, 1985.

Wilber, Ken. *Eye to Eye*. Boston: Shambhala, 1990.

Wolters, Clifton, trans. *The Cloud of Unknowing and Other Works*. Middlesex: Penguin Classics, 1987.

Zee, Anthony. *Fearful Symmetry*. New York: Macmillan Publishing Co., 1986.

Zee, Anthony. *An Old Man's Toy*. New York: Macmillan Publishing Co., 1989.

Zohar, Danah. *The Quantum Self*. New York: William Morrow & Co., 1990.

Zukav, Gary. *The Dancing Wu Li Masters*. New York: Bantam Books, 1989.

General Index

Acknowledgements

And last, but not least, grateful thanks to the following organizations and people for permission to reprint excerpts from these publications:

Masters of Time (pp. 6, 80, 224) © 1992 by John I. Boslough. Reprinted by permission of **Addison-Wesley Publishing Company**, Inc.

From *Science, Order and Creativity* by David Bohm and David Peat. Copyright © 1987 by David Bohm and David Peat. Used by permission of **Bantam Books**, a division of Bantam Doubleday Dell Publishing Group, Inc.

"Excerpts," from *A Brief History of Time* by Stephen W. Hawking. Copyright © 1988 by Stephen W. Hawking. Used by permission of **Bantam Books**, a division of Bantam Doubleday Dell Publishing Group, Inc.

From *Quantum Healing* by Deepak Chopra, M.D. Copyright © 1989 by Deepak Chopra, M.D. Used by permission of **Bantam Books**, a division of Bantam Doubleday Dell Publishing Group, Inc.

From *Recovering the Soul* by Larry Dossey, M.D. Copyright © 1989 by Larry Dossey, M.D. Used by permission of **Bantam Books**, a division of Bantam Doubleday Dell Publishing Group, Inc.

From *Genesis and the Big Bang* by Gerald Schroeder. Copyright © 1990 by Gerald Schroeder. Used by permission of **Bantam Books,** a division of Bantam Doubleday Dell Publishing Group, Inc.

From *A Garden Beyond Paradise* by Rumi, Jonathan Star & Shahram Shiva, translated by Jonathan Star. Translation copyright © 1992 by Jonathan Star. Used by permission of **Bantam Books**, a division of Bantam Doubleday Dell Publishing Group, Inc.

Selected excerpt from page 77 from *Godel, Escher, Bach: An Eternal Braid* by Douglas Hofstadter, Copyright © 1979 by Basic Books, Inc. Reprinted by permission of **Basic Books**, a division of HarperCollins Publishers, Inc.

Selected excerpts from *The End of Physics: The Myth of a Unified Theory* by David Lindley. Copyright © 1993 by David Lindley. Reprinted by permission of **BasicBooks**, a division of HarperCollins Publishers, Inc.

From *The Kabir Book* by Robert Bly. Copyright © 1971, 1977 by Robert Bly. Reprinted by permission of **Beacon Press**.

Aquinas, Thomas, *Summa Theologica of Thomas Aquinas* © 1952 **Benziger Publishing Company**. Reprinted by permission.

Blackwell Publishers, from Ian Stewart, *Does God Play Dice?*, © Ian Stewart 1989, Blackwell Publishers.

C.W. Daniel Company Limited, from John Davidson, *Subtle Energy*, © John Davidson 1987, The C.W. Daniel Company Limited.

C.W. Daniel Company Limited, from John Davidson, *The Secret of the Creative Vacuum*, © John Davidson 1989, The C.W. Daniel Company Limited.

P.C.W. Davies and J.R. Brown, editors, *The Ghost in the Atom.* © 1986 Cambridge University Press by arrangement with the British Broadcasting Corporation. Reprinted with permission of **Cambridge University Press**.

Reprinted from *Superstrings* by F. David Peat, © 1988. Used with permission of **Contemporary Books**.

Reprinted from *Einstein's Moon* by F. David Peat, © 1990. Used with permission of **Contemporary Books**.

Croom Helm Ltd., from Robert Slater and Mark Terry, *Tinnitus: A Guide for Sufferers and Professionals*, © 1987 Robert Slater and Mark Terry, Croom Helm Ltd.

Paul Davies, from Paul Davies, *Superforce*, © 1984 by Glenister Gavin Ltd., Simon & Schuster, Inc.

Paul Davies, from Paul Davies, *The Mind of God*, © 1992 by Orion Productions, Simon & Schuster, Inc.

Paul Davies, from Paul Davies and John Gribbin, *The Matter Myth*, © 1992 by Orion Productions and John Gribbin, Simon & Schuster, Inc.

David H. Freedman / © 1991 **Discover Magazine**. Reprinted with permission.

Hans Christian von Baeyer / © 1992 **Discover Magazine**. Reprinted with permission.

From *Quantum Reality: Beyond the New Physics* by Nick Herbert. Copyright © 1985 by Nick Herbert. Used by permission of **Doubleday**, a division of Bantam Doubleday Dell Publishing Group, Inc.

From *Science Matters: Achieving Scientific Literacy* by Robert M. Hazen and James Trefil. Copyright © 1991 by Robert M. Hazen and James Trefil. Used by permission of **Doubleday**, a division of Bantam Doubleday Dell Publishing Group, Inc.

From *Understanding the Present* by Bryan Appleyard. Copyright © 1992 by Bryan Appleyard. Used by permission of **Doubleday**, a division of Bantam Doubleday Dell Publishing Group, Inc.

From *Faster Than Light* by Nick Herbert. Copyright © 1988 by Nick Herbert. Used by permission of **Dutton Signet**, a division of Penguin Books USA Inc.

From *Elemental Mind* by Nick Herbert. Copyright © 1993 by Nick Herbert. Used by permission of **Dutton Signet,** a division of Penguin Books USA Inc.

Element Books, from M. O'C. Walshe translator and editor, *Meister Eckhart—Sermons and Treatises, Volume 1*, © 1979 by M. O'C Walshe, Element Books.

Element Books, from James G. Cowan translator, *Where Two Oceans Meet*, © James G. Cowan 1992, Element Books.

Element Books, from John Davidson, *The Robe of Glory*, © John Davidson 1992, Element Books.

Frog Ltd., from Andrew Harvey, *The Way of Passion*, © 1994 by Andrew Harvey, Frog Ltd.

HarperCollins Publishers, from Steven Weinberg, *The First Three Minutes*, © 1977, 1988 by Steven Weinberg, BasicBooks.

HarperCollins Publishers, from Aldous Huxley, *The Perennial Philosophy*, © 1944, 1945 by Aldous Huxley, copyright renewed 1973, 1974 by Laura A. Huxley, Harper & Row.

HarperCollins Publishers, from Dennis Overbye, *Lonely Hearts of the Cosmos*, © 1991 by Dennis Overbye, HarperCollins Publishers.

HarperCollins Publishers, from Michael Talbot, *The Holographic Universe*, © 1991 by Michael Talbot, HarperCollins Publishers.

HarperCollins Publishers, from Fritjof Capra and David Steindl-Rast, *Belonging to the Universe*, © 1991 by Fritjof Capra and David Steindl-Rast, HarperSanFrancisco.

Selected excerpt from pages 56 and 24 from *Hua Hu Ching: The Unknown Teachings of Lao Tzu* by Brian Walker, Copyright © 1992 by Brian Browne Walker. Reprinted by permission of **HarperCollins Publishers,** Inc.

From *Creating Health* by Deepak Chopra. Copyright © 1987 by Deepak Chopra. Reprinted by permission of **Houghton Mifflin Company**. All rights reserved.

Kyle Cathie Limited, from Karen Armstrong translator and editor, *The English Mystics of the Fourteenth Century*, © 1991 by Karen Armstrong, Kyle Cathie Limited.

Larson Publications, from Nevit Oguz Ergin translator, *Magnificent One*, © 1993 by The Society for Understanding Mevlana, Larson Publications.

From *Consciousness Explained* by Daniel Dennett. Copyright © 1991 by Daniel C. Dennett. By permission of **Little, Brown and Company**.

Macmillan Publishing Company, from Anthony Zee, *An Old Man's Toy*, © 1989 by Anthony Zee, Macmillan Publishing Company.

Macmillan Publishing Company, from Joe Rosen, *The Capricious Cosmos*, © 1991 by Joe Rosen, Macmillan Publishing Company.

McGraw-Hill, Inc., from Robert H. March, *Physics for Poets*, © 1992, 1978, 1970 by McGraw Hill, Inc., McGraw-Hill, Inc.

McGraw-Hill, Inc., from Edward Harrison, *Masks of the Universe*, © 1985 by Macmillan Publishing Company, Macmillan Publishing Company.

Mockingbird Books, from Raymond A. Moody, Jr., *Reflections on Life After Life*, © 1977 by Raymond A. Moody, Jr., Bantam Books.

Richard Morris, from Richard Morris, *The Edges of Science*, © 1990 by Richard Morris, Prentice Hall Press.

Newsweek, from Sharon Begley, "How Many Scientists Does It Take to Screw in a Quark," May 9, 1994, © 1994, Newsweek.

Reprinted by permission from *The Teachings of Rumi: The Masnavi* published by The **Octagon Press Ltd.**, London.

Omega Publications, from Hazrat Inayat Khan, *The Music of Life,* © 1983 by Sufi Order, Omega Publications.

Reprinted from *Beyond the Big Bang* by Willem B. Drees by permission of **Open Court Publishing Company,** La Salle, Illinois. © 1990 by Open Court Publishing Company.

By permission of **Oxford University Press,** from John D. Barrow, *Theories of Everything,* © John D. Barrow 1991, Oxford University Press.

By permission of **Oxford University Press,** from Roger Penrose, *The Emperor's New Mind,* © Oxford University Press 1989, Oxford University Press.

From *Dreams of a Final Theory* by Steven Weinberg. Copyright © 1992 by Steven Weinberg. Reprinted by permission of **Pantheon Books,** a division of Random House, Inc.

Reproduced by permission of **Penguin Books Ltd.**, from Clifton Wolters translator, *The Cloud of Unknowing and Other Works,* © Clifton Wolters 1961, 1978, Penguin Books Ltd.

Reproduced by permission of **Penguin Books Ltd.**, from Renée Weber, *Dialogues With Scientists and Sages: The Search for Unity,* © Renée Weber 1986, Arkana.

Princeton University Press, from Ahmed Ali translator, *Al-Qur'an,* © 1984 by Ahmed Ali, Princeton University Press.

Reprinted by permission of The **Putnam Publishing Group/Jeremy P. Tarcher, Inc.** from *The Self-Aware Universe* by Amit Goswami, Richard E. Reed, and Maggie Goswami. Copyright © 1993 by Amit Goswami, Richard E. Reed, and Maggie Goswami.

Random House, Inc., from H.L. Mencken editor, *A New Dictionary of Quotations,* © 1942 by Alfred A. Knopf, Inc., Alfred A. Knopf, Inc.

Random House, Inc., from John Updike, *Roger's Version,* © 1986 by John Updike, Alfred A. Knopf, Inc.

Random House, Inc., from Gia-Fu Feng and Jane English translators, *Tao Te Ching,* © 1972 by Gia-Fu Feng and Jane English, Vintage Books.

Routledge, Chapman & Hall, Inc., from Peter Gorman, *Pythagoras: A Life,* © Peter Gorman 1979, Routledge & Kegan Paul Ltd.

Science News, from Ivars Peterson and Carol Ezzell, "Crazy Rhythms." Reprinted with permission from *Science News,* © 1992, Science Service, Inc

Scientific American, from Larry Abbott, "The Mystery of the Cosmological Constant," May 1988, © 1988; also from Joseph F. Traub and Henryk Wozniakowski, "Breaking Intractibility," January 1994, © 1994; also from John Horgan, "Particle Metaphysics," February 1994, © 1994; also from David Albert, "Bohm's Alternative to Quantum Mechanics," May 1994, © 1994. Scientific American, Inc.

Reprinted with permission of **Scribner,** an imprint of Simon & Schuster from *Reading the Mind of God* by James Trefil. Copyright © 1989 James Trefil.

Reprinted with permission of **Scribner,** an imprint of Simon & Schuster, Inc. from *The Astonishing Hypothesis* by Francis Crick. Copyright © 1994 The Francis H.C. Crick and Odile Crick Revocable Trust.

From *The Tao of Physics* by Fritjof Capra. © 1975, 1983, 1991 by Fritjof Capra. Reprinted by arrangement with **Shambhala Publications, Inc.**, P.O. Box 308, Boston, MA. 02117.

From *Quantum Questions: Mystical Writings of the World's Great Physicists* edited by Ken Wilber. © 1984 Ken Wilber. Reprinted by arrangement with **Shambhala Publications, Inc.**, P.O. Box 308, Boston, MA. 02117.

Shambhala Publications, Inc., from Ken Wilber, *Eye to Eye*, © 1983, 1990 by Ken Wilber, Shambhala Publications, Inc.

Springer-Verlag, from Menas Kafatos and Robert Nadeau, *The Conscious Universe*, © 1990 Springer-Verlag New York, Inc.

Reprinted from *The Sufi Path of Love* by William C. Chittick by permission of the **State University of New York Press**. Copyright © 1983 State University of New York.

Trinity Press International, from John Polkinghorne, *Reason and Reality*, © John Polkinghorne 1991, Trinity Press International.

University of Chicago Press, from Arthur Arberry, translator, *Mystical Poems of Rumi*, © 1968 by A.J. Arberry.

University of Chicago Press, from Ivar Ekeland, *The Broken Dice and Other Mathematical Tales of Chance*, © 1993 by The University of Chicago.

From *Masks of God: Oriental Mythology* by Joseph Campbell. Copyright © 1962 by Joseph Campbell, renewed copyright © 1990 by Jean Erdman Campbell. Used by permission of **Viking Penguin**, a division of Penguin Books USA Inc.

B. Alan Wallace, from B. Alan Wallace, *Choosing Reality*, © 1989 by B. Alan Wallace, Shambhala Publications, Inc.

Warner Books, from Neil McAleer, *The Cosmic Mind-Boggling Book*, © 1982 by Neil McAleer, Warner Books.

Anthony Zee, from Anthony Zee, *Fearful Symmetry*, © 1986 by Anthony Zee, Macmillan Publishing Company.

Maypop Books, from Coleman Barks, trans, *We Are Three* © 1987 by Coleman Barks, Maypop Books

Index of First Lines of Rumi Quotes

Sacrifice your intellect for the love of the Friend 71
Since you are not a prophet, follow the Way! 99
So God compounded animality and humanity together 174
Speech is an astrolabe in its reckoning. 264
Sunlight fell upon the wall 120
That iniquitous man knows hundreds of superfluous matters in the sciences 47
That is the Ocean of Oneness, wherein is no mate or consort 105
That voice which is the origin of every cry and sound 123
That which God has decreed from all eternity 226
The Absolute Being works in nonexistence 109
The astronomer says, "You claim there is something other 59
The more awake one is to the material world 269
The paintings, whether aware or unaware, are present in the hand of the Painter 219
The poor body will not move until the spirit moves 291
The prophets and saints do not avoid spiritual combat 177
The surface of thought's stream 155
The universe displays the beauty of Thy Comeliness! 119
The use of words is that they set you searching and excite you 20
The workshop and treasure of God is in nonexistence 284
The world is kept standing through heedlessness 122
The world is snow and ice, and Thou art the burning summer 278
The world of creation possesses quarters and directions 140
There are a hundred thousand ranks 98
There is no doubt that this world is midwinter 268
Think of the soul as source 59
This at least is notorious to all men 256
This supreme joy has no resting place 253
This world and that world are forever giving birth 212
This world is a single thought of the Universal Intellect 155
Those persons who have made or are in the course of making their studies 65
Those who use the expression "if God wills" are the true lovers of God 228
Thou art the source that causes our rivers to flow 143
Though in the world you are the most learned scholar of the time 123
Though the worlds are eighteen thousand and more 126
Thy Attributes cannot be understood by the vulgar without analogy 106
To every image of your own imagination you say, "Oh, my spirit, my world!" 93
Waves of foam rise from the sea night and day xi
We are all darkness and God is light 267
We are like a bowl on the surface of the water 228
We have heard these melodies in Paradise 285
We're quite addicted to subtle discussions 89
What God most High has decreed in eternity 226
What sort of Beloved is He? 129